EXPOSED

EXPOSED

DESIRE AND DISOBEDIENCE

IN THE DIGITAL AGE

Bernard E. Harcourt

HARVARD UNIVERSITY PRESS

CAMBRIDGE, MASSACHUSETTS

LONDON, ENGLAND

2015

First printing

Library of Congress Cataloging-in-Publication Data

Harcourt, Bernard E., 1963–
Exposed : desire and disobedience in the digital age /
Bernard E. Harcourt.
pages   cm
ISBN 978-0-674-50457-8 (cloth)
1. Information technology—Social aspects.   2. Privacy,
right of.   I. Title.
HM851.H3664 2015
303.48'33—dc23   2015012788

To Isadora
To Léonard
To Mia

CONTENTS

The Expository Society   1

PART ONE *Clearing the Ground*   29

1. George Orwell's Big Brother   31
2. The Surveillance State   54
3. Jeremy Bentham's *Panopticon*   80

PART TWO *The Birth of the Expository Society*   105

4. Our Mirrored Glass Pavilion   107
5. A Genealogy of the New *Doppelgänger* Logic   141
6. The Eclipse of Humanism   166

PART THREE *The Perils of Digital Exposure*   185

7. The Collapse of State, Economy, and Society   187
8. The Mortification of the Self   217
9. The Steel Mesh   234

PART FOUR *Digital Disobedience*   251

10. Virtual Democracy   253

11. Digital Resistance   262

12. Political Disobedience   280

NOTES   285

ACKNOWLEDGMENTS   347

INDEX   349

EXPOSED

THE EXPOSITORY SOCIETY

EVERY KEYSTROKE, EACH MOUSE CLICK, every touch of the screen, card swipe, Google search, Amazon purchase, Instagram, "like," tweet, scan—in short, everything we do in our new digital age can be recorded, stored, and monitored. Every routine act on our iPads and tablets, on our laptops, notebooks, and Kindles, office PCs and smartphones, every transaction with our debit card, gym pass, E-ZPass, bus pass, and loyalty cards can be archived, data-mined, and traced back to us. Linked together or analyzed separately, these data points constitute a new virtual identity, a digital self that is now more tangible, authoritative, and demonstrable, more fixed and provable than our analog selves. Our mobile phones communicate and search for Wi-Fi networks even when cellular data is turned off. Our MetroCards and employee IDs leave traces with each swipe and tap. Every ATM withdrawal, web search, secured-building entry or elevator ride, every mobile payment leaves a mark that makes it possible for others to know our whereabouts at every moment, to track us at will, and to reconstitute our every action. In sum, today every single digital trace can be identified, stored, and aggregated to constitute a composite sketch of what we like, whom we love, what we read, how we vote, and where we protest.

The social media and web browsers we use—or accidentally visit— constantly collect a trove of our personal data. Our telecommunication companies record everything they can, as do other telecoms that, unbeknownst to us, route, switch, redirect, and retransmit our communications. Our intimate data are stockpiled by signals intelligence services in the United States and abroad, and by local law enforcement—but also by the retailers we use, by data brokers we've never heard of, by hackers, and simply by the curious among us using free network sniffers or stalking us on the web. Most of our digital information is available one way or another—for purchase by advertisers or for review by insurance companies, for supervision by our employers, for examination by the security apparatus, for capture by keystroke loggers, or for a quick peek on anonymous online message boards. Google and Facebook aggressively compete over who has more of our sensitive data to share with their users and to sell to advertisers. Free off-the-shelf sniffing programs allow anyone to read others' emails and see their web browsing on unsecured networks. Law enforcement agencies secretly collect, pool, and share as much of our digital information as possible. And the National Security Agency (NSA), the British Government Communications Headquarters, the French Direction Générale de la Sécurité Extérieure, the Chinese and Russian signals intelligence agencies, and practically every other intelligence service around the world share the ambition to know everything, to map the Internet universe, to be able to identify every end device connected to the Internet—in short, to know everything digital, everywhere, and at every moment.

Most of us are aware of this, although many of us put it out of our minds. We have read the *Guardian* articles and the *New York Times* and heard the investigative journalism on the radio. We have watched video clips of the congressional hearings. We've repeatedly seen the telltale advertisements popping up on the ribbon of our search screen, reminding us of our immediately past Google or Bing query. We've received the betraying emails in our spam folders. We've even scrutinized the top-secret NSA PowerPoint slides and other documents leaked by Edward Snowden. But it is one thing to know, and quite another to remember long enough to care—especially when there is the ping of a new text, the flag desktop notification of a new email, the flash of a new like on our Instagram photo, the doorbell noise of a new Facebook message, or just the simple desire to know how many people have seen our Snapchat story or commented on our blog post. It is quite another thing to pay attention in the face of the stimulating distractions and sensual pleasures of the new digital age—the constant news feeds and friend messages, the newest Vine or viral YouTube video, the access to every bit of information online, the ability to Google anything and everything. The anticipation, the desire for something new and satisfying, that sensation we get when we receive a bit of good news in our email in-box—how easily this distracts us from what we actually know about the breathtaking scope and ubiquity of these new forms of digital surveillance, data mining, profiling, and monitoring. We so easily get sidetracked by the most minor digital stimulus—and so often go there to avoid the emotional resistance of writer's block or the discomfort of a difficult thought or unpleasant interaction. How quickly, how instinctively we put our thumb on our smartphone, check email, read the Twitter feed, swipe to Facebook. Whatever. We've already put it out of our mind and are consumed with a new Snapchat, a viral wall post, or Assassin's Creed Unity. We ignore what we suspect or even know about being tracked and exposed. We put it out of our minds. But we do so at our peril.

. . .

The *Wall Street Journal* broke the story in May 2011, well before the name Edward Snowden meant anything to anyone.[1] The revelation did not draw much attention, though. It concerned those little icons on most websites—the thumbs-up of Facebook's like button, the little birdie of Twitter's tweet button, the multicolor Google+ widget, those small icons that line and populate websites, YouTube videos, news articles, travel websites, search ribbons, and so on.

It turns out that those little icons allow Facebook, Twitter, or Google to track our Internet browsing on the websites where the icons are placed, regardless of whether we are logged onto those social networks. As long as someone uses those social networks and has been logged onto them within the past month (and did not *actively* log out), their Internet surfing on *other sites* that carry those icons is tracked and reported back to Facebook, Twitter, or Google. In fact, you don't even need to be a user of social media—you can be tracked back from the other websites even if you *mistakenly* click onto one of those social media sites. And it turns out that those little icons are on lots of websites. Back in 2011, for instance, 33 percent of the top 1,000 most popular websites had the Facebook like button, 25 percent had the Google+ widget, and 20 percent had the Twitter tweet button. The icons are embedded in millions of websites today.[2]

The sequence is simple: logging onto any of those social media—Facebook, Twitter, Google+—will install software on your web browser that remains active even if you turn off your computer or shut down the browser. It only turns off if you affirmatively log out of the social media—if you intentionally click the "log out" button. Once activated, that software will then report back to the social media anytime you are on any other website that carries the little icon, regardless of whether you click or touch the little icon. Just being on a website with those like and tweet buttons embedded in them will allow those social media to track your Internet browsing.

The *Wall Street Journal* noted in passing, "Facebook says it still places a cookie on the computer of anyone who visits the Facebook.com home page, even if the user isn't a member."[3] So one's browsing history may be made available to others, even for those of us who do not have a Facebook account. The article goes on to report: "Until recently, some Facebook widgets also obtained browsing data about Internet users who had never visited Facebook.com, though Facebook wouldn't know their identity. The company says it discontinued that practice, which it described as a 'bug,' earlier this year after it was disclosed by Dutch researcher Arnold Roosendaal of Tilburg University."[4]

To be more exact, this is precisely how the tracking works. According to detailed communications dating from 2011 between reporters for *USA Today* and Facebook executives—Arturo Bejar, Facebook's engineering director, Facebook spokesmen Andrew Noyes and Barry

Schnitt, engineering manager Gregg Stefancik, and corporate spokes-woman Jaime Schopflin—these are the technical specifications:

- "The company compiles tracking data in different ways for members who have signed in and are using their accounts, for members who are logged-off and for non-members. The tracking process begins when you initially visit a facebook.com page. If you choose to sign up for a new account, Facebook inserts two different types of tracking cookies in your browser, a 'session cookie' and a 'browser cookie.' If you choose not to become a member, and move on, you only get the browser cookie.
- "From this point on, each time you visit a third-party webpage that has a Facebook Like button, or other Facebook plug-in, the plug-in works in conjunction with the cookie to alert Facebook of the date, time and web address of the webpage you've clicked to. The unique characteristics of your PC and browser, such as your IP address, screen resolution, operating system and browser version, are also recorded.
- "Facebook thus compiles a running log of all your webpage visits for 90 days, continually deleting entries for the oldest day and adding the newest to this log.
- "If you are logged-on to your Facebook account and surfing the Web, your session cookie conducts this logging. The session cookie additionally records your name, e-mail address, friends and all data associated with your profile to Facebook. If you are logged-off, or if you are a non-member, the browser cookie conducts the logging; it additionally reports a unique alphanumeric identifier, but no personal information."[5]

In case the minutiae got too tedious, it may be worth repeating here that, according to its engineers and spokesmen, Facebook links and ties all the information that it gleans from your web surfing back to "your name, e-mail address, friends and all data associated with your profile to Facebook."[6]

That was 2011. By 2014, this type of tracking software was old news, especially in the advertising business. Users were finding ways to disable cookies or clear them. Worse, cookies do not interface well with smartphones, and everyone is now spending more and more time on their mobiles. In the words of one tech expert, "The cookie is failing the advertising industry."[7] Facebook itself openly recognized the shortcomings of the cookie, acknowledging that "today's technology for ad serving and measurement—cookies—are flawed when used alone. Cookies don't work on mobile, are becoming less accurate in demographic targeting and can't easily or accurately measure the customer purchase funnel across browsers and devices or into the offline world."[8]

On September 29, 2014, Facebook launched a solution: "people-based marketing" via the new and improved Atlas product. "People-based marketing solves these problems," Facebook boasted.[9] The new technology is simple: take all of the data that Facebook has, including all the data it can mine through our smartphone apps, and share all that information with anyone who's willing to pay, so that they can then target users on all their other platforms at all times. In an appropriately titled article, "With New Ad Platform, Facebook Opens Gates to Its Vault of User Data," Vindu Goel of the *New York Times* explains that Atlas "will allow marketers to tap [Facebook's] detailed knowledge of its users to direct ads to those people on thousands of other websites and mobile apps."[10] By tapping into Facebook's trove of data, advertisers can then push products on all of the different devices and platforms accessed by the user, including video sites, game apps, Instagram, et cetera. Plus, Facebook can also then measure success and provide feedback to advertisers as to which ads are more effective. For instance, Facebook vaunts, "Instagram—as a publisher—is now enabled with Atlas to both measure and verify ad impressions. And for Atlas advertisers who are already running campaigns through Instagram, Instagram ads will be included in Atlas reporting."[11]

Facebook is the second-biggest digital advertising platform, after Google, but it has a particular advantage over most others in one important respect: users check Facebook on their mobile devices. As the *New York Times* explains, "The Facebook login is most useful on mobile devices, where traditional web tracking tools like cookies and pixel tags do not work. If a person is logged into the Facebook app on a smartphone, the company has the ability to see what other apps he or she is using and could show ads within those apps."[12] Or, as another tech expert explains: "As long as you are logged in to Facebook on your device, Atlas tracks activity even in apps that don't use a Facebook login."[13] Facebook boasts about this new technology. It is not hiding anything. On the contrary, it is trying to take over the digital advertising business—its revenue depends on it—so it is up front and explicit: "Atlas delivers people-based marketing, helping marketers reach real people across devices, platforms and publishers. By doing this, marketers can easily solve the cross-device problem through targeting, serving and measuring across devices. And, Atlas can now connect online campaigns to actual offline sales, ultimately proving the real impact that digital campaigns have in driving incremental reach and new sales."[14]

Social media has become, in the words of one analyst describing Google's Gmail and other services, "a massive surveillance operation that intercepts and analyzes terabytes of global Internet traffic every day, and then uses that data to build and update complex psychological profiles on hundreds of millions of people all over the world—all of it in real time."[15]

. . .

The Associated Press (AP) broke this other story in 2014, leading Sen. Patrick Leahy of Vermont, chairman of the Senate Appropriations Committee's State Department and Foreign Operations Subcommittee, to hold hearings and substantiate the details in the spring of 2014. The episode, which dates back to 2009, involved another trove of data: a large database consisting of half a million telephone numbers of cell phone users in Cuba—or, more precisely, a trove of data *and* a few entrepreneurial officials and subcontractors at the United States Agency for International Development (USAID), the agency that delivers billions of dollars of aid and humanitarian assistance to needy countries.[16]

It all started when an employee at Cubacel, the Cuban state-owned cell phone provider, surreptitiously slipped half a million cell phone numbers to a Cuban engineer living in Spain. That Cuban expat then turned the database over, "free of charge" according to the documents reviewed by the AP, to officials at USAID and to a for-profit Washington, D.C., company, Creative Associates International. (Creative Associates had received millions of dollars in contract business from USAID.) A manager there got, well, as the company's name would suggest, creative: with an associate in Nicaragua, she hatched the idea of sending bulk text messages from different countries to the Cuban cell phone users. It was a creative way to circumvent the strict state control of the Internet in Cuba. A way to kick-start a basic type of social media, on the model of Twitter.[17]

The idea was to surreptitiously set up from scratch a social network for Cubans, a rudimentary "Cuban Twitter," with the long-term goal of fomenting political dissent. Top USAID administrators deny that was the intent—but the documentary evidence belies any such disavowals.[18] As the AP reported, after carefully reviewing more than 1,000 pages of documents: "Documents show the U.S. government planned to build a subscriber base through 'non-controversial content': news messages on soccer, music and hurricane updates. Later when the network reached a critical mass of subscribers, perhaps hundreds of thousands, operators would introduce political content aimed at inspiring Cubans to organize 'smart mobs'—mass gatherings called at a moment's notice that might trigger a Cuban Spring, or, as one USAID document put it, 'renegotiate the balance of power between the state and society.'"[19]

They gave the undertaking the name ZunZuneo, which means, in Cuban slang, a hummingbird's tweet. They then drew on user reactions and responses to start profiling the Cuban subscribers. One USAID contractor was tasked with categorizing the subscribers as either "pro-revolution," "apolitical," or "anti-revolution" based on their answers to prompts.[20] According to the AP, this contractor "collected a sample of more than 700 responses and analyzed them according to two variables. The first was the level of interest in the messages received, and the second was the political nature of the response. She wrote in her report that 68 percent of responses showed mild interest in the texts."[21]

In another operation, "USAID divided Cuban society into five segments depending on loyalty to the government. On one side sat the 'democratic movement,' called 'still (largely) irrelevant,' and at the other end were the 'hard-core system supporters,' dubbed 'Talibanes' in a derogatory comparison to Afghan and Pakistani extremists."[22] The key objective of the overarching project, according to the AP, was to "move more people toward the democratic activist camp without detection."[23] A close parsing of the USAID documents reveals that "their strategic objective in Cuba was to 'push it out of a stalemate through tactical and temporary initiatives, and get the transition process going again toward democratic change.' "[24]

At one point, in March 2011, ZunZuneo had about 40,000 subscribers. None of the subscribers had any idea that the social media was created, sustained, or fed by USAID operatives. None of them realized that their texts were profiled by USAID contractors to determine their political allegiance. None of them realized that the messaging was intended to motivate them politically.

To hide all this, USAID set up "a byzantine system of front companies using a Cayman Islands bank account, and recruit[ed] executives who would not be told of the company's ties to the U.S. government," according to the AP investigation.[25] They had a British company set up a corporation in Spain to run ZunZuneo. They set up a companion website to the cell phone texting service so that the cell phone users could subscribe, offer feedback, and themselves send texts for free.[26] And, the documents reveal, they discussed how to make this all seem legitimate: "Mock ad banners will give it the appearance of a commercial enterprise," a document reads.[27] Most important, as the AP reported, " 'there will be absolutely no mention of United States government involvement,' according to a 2010 memo from Mobile Accord, one of the project's contractors. 'This is absolutely crucial for the long-term success of the service and to ensure the success of the Mission.' "[28] The ZunZuneo team even hired a Havana-born satirical artist to write Twitter-like messages that had a Cuban flavor.[29]

ZunZuneo shut down in mid-2012 when the USAID money dried up. After the program was exposed, Senator Leahy of Vermont began holding hearings in Congress in the spring of 2014. The Senate hearings confirmed the story.

. . .

Launched in 2007, the PRISM program allows the NSA to access data from Google, Facebook, Microsoft, Yahoo, Paltalk, YouTube, Skype, AOL, Apple, and more—for a mere $20 million per year, a trifling sum for an intelligence program.[30] In conjunction with other software, such as the XKeyscore program, PRISM "allows officials to collect material including search history, the content of emails, file transfers and live chats"; they can extract a person's email contacts, user activities, and webmail, as well as all contacts listed in the to, from, CC, and BCC lines of emails.[31] Using other programs and tools, such as DNI Presenter, they can "read the content of stored emails," "read the content of Facebook chats or private messages," and "learn the IP addresses of every person who visits any website the analyst specifies."[32] Moreover, the NSA has "developed methods to crack online encryption used to protect emails, banking and medical records" giving it access to all our private and legally protected information.[33] The possibilities and the amounts of data are simply staggering. The *Washington Post* reported back in 2010 that "every day, collection systems at the [NSA] intercept and store 1.7 billion emails, phone calls and other type of communications."[34]

The PRISM program gives the government access to an individual's emails, photos, videos, attachments, VoIP, and so on. At practically no cost, the government has complete access to people's digital selves. And although the NSA denied that it has immediate access to the data, the NSA documents leaked by Edward Snowden state that the agency "claims 'collection directly from the servers' of major US service providers."[35] Bart Gellman of the *Washington Post*, after fully reinvestigating the PRISM program, affirms that "from their workstations anywhere in the world, government employees cleared for PRISM access may 'task' the system and receive results from an Internet company without further interaction with the company's staff."[36] The trajectory of cooperation is illustrated in a classified NSA training slide leaked by Snowden in June 2013 (see Figure 1).

As Glenn Greenwald of the *Guardian* reported, "The Prism program allows the intelligence services direct access to the companies' servers. The NSA document notes the operations have 'assistance of communications providers in the US.'"[37] The *Guardian* explained, based on an NSA document dating to April 2013: "The Prism program allows the NSA, the world's largest surveillance organisation, to obtain targeted communications without having to request them from the service providers and without having to obtain individual court orders. With this program, the NSA is able to reach directly into the servers of the participating companies and obtain both stored communications as well as perform real-time collection on targeted users."[38] This allows the NSA to obtain all kinds of data—as illustrated in another classified NSA training slide leaked by Snowden to the *Guardian* in June 2013 (see Figure 2).

According to the *Guardian*, there is no warrant requirement or the need for individual authorization under the Foreign Intelligence Surveillance Act for any of these collection and analysis activities so long as the analyst searching the communications had "reasonable suspicion that one of the parties was outside the country at the time the records were collected by the NSA."[39] The program is apparently leading to exponential rates of growth of search queries. "The document highlights the number of obtained communications increased in 2012 by 248% for Skype—leading the notes to remark there was 'exponential growth in Skype reporting; looks like the word is getting out [among intelligence analysts] about our capability against Skype,'"

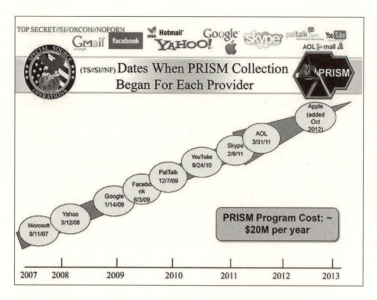

FIGURE 1 Top-secret NSA PowerPoint slide on PRISM program history (2013)
Source: "NSA Slides Explain the PRISM Data-Collection Program," *Washington Post,*
June 6, 2013.

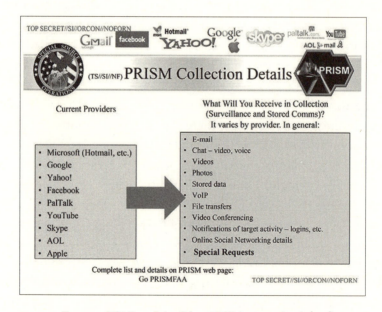

FIGURE 2 Top-secret NSA PowerPoint slide on PRISM program details (2013)
Source: "NSA Slides Explain the PRISM Data-Collection Program," *Washington Post,*
June 6, 2013.

the *Guardian* reports. "There was also a 131% increase in requests for Facebook data, and 63% for Google."[40]

IN OUR DIGITAL FRENZY to share snapshots and updates, to text and videochat with friends and lovers, to "quantify" ourselves, we are exposing ourselves—rendering ourselves virtually transparent to anyone with rudimentary technological capabilities. We are exhibiting ourselves through petabytes of electronic traces that we leave everywhere, traces that can be collected, linked together, and amalgamated, traces that, paradoxically, although they are virtual, have become more tangible, more provable, more demonstrable, and more fixed than our analog selves. Ernst Kantorowicz spoke of the king's two bodies, but the metaphor is more applicable today to the subject's—or, rather, the liberal democratic citizen's—two bodies: the now permanent digital self, which we are etching into the virtual cloud with every click and tap, and our mortal analog selves, which seem by contrast to be fading like the color on a Polaroid instant photo.

For many of us, we have brought this upon ourselves willingly, enthusiastically, and with all our passion: through our joyful and fulfilling embrace of social media and online shopping, through our constant texting to loved ones and our Google searches. Many of us give ourselves away and expose our innermost being with all our predilections and lust; our yearning for more attention and the costlessness of publicity combine in this toxic way and lead many of us to share our most intimate details on Facebook, to display our resumés and accomplishments on personal websites, to share our personal photos and travels on Instagram and Tumblr, to avow our political faith on virtual walls and digital protest sites. Not everyone, of course, and not always so willingly. By November 2014, Facebook could boast a community of 1.35 billion people, but there remain a number of us who continue to resist.[41] And there are many more who are ambivalent about the loss of privacy or anonymity, who are deeply concerned or hesitant. There are some who anxiously warn us about the dangers and encourage us to maintain reserve.[42] There are many of us who are confused and torn between, on the one hand, the desire to share our lives with loved ones and friends, to videochat on Skype from any-place on earth, to freely share photos on Flickr, and, on the other hand, the discomfort of seemingly undressing ourselves in public or entering our private data on commercial websites. And yet, even when we hesi-tate or are ambivalent, it seems there is simply no other way to get things done in our new digital age. No other way to reserve the hotel room or seat on the plane, to file the IRS form, to recall the library book, or to send money to our loved one in prison. No other way to do it but online. Even when we do not will it, so many of us hesitantly ex-pose ourselves despite all our reservations and care.

And by exposing ourselves, we make it so easy, so tempting, so cheap to watch us, to monitor us, to target us—as well as to track us, to detain us, and, for some, to extract and punish us. We make ourselves virtually transparent for everyone to see, and in so doing, we allow ourselves to be shaped in unprecedented ways, intentionally or unwittingly. Our selves and our subjectivity—the very subjectivity that embraces the digital apps, platforms, and devices—are themselves molded by the rec-ommender algorithms and targeted suggestions from retailers, the spe-cial offers from advertisers, the unsolicited talking points from political parties and candidates. Through the flood of suggestions and recom-mendations, through our own censorship and self-governance in the face of being watched, we are transformed and shaped into digital sub-jects. We are brought into being through the processes of digital expo-sure, monitoring, and targeting that we embrace and ignore so readily. And we give ourselves up to new forms of subjectivity and social order, marked by unprecedented restrictions on privacy and anonymity and by seemingly unlimited levels of monitoring and surveillance.

Our digital cravings are matched only by the drive and ambition of those who are watching. One data broker, Acxiom, boasted in 2014 that it had 3,000 points of data on practically every consumer in the United States.[43] The Global Access Operations unit of the NSA collected data, in a single month in 2013, on more than 97 billion emails and more than 124 billion telephone calls from around the globe.[44] Facebook's ef-fort to become the biggest digital advertising platform means it collects gargantuan amounts of personal data. Google's ambition to map every street and vacuum up every bit of Wi-Fi traffic with its Street View cars and to digitize every book in the world is colossal. The NSA has set out to map, know, and tap into the *entire* Internet universe: according to NSA documents leaked by Snowden and viewed by *Der Spiegel,* the NSA and the British signals intelligence agency are working together to, in the words of the secret document, "map the entire Internet—[a]ny device, anywhere, all the time."[45] The program, called Treasure Map, seeks to identify and make visible "every single end device that is connected to the Internet somewhere in the world—every smartphone, tablet and computer."[46] Today, the drive to know everything, everywhere, at every moment is breathtaking.

WE LIVE TODAY in a new political and social condition that is radically transforming our relations to each other, our political community, and ourselves: a new virtual transparence that is dramatically reconfiguring relations of power throughout society, that is redrawing our social landscape and political possibilities, that is producing a dramatically new circulation of power in society. A new expository power constantly tracks and pieces together our digital selves. It renders us legible to others, open, accessible, subject to everyone's idiosyncratic projects—whether governmental, commercial, personal, or intimate. It interpellates us into digital subjects born into a new culture "that allows an airline or a politician to know more about us than our mothers or our lovers do."[47] And it does so with our full participation. There is no conspiracy here, nothing untoward. Most often we expose ourselves for the simplest desires, the pleasures of curiosity, a quick distraction—those trifling gratifications, that seductive click the iPhone "shutter" makes, the sensual swoosh of a sent email. That, and the convenience and apparent costlessness with which we can shop online, renew a subscription, deposit a check via our mobile phone, carry a library on our e-reader. For those of us who hesitate at first, the allure and efficiency of costless storage on Dropbox, of gratis transfers of megabytes of data on WeTransfer.com, of free calendaring have made exposure practically irresistible. We know, of course, that none of this is "free" and that we pay for it by giving complete access to our personal data.[48] We pay for it also with our attention and distraction.[49] But those costs do not feel expensive or prohibitive, or sufficiently so, at the moment we have our finger on the mouse.

In the wake of the two grand jury decisions to refuse to indict police officers in the homicides of Michael Brown in Ferguson and Eric Garner in Staten Island, a protest was organized for Saturday, December 13, 2014, in New York City. The organizers of the Millions March set up a Facebook page, where, by the night before, more than 45,000 people had RSVPed. It was posted as a public event on Facebook, so everyone and anyone could see who had signed up to attend—providing everyone and anyone, including the social media unit of the New York City Police Department, a costless, pristine list of all the individuals who feel so strongly about the problem of police accountability that they are willing to identify themselves publicly.

It takes little imagination to think of the ways that such a list could be exploited: As a background check during a police-civilian encounter or stop-and-frisk. As a red flag for a customs search at the airport, or a secondary search at a random checkpoint. As part of a larger profile for constructing a no-fly list, or for attributing a lower priority to a 911 emergency call. For more aggressive misdemeanor arrests in neighborhoods that have higher concentrations of protesters. As part of a strategy to dampen voter turnout in certain precincts. For a cavity search in case of arrest. There are myriad creative ways to misuse the data; our imagination is the only limit. And with a single click, a prying eye can learn everything about each of the digital selves that signed up for the protest on Facebook; using a simple selector, an intelligence analyst could collect all of the digital information about any one of those signatories, read all their emails, attachments, wall posts, and comments, decipher their political opinions and engagements, scan their photos and texts, target their videochats, track them by cell phone location—in sum, follow their every movement and digital action throughout every moment of their day. Once we are identified, we can be relentlessly monitored across practically every dimension of our daily routine—by means of the GPS on our phones, our IP addresses and web surfing, our Gmail contacts, MetroCards, employee IDs, and social media posts.

We face today, in advanced capitalist liberal democracies, a radically new form of power in a completely altered landscape of political and social possibilities. Earlier models of sovereign power, characterized by dramatically asymmetric and spectacular applications of brute force, do not quite capture this moment. We have moved far beyond the juridical relations of ruler to subject that characterized Thomas Hobbes's *Leviathan,* we no longer depend on the éclat of spectacular ordeals and tortures—even if residues and traces remain.[50] For the most part, there is no longer as much need today for the stunning display, nor for the snitch or *mouchard* of the ancien régime, nor for the Stasi informant. There is hardly any need for illicit or surreptitious searches, and there is little need to compel, to pressure, to strong-arm, or to intimidate, because so many of us are giving all our most intimate information and whereabouts so willingly and passionately—so voluntarily.

There is also little need today to discipline us. We are no longer forced to follow strict timetables and work routines, to walk in straight lines, stand at attention, or stay seated behind rows of school desks. No need to confine us to observation and examination. No, we have been liberated from all that. We are free to keyboard on our laptops in bed or dictate an email at the beach, to ask Siri a question at the dinner table, to text while riding a bike. Although the digital transparence at times feels panoptic, there is in truth little need for bars and cells and watchtowers—there is little need to place the subject in a direct line of sight. There is little need to *extract* information, because we are giving it out so freely and willingly, with so much love, desire, and passion—and, at times, ambivalence or hesitation.

We have also moved beyond biopower or securitarian logics—those modalities live on, yes, but do not fully account for the new forms of power circulating today. The new recommender algorithms that profile us do not concern themselves with populations or even groups. They do not seek an equilibrium, nor to maximize outcomes under cost constraint, in large part because the cost analysis is entirely different today: those who govern, advertise, and police are dealing with a primary resource—personal data—that is being handed out for free, given away in abundance, for nothing. And rather than optimize at reduced cost levels, those who are mining the data can seek to construct perfectly unique matches for each person's individuality. They can seek to know every device and every user, at a paltry price. The fact that the PRISM program only costs the NSA $20 million is telling: we, digital subjects, are providing the information for a pittance. Virtual transparence is now built into the technology of life, into the very techniques of living, and it makes possible an individually targeted gaze that pierces through populations.

No, we are not so much being coerced, surveilled, or secured today as we are *exposing* or *exhibiting* ourselves knowingly, many of us willingly, with all our love, lust, passion, and politics, others anxiously, ambivalently, even perhaps despite ourselves—but still, *knowingly* exposing ourselves. The relation of power is inverted: we digital subjects, we "data doubles," we *Homo digitalis* give ourselves up in a mad frenzy of disclosure.[51] Many of us exhibit our most intimate details in play, in love, in desire, in consumption, in the social, and in the political throughout our rich digital lives—through our appetites, in our work, for our political convictions, to become ourselves. Even those of us who do not partake of the seductive world of social media most often have no alternative but to share our intimate lives and political views in emails, texts, and Skype conversations—knowing that we are exposing ourselves.

For many of us, our digital existence has become our life—the pulse, the bloodstream, the current of our daily routines. For adolescents and young adults especially, it is practically impossible to have a social life, to have friends, to meet up, to go on dates, unless we are negotiating the various forms of social media and mobile technology. The teens that danah boyd interviews give voice to this: if you're not on social media, one of the girls explains, "you don't exist."[52] Social media communications are necessary for most youths today in advanced capitalist democracies, necessary to have a fulfilling social life.

Writing about one of her informants, a sixteen-year-old girl, boyd observes: "For Heather, social media is not only a tool; it is a social lifeline that enables her to stay connected to people she cares about but cannot otherwise interact with in person."[53] For many of us, these digits have become our social lives, and to fit in, we have to be wired, plugged in, online. It's not so much a question of choice as a feeling of necessity.

GUY DEBORD SPOKE of the "society of the spectacle."[54] Michel Foucault described instead the panoptic society, the disciplinary society—or what he referred to as the "punitive society."[55] "Our society is one not of spectacle," Foucault declared, "but of surveillance."[56] Gilles Deleuze went further and prefigured the "societies of control."[57] But it seems as if today we live, rather, in a society of *exposure* and *exhibition*. We live in what I would call the *expository society*.

In this expository society, so many of us—or at least those of us on this side of the digital divide, enmeshed in a constant digital pulse—have become dulled to the perils of digital transparence. Dulled by decades of blind faith in the virtues of entrepreneurialism, self-interest, and self-centeredness, and in the illusion of free markets. Dulled by the deafening simplicity of that neoliberal mantra: "Would you prefer to have the government running innovative companies or would you rather have the private sector running them?"—as if those were the only two options or, even, real choices.[58] Dulled by decades of militaristic homeland security, national "terror" alerts, and massive overincarceration—and the belief that these have made us so much safer. Dulled by the mortification of the analog self that the loss of privacy, autonomy, and anonymity have brought upon us. In short, dulled into not caring because there is "nothing to hide" and "no place to hide"—perhaps the ultimate stage of mortification.[59] Numb to the risk of digital transparence, or simply overwhelmed by the seductive pleasures of the red jelly beans and green chiclets of Candy Crush, the new emoticon of a doll or a smoking pipe, the paper-crumpling "noise" a document makes when it is trashed, the surprisingly perfect book recommendations by Amazon.com, or the color options of the new iPhone. Simply distracted by all these—while the power and scope of the commercial interests that seduce us, profile and track us, watch our every e-motion, and compile every slightest detail of our intimate lives grow far more awesome, day by day, than the East German government could ever have dreamed.

And as we play with our Xbox consoles, Android phones, and Apple Watches, our digital lives begin to converge with a form of electronic monitoring that increasingly resembles correctional supervision. This is a double movement, a pas de deux: on the one hand, our daily routines gravitate from the analog to the digital, with all of the accompanying data collection and mining by social media, advertisers, commercial interests, and government surveillance; on the other hand, our punishment practices, at least in the West, themselves also eerily gravitate from the analog to the digital, from the brick-and-mortar and iron bars of the prison to electronic monitoring and GPS tracking. The costs of massive overincarceration, especially during recent times of economic and financial crisis, have pushed us, in advanced capitalist liberal democracies, to turn increasingly to supervised parole and probation, to the electronic bracelet and the CCTV, to methods of digital monitoring of supervised correctional populations in the free world. Throughout the United States, the use of GPS monitoring has grown exponentially during the twenty-first century, and other Western European countries have also turned increasingly to open supervision. The major reform of the French penal code in 2014, for instance, centered with much fanfare on replacing physical detention with *le suivi*—digital monitoring in the free world.[60]

Some of us are forced to wear electronic ankle bracelets, others lustfully strap Apple Watches onto their wrists, but in both cases, all of our daily motions, activities, and whereabouts become easily accessible to those with rudimentary technology—that is, when we are not actively broadcasting our heartbeat to our loved ones, living reality on camera, or tweeting our lives minute by minute. The Samsung smart TV that we proudly display in our den records and shares our private conversations as if it were an electronic monitoring device—or a telescreen that we gleefully hooked to the wall.[61] In these ways, ordinary life is uncannily converging with practices of punishment: The see-throughness of our digital lives mirrors the all-seeingness of the penal sphere. Eventually there may be no need for punishment since we will be able to monitor every movement and action—and perhaps intervene remotely with similar know-how to the Google driverless car or drone technology. No need to distinguish ordinary life from the supervised correctional condition, since we will be watched, tracked, analyzed, and known at every moment of the day.

"Not to punish less, but to punish better": we have now overcome discipline, freed ourselves of the institutional straitjacket, reached a privileged space of utter freedom where we get to do everything we desire—to tweet, to write in emoticons, to work remotely from our beds, to text and sext, to play Candy Crush on the subway or in the classroom, to stalk our friends and lovers on Facebook.[62] It is a free space where all the formerly coercive surveillance technology is now woven into the very fabric of our pleasure and fantasies. In short, a new form of expository power embeds punitive transparence into our hedonist indulgences and inserts the power to punish in our daily pleasures. The two—pleasure and punish—can no longer be decoupled. They suffuse each other, operate together. They have become inextricably intertwined.

WE ARE INUNDATED today with talk of "datafication": with the idea that amassing large data sets, mining and analyzing them will reveal new truths about society and ourselves that we would never have known before and which will allow us to find solutions to problems that we might never have discovered. All we have to do, we are told, is "let the data speak." We are surrounded by a kind of veneration and mystification of the new digital mediums, the connected things, the Internet of Things. And, no doubt, there exist situations where the analysis of large data streams has identified or brought to our attention significant events before we would otherwise have noticed them. We are often reminded of how Google searches can forecast flu epidemics.[63]

New studies reveal that digital monitoring may have a dampening effect on police use of force—although the videos of excessive force continue to stream in.[64] At times the new digital medium allows us to self-present more carefully and may even momentarily enhance our self-esteem.[65]

Today, the dominant self-image of datafication is that it can reveal truth and solve problems. But the digital realm does not so much give us access to truth as it constitutes a new way for power to circulate throughout society. It is less a means to truth than an exercise of power. It does not so much reveal truths about society and ourselves as produce them. Digital capabilities have fundamentally shifted the way power flows in our advanced capitalist democracies, producing a new digitized way of life with a unique recommender rationality that dominates this side of the digital divide—with its rich circuit of texts, tweets, and emails, digital photos, scans and PDFs, Skype calls, Facebook posts, Google searches and Bings, pings and Snapchats, Venmo payments, Vimeos and Vines. Embedded in all these platforms, there is a technology of virtual transparence that allows for pervasive data mining, digital profiling, facial recognition, Amazon recommendations, eBay special offers, Netflix algorithms, and NSA surveillance. It enables a new circulation through which we expose ourselves and our most intimate desires, inescapably, to the technological knowhow of the market and the state, and in the process become marketized subjects in marketized democracies, increasingly oriented by our digital rankings, ratings, and scores as consumers, less and less connected to our political status as citizens and private analog subjects.

The expository society—such is our new political, social, and intimate condition. Not new in the sense that pervasive surveillance of citizens is in any way novel in liberal democracy. No, collecting intelligence on one's own has been a core function of the security apparatus of democratic regimes since their inception. In fact, the surveillance practices that predated the republican revolutions were inscribed into the very fabric of liberal democracies at their birth during moments of crisis and survival.[66] Creating networks of informants, compiling biometric databases (whether physical descriptors, fingerprints, or DNA), maintaining files (from the crude mug book filled with head shots to refined facial recognition techniques)—the fact is, domestic security agencies have depended on data gathering since the early days of the Republic. State surveillance of our ordinary lives, even illegal spying on ordinary fellow citizens, is by no means new in liberal democracy.

Not new either in the sense that these awesome digital intelligence capabilities have just been discovered or that the data-mining practices were previously unknown. No, astute observers sounded the alarm bell many years ago, minutely describing these new digital capabilities in best-selling books that read more like dystopian science fiction than journalistic reporting—all of that long before Edward Snowden even imagined leaking any secret documents.[67]

Nor, finally, is anything new about the desire to know. In a previous time—in the actuarial age of the early twentieth century, for instance—the same drive to know prevailed: the security apparatus tried to collect as much data as possible and predict behaviors using actuarial instruments. Parole boards rummaged through an inmate's jacket, collecting facts and predicting the likelihood of reoffending if released; sentencing judges culled every detail from the presentence report to estimate the future dangerousness of a convict; insurance companies collected background information to better tabulate life expectancy. But the tools were crude, rough estimates, with lots of error. Predicting future dangerousness, even among high-risk offenders, was a tricky business. We had big dreams of perfect surveillance, but we didn't have the means. We do now, and the effects are dramatic. Not just for the cosmopolitan elites, who are glued to their latest iPhones and iWatches, but across the socioeconomic spectrum in advanced capitalist democracies.[68]

New, then, in that power circulates today in a manner radically different from what we had grown accustomed to in the recent past. Fed by corporate surveillance and data-mining programs of social media, Internet, and retail giants, and replenished by our own curiosity and pleasure—retweeted, friended, shared, and reposted—the new digital capacities expose us and shape our subjectivity in unprecedented ways. One website, called Insecam, has taken upon itself to broadcast the live feed from private security cameras of anyone who has, either intentionally or inadvertently, failed to change the default password on her security camera. The result is that "Insecam is streaming over 11,000 cameras in the US, nearly 2,500 in the UK, six in Tanzania and others everywhere in between," with "footage of people hanging out in living rooms, kids sleeping in their beds, garages, neighborhoods, businesses and more."[69] Programs such as XKeyscore can instantaneously give an intelligence analyst access to all the content that a user types in an email, the websites she visits, the searches she performs, as well as all of her metadata—in short every keystroke, every touch, every click.[70] They provide access, as Glenn Greenwald documents, to "nearly everything a typical user does on the internet."[71] And they achieve this in massive quantities. In a single one-month period in 2012, the XKeyscore program "collected over 41 billion records."[72] In 2012, the NSA was processing "more than twenty billion communications events (both Internet and telephone) from around the world *each day*."[73] The data-mining capabilities exceed the basic level of metadata, allowing access to real-time content—though metadata alone is extraordinarily powerful. In the words of Gen. Michael Hayden, former director of both the NSA and the CIA: "We kill people based on metadata."[74] Google and Facebook follow our every digital move to accumulate troves of our personal data; as Eben Moglen reminds us, "If you have a Facebook account, Facebook is surveilling every single moment you spend there"—and elsewhere.[75] The digital age has effectively inserted the surveillance capability into our everyday pleasures.

We can no longer afford to ignore this, wittingly or unwittingly. We can no longer disregard the unique ways in which digital technologies are reshaping relations of power in society. With the revelations by Edward Snowden, Wikileaks, and others, the investigative reporting of Glenn Greenwald, Laura Poitras, Ewen MacAskill, and so many others, the critical writings and engagements of Julie Cohen, Kevin Haggerty, Larry Lessig, Eben Moglen, Torin Monahan, Frank Pasquale, Daniel Solove, and more, we have been pushed out of an analog political era and forced to face an altered political space that has radically transformed our social relations. Siva Vaidhyanathan writes in *The Googlization of Everything* that "to search for something on the Web using Google is not unlike confessing your desires to a mysterious power."[76] The aim of this book is to unpack that mysterious power, to lay it bare, to render it, too, transparent, and to decipher how this new form of expository power circulates today. To trace the emergence of this new way of life—of exhibition and watching. To identify and characterize its central features. To excavate the forms of complacency that have allowed it momentum. To grasp the mortification of the self that it is producing. To see the evolving convergence with electronic monitoring and punitive practices. These are the tasks that this study has set for itself. To be more precise:

1. To trace the emergence of a new architecture of power relations throughout society, to excavate its antecedents, to explore what it is constructing. The new expository power circulates by means of a new logic, a new form of rationality. It is a data-driven, algorithmic searchability that constantly seeks to find, for each and every one of us, our perfect match, our twin, our look-alike, in order to determine the next book we want to buy, the person we would like to resemble, the perfect answer to our question, the conspiracy we intend to join. This new form of rationality—this digital *doppelgänger* logic—has a genealogy, with precursors in the actuarial methods of the early twentieth century and in the methods of systems analysis from the Cold War. With its recommender algorithms and tracking technologies, it is a unique logic that is allowing us to construct, for ourselves, the equivalent of a mirrored-glass and steel pavilion—a mesh of wireless digits, with "mirror reflections," "uncanny effects," virtual transparence, and "phenomenal opacity."[77] An encasing that combines both the visibility of the crystal palace and the steel cage of electronic monitoring.
2. To document its effects on our political relations, on our conceptions of self, and on our way of life. Digital technology is

breaking down whatever semblance of boundaries existed—
if they ever existed at all—between the state, the economy,
and private existence, between what could be called governing,
exchanging, and living.[78] The consequences are far-reaching.
By breaking down the boundaries, the aspiration to any liberal
limits, or checks and balances, evaporates as well. The liberal
ideal—that there could be a protected realm of individual autonomy—no longer has traction in a world in which commerce
cannot be distinguished from governing or policing or surveilling or just simply living privately. The elision, moreover, fundamentally reshapes our subjectivity and social order: the massive
collection, recording, data mining, and analysis of practically
every aspect of our ordinary lives begins to undermine our
sense of control over our destiny and self-confidence, our sense
of self. It begins to shape us, at least many of us, into marketized subjects—or rather subject-objects who are nothing
more than watched, tracked, followed, profiled at will, and
who in turn do nothing more than watch and observe others.
In the process, our digital lives eerily begin to resemble the
electronically monitored existence of the carceral subject
under supervision.

3. To explore how to resist and disobey.

But before proceeding a step further, it is crucial first to clear the
ground conceptually—or at least to winnow the metaphors that so
many of us have instinctively turned to. The stream of revelations—
about the collection and mining of our personal information by social
media, retail, and tech companies, about the widespread monitoring
of our digital lives by the NSA, about the cooperation between domestic and foreign signals intelligence agencies and large multinational corporations—have provoked an array of criticisms drawing on
a range of metaphors. Some critics invoke George Orwell's novel *1984*
and draw parallels between our current political condition and the
haunting menace of a Big Brother.[79] Others contest the analogy with
Orwell and turn instead to Franz Kafka, especially to the night-

marish, suffocating, bureaucratic "black box" at the heart of Kafka's novel *The Trial*.[80] Still others have converged on the idea that we are now facing a new surveillance state, while others gravitate to the writings of Jeremy Bentham and Michel Foucault on the panopticon.[81] Some critics draw on all of these concepts at once.[82]

Big Brother, the surveillance state, the panopticon—these are all, to be sure, compelling analogies and frightening metaphors. In certain respects, each works well and sheds some light on our situation. But they also tend to distort our understanding of the unique aspects of digital exposure. They are, in a sense, ready-made, off-the-shelf, and in that respect they have the potential to hide important dimensions of our new situation. What we need is to carefully reexamine each of these concepts, and winnow them, to retain what might be enlightening while discarding what is inapposite.

These metaphors have significant effects on how we interpret, characterize, and eventually respond to the current condition. To give but one illustration: the controversy over whether the NSA or the private telecommunications companies should store our bulk telephony metadata under what is known as the Section 215 program.[83] The metadata collection program, revealed by Edward Snowden in June 2013, triggered a robust public debate, spawned a number of federal court decisions, and resulted in the passage of the USA FREEDOM Act on June 2, 2015.[84] The solution to the privacy problems enacted in the legislation—a solution urged by President Obama's advisers in their report *Liberty and Security in a Changing World*—was to have the private telecommunications companies gather, hold, and store the data themselves, rather than have the NSA do it.[85] This solution only makes sense, however, if the problem we are facing is that of a surveillance state. It does little to address the problem if it involves a larger amalgam of corporate, intelligence, and security interests, or, to use the words of James Bamford, a "surveillance-industrial empire" that includes those very telecommunication companies, as well as social media, retailers, and intelligence services, among others.[86] If the problem goes beyond a surveillance state, this solution—including *paying* the telecoms to hold the data—is nothing more than a fig leaf. It would be, in effect, the unfortunate consequence of a misleading metaphor, the product of not yet fully understanding our current digital condition.

No, we need to reexamine the metaphors closely. Metaphors matter. They shape the way we understand the present. They affect the way we respond. They are powerful devices. And they need to be rethought. For, as it will soon become clear, our new political condition cannot entirely or properly be described through the lenses to which we have become accustomed. We are not just facing an Orwellian Big Brother. Nor are we wrestling primarily with a surveillance state. And Foucault's notion of panopticism also misses important aspects of our contemporary situation. None of those tried-and-true models fit our new condition well. They were born in an analog age and are not entirely adequate to the digital era. It is time, preliminarily, for some rethinking. Let me begin, then, by clearing the conceptual brush of some ambiguities and imprecisions.

PART ONE

Clearing the Ground

ONE

GEORGE ORWELL'S BIG BROTHER

IN THE WAKE OF THE Edward Snowden revelations, interest in George Orwell's novel *1984* soared, leading to an exponential rise in sales. Less than a week after the first leaks were revealed by the *Guardian* in June 2013, the *Los Angeles Times* reported that sales of Orwell's book had increased by nearly 6,000 percent.[1] One edition in particular, which includes a foreword by Thomas Pynchon, increased in Amazon sales by nearly 10,000 percent, becoming "the third hottest book on Amazon," according to its "Movers and Shakers" list.[2]

Editorialists across the globe instantly drew the connection. Alan Rusbridger, the *Guardian*'s editor in chief, announced that the "NSA surveillance goes beyond Orwell's imagination," while the newspaper quickly assembled letters to the editor under the headline "Orwell's Fears Refracted through the NSA's Prism."[3] At the *New York Times*, Nicholas Kristof jumped on Twitter to ask, "I wonder if NSA has to pay a licensing fee for its domestic surveillance program to the George Orwell estate. #1984."[4] Even the *International Business Times* ran a story that read "NSA PRISM: 3 Ways Orwell's '1984' Saw This Coming."[5]

President Obama stoked the fire, alluding to Orwell's novel just two days into the revelations: "In the abstract, you can complain about

Big Brother and how this is a potential program run amok, but when you actually look at the details, then I think we've struck the right balance."[6] Edward Snowden refreshed the allusion in his Christmas 2013 greeting, speaking directly into the TV camera from Moscow, telling anyone who would listen that the NSA surveillance capabilities vastly surpass Orwell's fears. "The types of collection in the book—microphones and video cameras, TVs that watch us—are nothing compared to what we have available today. We have sensors in our pockets that track us everywhere we go," Snowden said. "Think about what this means for the privacy of the average person."[7] Glenn Greenwald also invoked Orwell's *1984,* emphasizing that "the echoes of the world about which he warned in the NSA's surveillance state are unmistakable."[8] And the metaphor continued to persist, with District of Columbia federal district court judge Richard Leon—who had been appointed by President George W. Bush—striking down the bulk telephony metadata program in *Klayman v. Obama,* calling the NSA technology "almost-Orwellian"—technology that, in his words, "enables the Government to store and analyze phone metadata of every telephone user in the United States."[9]

These constant references to *1984* have fueled a rich debate about whether the metaphor is apt—whether Orwell's dystopian vision accurately captures our political condition today, whether it exceeds or minimizes it, whether it is a useful lens with which to analyze the present.

For some, the similarities are practically perfect. Richard Jinman, in the British news magazine *The Week,* highlights five matches: the environment of perpetual war in the fictional Oceania and the United States' own perpetual "war on terror"; the "charismatic leader" of Big Brother and the popular president Barack Obama; the two "anti-heroes," Winston Smith and Edward Snowden; the "beautiful heroines," Julia and Snowden's girlfriend, Lindsay Mills; and "the sinister technology" and "the sinister language" of doublespeak present in government and corporate responses to the Snowden leaks.[10]

Ian Crouch at the *New Yorker* also highlights a number of symmetries between *1984* and the Snowden affair.[11] Crouch notes that Snowden is much like Orwell's Winston, especially had the latter "been a bit more ambitious, and considerably more lucky, and managed to defect from Oceania to its enemy Eastasia and sneak a message to the telescreens back home." Second, the technological capabilities of NSA's surveillance apparatus mirror those in Oceania in *1984*—especially the "safe operating assumption" that most of our daily, non-face-to-face communications are capable of being recorded. This, despite the fact that the "technological possibilities of [the NSA's] surveillance and data collection and storage surely surpass what Orwell imagined," Crouch adds.[12] Third, today's language and words "are manipulated by the three branches of government to make what might seem illegal legal—leading to something of a parallel language that rivals Orwell's Newspeak for its soulless, obfuscated meaning."[13] Crouch adds: "Indeed, there has been a hint of something vaguely Big Brotherian in Obama's response to the public outcry about domestic surveillance, as though, by his calm manner and clear intelligence, the President is asking the people to merely trust his beneficence—which many of us might be inclined to do."[14] Nonetheless, Crouch does point out a key difference with *1984:* "We are far from the crushing, violent, single-party totalitarian regime of Orwell's imagination." That is, our own political regime does not resemble the "boot stamping on a human face—forever" that Orwell described.[15]

Along the latter lines, Robert Colls at Salon.com argues that the world depicted in Orwell's novel differs in key respects from the particularities of the current NSA surveillance programs. For one thing, Colls suggests, the means by which Orwell's dystopian society was monitored were far more visible and more encompassing than the NSA's surveillance apparatus. The figure of Edward Snowden himself—an individual employee, working at a private consulting firm, with access to bulk data—is different from the coercive figure of Big Brother, who listens in every possible way to every possible conversation (though Colls is careful to note that Snowden "leaves no one in doubt that in slightly less serious hands this infinity of tapping could disfigure not only other people's lives but other people's countries too"). But "the decisive difference," Colls asserts, between the reality of

Snowden and Orwell's depiction of Winston Smith "is that Snowden is not really alone." Edward Snowden has forms of public support that were not possible for Winston Smith:

> When the *New York Times, Der Spiegel,* the *Guardian,* the American Civil Liberties Union, and some very influential legal and political opinion, including that of former President Jimmy Carter on the left and libertarian lawyer Bruce Fein on the right, and influential Democrats and Republicans like Patrick Leahy and Jim Sensenbrenner in the middle—when they and head of US Intelligence James Clapper all voice their various and diverse levels of support for a debate started by a man whom the other side calls an out and out traitor, then you know that unlike Winston Smith, Edward Snowden will not be a traitor forever. The 9th December statement by the giant American information companies is just as significant. Apple, Google, Microsoft, Facebook, and the rest are American brands as well as American corporations. Money counts.[16]

Clearly, opinions are divided on the relevance of Orwell's novel. Some emphasize the similarities with our present condition, others the stark differences. In this case, though, it may well be the differences that make the novel so relevant.

WHAT MAKES ORWELL'S NOVEL *1984* such a precious metaphor today is precisely that Orwell was prophetic in so many ways, yet he got the most crucial dimension flat wrong—namely, the role that desire would play in enabling digital exposure today. It is precisely in the jarring contrast between the myriad aspects of the novel that are so prescient and this one thing that is so off—the place of pleasure—that the *1984* metaphor is so telling. What it highlights and brings out is the central operating mechanism of our own dystopian present: our fancies, our predilections, our simplest desires, all recommended, curated, and cultivated in this digital age.

Across a range of dimensions, Orwell was clairvoyant: on the technologies and pervasiveness of surveillance, on the very idea of thoughtcrime, on the form of punishment, especially on *what* and *how* we punish—or at least punished until recently. On everything analog, on the bars and the cinder blocks, on the physical, tangible reality of punishment, Orwell got it. But he misfired at the core of the digital. Orwell got one key feature wrong, marvelously wrong—marvelously, because it sheds so much light on our current condition. In *1984*, the fundamental political strategy of oppression—of Big Brother, of O'Brien, of what the fictitious traitor Emmanuel Goldstein described in his splendid book *The Theory and Practice of Oligarchical Collectivism*—was to *crush* and *eradicate* desire. With their Junior Anti-Sex Leagues that "advocated complete celibacy for both sexes" and their drive to abolish the orgasm, the central tactic was to hollow out the men and women of Oceania, to neutralize—in effect, to neuter—all their drives and passions, to wear them down into submission, with the smell of boiled cabbage and old rag mats, coarse soap, blue uniforms, and blunt razors. To eviscerate all desire and fill them instead with hate—with the "Hate Week" and the "Two Minutes Hates."

Today, by contrast, everything functions by means of "likes," "shares," "favorites," "friending," and "following." The drab blue uniform and grim gray walls in *1984* have been replaced by the iPhone 5C in all its radiant colors—in sharp pink, yellow, blue, and green. "Colorful through and through" is our marketing slogan today, and it is precisely the desire for color-filled objects, for shared photos on Instagram, for vivid emoticons, for more "likes," clicks, and tweets, that make us expose our most intimate desires and deliver ourselves to the surveillance technologies.[17]

It is almost as if someone had learned from Orwell's greatest error: it is far easier to tame men and women by means of their passions—even passions for the simplest things, like real tea, real coffee, and real sugar for Winston and Julia, or for us, lattes, frappes, and free Wi-Fi—than it is by trying to quash desire and lust, by trying to defeat "the Spirit of Man."[18] Orwell's misfire, I would say, is so terribly revealing because it is precisely through our passions, desires, and interests—

nurtured, curated, and fed by Facebook, MySpace, Google, Instagram, Flickr, YouTube, and Snapchat—that we, at least in the United States, have become so transparent and so exposed today to pervasive (one might even say Orwellian) surveillance, not only by the government and the police, but by private corporations, social media, retailers, nonprofit entities, foreign countries, and each other. It is our passions— in tension with our ambivalences, anxieties, and discomforts, to be sure—that feed the expository society.

So, plainly, this is not to deny that there is much that is prescient in the novel. Written in 1949, Orwell's 1984 accurately prefigured a range of deep transformations that would take place in the area of crime and punishment, and in surveillance, over the course of the second half of the twentieth century.

Orwell's novel foreshadowed a fundamental transformation in what we punish: instead of punishing people solely for the commission of acts, we began to punish people—and we punish them today—for *who they are*. As O'Brien exclaimed to Winston, "We are not interested in those stupid crimes that you have committed. The Party is not interested in the overt act: the thought is all we care about" (253).[19] The thought, or more precisely, the person: the habitual offender, the sexual predator—in short, the dangerous individual. The invention of actuarial methods at the dawn of the twentieth century and the steady rise in their use over the past fifty years fundamentally transformed what we punish in the twenty-first century—not just the act, but the character of the offender, who he is, his age, his schooling, his employment record, his marital history, his prior treatments and incarcerations. And while I trace this shift back to the implementation of the actuarial in American criminal justice rather than to discipline or governmentality, I agree fundamentally with Foucault, who would write, twenty-five years after Orwell, that judges have "taken to judging something other than crimes, namely, the 'soul' of the criminal."[20] To be sure, there remains today the requirement of an overt act in order to be convicted of a crime—though that, effectively, was also true in Orwell's

1984. Lord knows that Winston engaged in overt acts. Buying that coral paperweight would surely have been considered a "voluntary act" in both Blackstone's *Commentaries on the Laws of England* and the American Law Institute's Model Penal Code.[21] Yet *what* we punish has changed. Ian Crouch perhaps disagrees, emphasizing that the NSA "is primarily interested in overt acts, of terrorism and its threats, and presumably—or at least hopefully—less so in the thoughts themselves."[22] But he fails to grasp that the mere trigger for culpability may not reflect the object of our punishment practices, and also how easily thought crimes with overt acts, surprisingly even in the terrorism context, can be the product of entrapment.[23]

Orwell's novel foreshadowed as well a fundamental transformation in *how* we punish—or, to be more exact, how we punished several decades after Orwell wrote, though perhaps no longer really today: namely, by means of a form of subjective rewiring intended to completely transform the offender into a different being. "No!" O'Brien exclaimed when Winston said he assumed he was going to be punished. "Shall I tell you why we have brought you here? To cure you! To make you sane! Will you understand, Winston, that no one whom we bring to this place ever leaves our hands uncured? . . . We do not merely destroy our enemies," O'Brien adds, "we change them" (253). That is precisely what prefigured the rehabilitative model of the 1960s and early 1970s—so thoroughly and excruciatingly critiqued by the likes of Thomas Szasz, R. D. Laing, David Cooper, Foucault, and others.[24]

Orwell also captured brilliantly the psychoanalytic transference that would be necessary to effect such a cure. At an important juncture in his treatment, when Winston no longer knows what $2 + 2$ equals, when he is practically sure he is about to be sent to his death, Winston begins to love his psychotherapist-torturer, O'Brien—his "tormentor," "protector," "inquisitor," and "friend" (244). "If he could have moved," Orwell writes, Winston "would have stretched out a hand and laid it on O'Brien's arm. He had never loved him so deeply as at this moment. . . . In some sense that went deeper than friendship, they were intimates" (252). Winston honestly believes that O'Brien feels the same way toward him. Winston has been in O'Brien's care, you will recall. "You are in my keeping," O'Brien explains. "For seven years I have watched over you. Now the turning point has come. I shall save you, I shall make you perfect" (244). This exact psychotherapeutic relationship would shape our punishment practices in the 1960s—ending, somewhat brutally, with the "nothing works" of the 1970s.

What we punish and *how* we punish, or punished—these are some of the remarkable insights Orwell had. Orwell even prefigured Foucault's return to Bentham's panopticon. Recall the picture fixed to the wall of Winston and Julia's little refuge above the antique store. The picture that hid the telescreen. The picture, in Orwell's words, of "an oval building with rectangular windows, and a small tower in front" (97). A picture, indeed, of a panopticon that repeats, in an iron voice, "You are the dead." The telescreen itself is also deeply panoptic: "There was of course no way of knowing whether you were being watched at any given moment," Orwell writes (3). "How often, or on what system, the Thought Police plugged in on any individual wire was guesswork. It was even conceivable that they watched everyday all the time" (3). That is, after all, Foucault's description of the panoptic prison: "The inmate must never know whether he is being looked at at any one moment; but he must be sure that he may always be so."[25] In effect, Winston's apartment was a Foucauldian cell.

And that totalizing surveillance—the brutally omniscient and omnipresent nature of the capabilities—that too was prescient. O'Brien knew *everything* about Winston, even the fact that he had put that speck of white dust on his journal to verify that no one would touch it. "For seven years the Thought Police had watched him like a beetle under a magnifying glass," Orwell writes. "Even the speck of whitish dust on the cover of his diary they had carefully replaced" (276). O'Brien had seen and read everything Winston had written in his journal. He had seen Winston's every bodily movement in his cubicle at work—even the hidden gesture, when he slipped the photo of the three heroes turned traitors down the memory hole (78–79). Orwell imagined "total information awareness" well before we did, and his vivid imagination would come true in today's computer algorithms and programs, like XKeyscore and radio-frequency technology, that can capture and record every single keystroke and mouse click we perform on our computers.[26]

Total awareness accompanied by constant messaging: Winston was surrounded by telescreens—we today by digital billboards, smart TVs, electronic tickertapes, and LED displays streaming videos and announcements from CNN, MSNBC, and Fox News, and all their commentators and Skyped-in blogging heads from across the Internet. Winston was surrounded by posters of Big Brother or Eurasian soldiers—we today by advertisements and commercials trying to sell products, to shape our desires, to seduce us at every turn.[27] Cable news coming at us every moment, 24/7, interrupted every other minute by adverts, product placements in TV shows, a barrage of marketing and endless political propaganda.

And perhaps the most haunting idea connected with punishment: that Big Brother would do everything humanly possible to avoid creating martyrs. What "the German Nazis and the Russian Communists" never understood, O'Brien tells Winston, was precisely that a government must avoid turning those it punishes into martyrs (254). And so, before eliminating, before executing, before disappearing people, the state had to first cure, to first hollow out the individual—that was the only way to make sure that the dead would not "rise up against us" (254). The only way to make sure that "posterity will never hear of you" and that "you will be lifted clean out from the stream of history" (254). It should come as no surprise that, today as well, we sometimes make sure that certain images or photographs are not broadcast, certain events not recorded, certain deaths not publicized. And certain bodies dumped at sea.

YET DESPITE THESE PRESCIENT INSIGHTS, it almost feels as if whoever designed our new surveillance age—as if there were such a person—learned from the error of *1984*. For in contrast to today, everything O'Brien and Big Brother did in Orwell's novel worked by quashing, destroying, liquidating desire and passion. Everything turned on the elimination of human pleasure. We see this well with the Junior

Anti-Sex League and the central effort, as O'Brien explains, to eradicate the sexual instinct: "Procreation will be an annual formality like the renewal of a ration card. We will abolish the orgasm. Our neurologists are at work upon it now" (267). Big Brother also intended to eliminate art and literature: "There will be no art, no literature, no science" (267). All those forms of pleasure would be vaporized. "The whole literature of the past will have been destroyed. Chaucer, Shakespeare, Milton, Byron—they'll exist only in Newspeak versions, not merely changed into something different, but actually changed into something contradictory of what they used to be" (53). Eradication of desire and pleasure is the fundamental strategy of Big Brother: to annihilate *jouissance,* to make it a crime. Sex, "the natural feeling," had to be "driven out of them" (68). Desire had to be transformed into thoughtcrime—or as Winston reflected, "Desire *was* thoughtcrime" (68; emphasis added).

Now, part of this was about eradicating empathy—a task that may be essential for an authoritarian regime. As the philosopher Martha Nussbaum suggests, the effort was aimed in part at "extinguishing compassion and the complex forms of personal love and mourning that are its sources, and replacing them with simple depersonalized forms of hatred, aggression, triumph, and fear."[28] That was certainly a key aspect of *1984.* Winston repeatedly emphasizes the love and friendship of an earlier age: "Tragedy, [Winston] perceived, belonged to the ancient time, to a time when there were still privacy, love, and friendship, and when the members of a family stood by one another without needing to know the reason" (30). A central goal of Big Brother was to annihilate in individuals those feelings of compassion and love. "Never again will you be capable of ordinary human feeling," O'Brien explains. "Never again will you be capable of love, or friendship, or joy of living, or laughter, or curiosity, or courage, or integrity. You will be hollow. We shall squeeze you empty" (256). Empty of love. Empty of friendship.

But also, importantly, empty of *desire,* empty of *jouissance,* empty of *pleasure and attraction.* That is central to the political strategy of Big Brother in *1984:* to eliminate the sense of ecstasy—of all kinds. "Sexual

intercourse was to be looked on as a slightly disgusting minor operation, like having an enema" (65). The Party's "real, undeclared purpose
was to remove all pleasure from the sexual act. Not love so much as
eroticism was the enemy, inside marriage as well as outside it" (65).
And not just because sexual deprivation can be channeled into obedience, hatred, and fear—as some have suggested.[29] But because *all*
desire had to be extracted and replaced with hatred and allegiance.
Big Brother exterminates all desire and then fills the emptiness with
itself. As O'Brien explains to Winston: "We shall squeeze you empty,
and then we shall fill you with ourselves" (256). And so, as they were
hollowing out souls, O'Brien and Big Brother were folding everything
back onto hate, with "Hate Week," "the Hate Song," and the "atrocity
pamphlets" (148).

How radically different our situation is today. We do not sing hate,
we sing praise. We "like," we "share," we "favorite." We "follow." We
"connect." We get "LinkedIn." Ever more options to join and like and
appreciate. Everything today is organized around friending, clicking,
retweeting, and reposting. We share photos on Instagram so that others
will like them, and constantly check to see how many people have—and
we form judgments of ourselves based on how many likes we get. We
make Snapchat "stories" and continually check to see who has viewed
them—and to figure out how to make them more popular. We tweet
to our followers, hoping that they will retweet or "favorite" our post.
We write blogs, hoping they will get trackbacks, be linked to, and become embedded. We are appalled by mean comments—which are
censored if they are too offensive (which is not to diminish the fact
that they often are nasty). But we privilege the positive. Many of us
don't even want people to be able to "dislike"! Recall that there was
tremendous contestation and consternation around the very question
of whether Facebook would allow a "dislike" button on its pages—and
that Facebook ultimately rejected the idea. What we seem to want is
more and more "likes," not "dislikes."

We want to be loved, we want to be popular, we want to be de-
sired—and we want to desire. Those very instincts drive the digital
condition. They are what lead so many of us to give away our personal,

even intimate information so freely and enthusiastically. They are also what have made us virtually transparent to surveillance today. They are how come our information can be collected and analyzed so easily. They are precisely what make it possible for the government and police to stalk the Twitter and Facebook pages of us all—especially of the usual suspects.[30]

. . .

Facebook is keenly aware that its business model depends on users enjoying being on Facebook. It is also acutely aware that people's emotions are contagious: whether someone likes using Facebook may be related to being surrounded by other people who like using Facebook. It is for that reason, not surprisingly, that Facebook has a "like" button but not a "dislike" button. Facebook *wants* people to like being on Facebook—that's good for its business. And like any savvy corporation, Facebook is constantly trying to figure out new and creative ways to make users enjoy Facebook even more.

In January 2012, one of Facebook's scientific researchers conducted an experiment on Facebook users.[31] The idea was to test whether their good mood and positive attitude—and their enjoyment of Facebook—could be affected by the mood of their Facebook friends. The motivation—or rather the worry—was simple: "We were concerned that exposure to friends' negativity might lead people to avoid visiting Facebook."[32]

The Facebook researcher, Adam D. I. Kramer of the Core Data Science Team at Facebook, Inc., in Menlo Park, and two of his colleagues at Cornell University, in Ithaca, New York, ran the experiment on 689,003 Facebook users.[33] (The users, incidentally, did not know they were being experimented on, nor had they given informed consent. Facebook's view was that they had implicitly consented when they originally agreed to Facebook's terms of service. Facebook apparently "argued that its 1.28 billion monthly users gave blanket consent to the company's research as a condition of using the service.")[34]

To test emotional contagion, Kramer and his colleagues manipulated the news feeds that users received by increasing or decreasing the amount of positive or negative emotional content. Based on those manipulations, they then tried to determine whether a user's own messages and posts on Facebook would also change in emotional content. The idea was to test, in the study's words, "whether exposure to emotional content led people to post content that was consistent with the exposure—thereby testing whether exposure to verbal affective expressions leads to similar verbal expressions, a form of emotional contagion."[35]

Kramer and his colleagues measured emotional content by means of linguistic data-mining software that could determine how many positive or negative emotional words were in the messages they were receiving and sending.[36] This served to measure both the input and the output. As the authors explained: "Posts were determined to be positive or negative if they contained at least one positive or negative word, as defined by Linguistic Inquiry and Word Count software (LIWC2007) word counting system, which correlates with self-reported and physiological measures of well-being, and has been used in prior research on emotional expression. LIWC was adapted to run on the Hadoop Map/Reduce system and in the News Feed filtering system."[37]

So what the researchers did, effectively, was to manipulate the sensory input of Facebook users in order to see if that had any effect on their sensory output—in other words, to limit certain kinds of news feed posts, based on their emotional content, and increase others, in order to see the effect on users' emotions. The researchers ran "two parallel experiments," one for positive emotion and one for negative emotion: "one in which exposure to friends' positive emotional content in their News Feed was reduced, and one in which exposure to negative emotional content in their News Feed was reduced."[38] Or, to be more exact, "when a person loaded their News Feed, posts that contained emotional content of the relevant emotional valence, each emotional post had between a 10% and 90% chance (based on their User ID) of being omitted from their News Feed for that specific viewing."[39]

To determine whether there was any contagion effect, the Facebook researcher and his colleagues then examined the posts the subjects made after being exposed to more or less emotional content, to see whether those posts were more or less positive. In technical terms, the dependent variable—the thing that might have been affected by the treatment—was the posts of the Facebook users: "Dependent variables were examined pertaining to emotionality expressed in people's own status updates: the percentage of all words produced by a given person that was either positive or negative during the experimental period."[40] All in all, the researchers analyzed more than 3 million Facebook posts, which contained "over 122 million words, 4 million of which were positive (3.6%) and 1.8 million negative (1.6%)."[41]

What the Facebook researcher and his colleagues found was not entirely surprising but very significant. As they reported in the *Proceedings of the National Academy of Sciences,* under the heading "Significance": "We show, via a massive (N = 689,003) experiment on Facebook, that emotional states can be transferred to others via emotional contagion, leading people to experience the same emotions without their awareness."[42] Or, in slightly more detail:

> The results show emotional contagion. As Fig. 1 illustrates, for people who had positive content reduced in their News Feed, a larger percentage of words in people's status updates were negative and a smaller percentage were positive. When negativity was reduced, the opposite pattern occurred. These results suggest that the emotions expressed by friends, via online social networks, influence our own moods, constituting, to our knowledge, the first experimental evidence for massive-scale emotional contagion via social networks, and providing support for previously contested claims that emotions spread via contagion through a network.[43]

We are happier when we get good news. And we are happier when we are surrounded by people who are getting good news. Happiness is contagious. And this is significant at the scale of Facebook. In the words of the Facebook researcher and his colleagues: "After all, an effect size of $d = 0.001$ at Facebook's scale is not negligible: In early 2013, this would have corresponded to hundreds of thousands of emotion expressions in status updates per day."[44]

With that in mind, it is worth noting that Facebook constantly modifies the algorithms it uses to sort and discriminate between items in news feeds. Facebook is frequently adjusting the models, for instance, to send users more "high-quality content" or more "relevant" news, or, in its own words, "to weed out stories that people frequently tell us are spammy and that they don't want to see."[45] The reality is, of course, that Facebook always screens and filters the news that users receive—for the simple reason that there is way too much news to share. Usually, when a user logs on to his or her Facebook account, there are around 1,500 feeds that typically could be displayed, but room for only about 300.[46] Facebook decides what goes through and what does not. As the Facebook researcher acknowledged and explained:

> Because people's friends frequently produce much more content than one person can view, the News Feed filters posts, stories, and activities undertaken by friends. News Feed is the primary manner by which people see content that friends share. Which content is shown or omitted in the News Feed is determined via a ranking algorithm that Facebook continually develops and tests in the interest of showing viewers the content they will find most relevant and engaging.[47]

Now, the algorithms that Facebook uses are intended to promote user satisfaction, to improve the experience, to create "a more alluring and useful product."[48] Chris Cox, the chief product officer of Facebook, explains, "We're just providing a layer of technology that helps people get what they want. . . . That's the master we serve at the end of the day."[49] Kramer explained, after the research was published: "The goal of all of our research at Facebook is to learn how to provide a better service."[50] I take it that means a service we are going to "like" more. The digital equivalent of the perfect hallucinogen, "soma," from

Aldous Huxley's *Brave New World*—a magical substance, without side effects or hangovers, that is perfectly satisfying. An experience one step closer to that perfect state of happiness in which we consume more, we communicate more, we expose ourselves more.

DURING WINSTON'S INTERROGATION, O'Brien asks him what principle could possibly defeat the Party, and Winston responds, naively, "I don't know. The Spirit of Man" (270). In another passage, Winston refers to "simple undifferentiated desire." That, he explains to Julia, is the only "force that would tear the Party to pieces"—"not merely the love of one person, but the animal instinct, the simple undifferentiated desire" (126). Those desires, those passions, that lust: those are the only things that could bring down Big Brother.

Now, sadly—the novel is, after all, a dystopia—Big Brother wins in the end, in Orwell's account. Nothing brings it down. The "spirit of man" does not defeat the Party. To the contrary. On my reading, in fact, Winston is liquidated right after the close of the book. The metaphor in the last sentence of the penultimate paragraph is no mere literary flourish: "The long-hoped-for bullet was entering his brain" (297). Big Brother would reduce the traitor to complete obedience first, but then kill him. O'Brien had often said so. He was honest in this respect. "How soon will they shoot me?" Winston had asked. " 'It might be a long time,' said O'Brien. 'You are a difficult case. But don't give up hope. Everything is cured sooner or later. In the end we shall shoot you' " (274). Winston would be shot and then forgotten. "You will be lifted clean out from the stream of history. We shall turn you into gas and pour you into the stratosphere. Nothing will remain of you: not a name in a register, not a memory in a living brain. You will be annihilated in the past as well as in the future. You will never have existed" (254).

Political disobedients are indeed disappeared in *1984*. Winston was lobotomized (257) and he would be killed—but only after he had reached the point at which he would obey completely and love Big Brother, which, as we know, only happens in the last sentence of the book, in Winston's last thoughts: "He had won the victory over himself. He loved Big Brother" (298). It is only then that Winston can be eliminated, can be killed, can be wiped from history. It is only then that all traces of Winston Smith can be extinguished. Big Brother does prevail in *1984*.

But somehow, we still believe today—or hope—that the "spirit of man" and "simple undifferentiated desire" are the things that *might have* defeated Big Brother. We tend to think that Winston was right—or at least, there is a lesson we think we can learn from *1984*, namely, that surveillance cannot succeed, that populations cannot be mastered by *eliminating* desire. That desire, passion, lust, and love will save us from oppression.

Unlikely. The truth is, today it is those very passions, shaped and cultivated in the digital age, that render us virtually transparent. This is perhaps the dark side of desire and pleasure. It is far easier to master a population with their passions, through their desires, than against them—an insight that traces back at least to eighteenth-century liberal thought and forward to the more modern writings of Gilles Deleuze, Félix Guattari, and Wilhelm Reich.[51] Our desires and passions can enslave us. Petty desires at that—those are all that Winston and Julia really wanted to satisfy in that basic little apartment above the antique dealer. They just wanted the good taste and smell of real coffee, a little perfume and some rouge, to hold each other, to have sex. How, one might ask, has Starbucks managed to dominate five continents today? With decent coffee and, now, free Wi-Fi—so rare even in the most urbane spaces.

That is primarily how surveillance works today in liberal democracies: through the simplest desires, curated and recommended to us. The desire to play our favorite videogame, Facebook a viral Vine, Instagram a selfie, tweet a conference, look up the weather, check-in online, Skype our lover, instant-message an emoticon. The impulse to quantify ourselves, to monitor all our bodily vitals and variations and changes. And in the process, we expose ourselves. As Siva Vaidhyanathan reminds us, "We are not Google's customers: we are its product. We—our fancies, fetishes, predilections, and preferences—are what

Google sells to advertisers. When we use Google to find out things on the Web, Google uses our Web searches to find out things about us."[52] For the most part, we give ourselves up so willingly and freely today out of desire, lust, friendship, and love.

This is one of the central lessons—a brilliant insight—of Luc Boltanski and Eve Chiapello's *The New Spirit of Capitalism:* the revolutionary spirit of 1968 was tamed, mastered, disciplined, turned back to corporate profit by tapping the creative energy and artistic potential of the young militants, not by restricting them.[53] Engineers today at Google and Apple can wear jeans and tie-dyed T-shirts and flip-flops, can have entirely creative and artistic lives, have a gym to work out in, drink cappuccinos and frappes and lattes on demand, telecommute, play video games at their computers, re-create *Star Wars* in their own cubicle—the fact is, all that creativity, all that energy, all that passion is turned back into and reinvested in the capitalist enterprise to make a profit, and to mold desire into corporate returns.

Julia, when she wants to be happy, puts on makeup, heels, a dress: "She had painted her face.... Her lips were deeply reddened, her cheeks rouged, her nose powdered" (142). She had found perfume. "In this room, I am going to be a woman, not a Party comrade," Julia exclaims (142). Big Brother's mistake in *1984* was to go after *that*—after the simplest desires of the self and the desire for others. That is precisely what we have learned *not* to do today. Instead, we feed people's desires, we fuel and shape them with recommendations and targeted solicitations, and we lose all of our resistance. We give ourselves up, some joyfully, others anxiously, to the watching eyes of Amazon.com, Netflix, and the NSA.

IN THEIR WILDLY imaginative book *Anti-Oedipus,* Gilles Deleuze and Félix Guattari explore these deep regions of desire.[54] They set out to imagine desire in its full potentiality. They dream of desire as a machine—a desiring and desire-producing machine that seeks to lock onto other desiring machines. Desire, they argue, does not seek out an object for its own sake, does not try to fill a void, but seeks a whole assemblage of interrelated sensations, experiences, pleasures, and symbols. We are desiring machines, they argue, that glom on to other machines of desire. "Desire," they tell us, "is a machine, a synthesis of machines, a machinic arrangement—desiring-machines."[55]

The unconscious, Deleuze emphasizes, is not some kind of theater in which Hamlet or Oedipus dominates, but rather a factory.[56] It is productive. It is pure production. It is a desiring machine that wants not simply the mother or the father, but the entire world. It is outward-looking and seeks an assemblage of things, not a simple object.[57] Assemblages of icons and images and experiences. "Flags, nations, armies, banks get a lot of people aroused," they write.[58] These can be all-encompassing and all-defining, and they become coextensive with the social. As Deleuze and Guattari emphasize, *"There is only desire and the social, and nothing else."*[59] It is coextensive with the social, and infused with the sexual. Sexuality, they maintain, is pervasive. "The truth is that sexuality is everywhere: the way a bureaucrat fondles his records, a judge administers justice, a businessman causes money to circulate; the way the bourgeoisie fucks the proletariat; and so on."[60] Even Hitler, they suggest, "got the fascists sexually aroused."[61]

Indeed, these assemblages—libidinal, phantasmal, productive—can serve despicable ends. They can be fascistic, genocidal. "Even the most repressive and the most deadly forms of social reproduction are produced by desire within the organization," Deleuze and Guattari remind us.[62] For this reason, it is desire that we must focus on and explain.[63] These instances of social obscenity demand, they write, "an explanation that will take their desires into account, an explanation formulated in terms of desire: no, the masses were not innocent dupes; at a certain point, under a certain set of conditions, they *wanted* fascism, and it is this perversion of the desire of the masses that needs to be accounted for."[64] Wilhelm Reich was on the right path, but he failed to push the analysis beyond the traditional dualisms of rational and irrational and toward the concepts of desiring machines and the production of desire.[65] The important point is that the key to understanding our commitments—or what Deleuze called our "investments"—is precisely through desire.[66]

Deleuze and Guattari's ideas may seem extreme to many readers, but they somehow capture best the relation of desire to expository power in the digital age: the sense that we are desire-producing machines that seek an entire assemblage of sociality, sexuality, and pleasure, and that we are, today, stuck to these other machines: iPhones, Droids, iPads, the Internet, Facebook, Instagram. They also capture well the idea of our disembodied and abstracted physical selves, and how these are reconstituted through digital surveillance and turned into digital assemblages—as Kevin Haggerty and Richard Ericson chart in their important work on the "surveillant assemblage."[67] These libidinal and phantasmal assemblages attach to digital devices. "Desire is a machine, and the object of desire is another machine connected to it," Deleuze and Guattari wrote.[68] How appropriate today, with everyone glued to their smartphones as they walk down the street, sitting with their laptops in the quad, connected to the Internet. This is the production of desire in a machinelike way, as a process, in a cycle of desire production—in a schizoid frenzy of reproduction.

Inspired by the surrealists and the theater of the absurd, by R. D. Laing and Wilhelm Reich, but surpassing them deliberately, Deleuze and Guattari prefigure digital lust. That frenzied desire for digital updating, constantly rechecking social media to check the number of likes, or shares, or comments. Addictively checking Gmail to see if there is any good news—like a Pavlovian dog, or a mouse nibbling on an electrical wire. Digital stimulation taps into our lateral hypothalamus, triggering the kind of addictive "seeking" that absorbs us entirely, annihilates time, and sends us into a constant, frenzied search for more digital satisfaction.[69] "The desiring-machines pound away and throb in the depths of the unconscious," Deleuze and Guattari write.[70] These devices keep "pinging" us, "flagging" us, stimulating our cortex, poking at our reflexes. The digital space itself is precisely the machine that we are connected to: it is full of life and energy, of color and movement, of stimulation and production. Deleuze and Guattari spoke of machines in a metaphorical way. By happenstance, perhaps, they fixated on this idea of desire as a bodiless or disembodied automaton generating wants and drives that are not directed at any particular objects but are churning forward—like a machine: "Desire does not lack anything; it does not lack its object. It is, rather, the *subject* that is missing in desire, or desire that lacks a fixed subject; there is no fixed subject unless there is repression. Desire and its object are one and the same thing: the machine, as a machine of a machine."[71]

In his preface to *Anti-Oedipus,* Michel Foucault suggested that these radically new ideas form an "anti-oedipal . . . life style, a way of thinking and living."[72] Foucault identified, in this work, a liberatory ethic. He in fact described the book as "a book of ethics, the first book of ethics to be written in France in quite a long time."[73] A book of ethics that could help us in our tactical, strategic engagements against fascisms of all kinds, "the fascism in us all, in our heads and in our everyday behavior, the fascism that causes us to love power, to desire the very thing that dominates and exploits us."[74] To Foucault, *Anti-Oedipus* could teach us: "How does one keep from being fascist, even (especially) when one believes oneself to be a revolutionary militant? How do we rid our speech and our acts, our hearts and our pleasures, of fascism? How do we ferret out the fascism that is ingrained in our behavior?"[75] Foucault retitled *Anti-Oedipus* an *"Introduction to the Non-Fascist Life"* and drew from it maxims and guides for everyday life—the last of which was "Do not become enamored of power."[76]

But how much have we learned from this book of ethics, from this tract against fascism? Has it served to liberate us—or rather, does it merely describe our newfound dependencies on these digital desiring machines? A telling description, perhaps. More than others, Deleuze and Guattari plumbed the depth of capitalism as desire—of capitalism as this "process of production [that] produces an awesome schizophrenic accumulation of energy or charge."[77] Consumerism, they suggested, is what "liberates the flows of desire."[78]

More a series of questions and wild hypotheses than a manual to liberate us and free us from the drive to fascism, *Anti-Oedipus* pushes us to rethink our current situation. "The question posed by desire is not 'What does it mean?' but rather *'How does it work?'* How do these machines, these desiring-machines, work—yours and mine? With what sort of breakdowns as a part of their functioning? How do they pass from one body to another? ... What occurs when their mode of operation confronts the social machines? ... What are the connections, what are the disjunctions, the conjunctions, what use is made of the syntheses?"[79]

Perhaps Deleuze and Guattari are right. We have become desiring machines, and we lock onto other desiring machines. Those other machines, we know them all too well today. We are glued to them. Inseparable. And we give ourselves up to them—in the process, giving ourselves away. "Everything revolves around desiring-machines and the production of desire."[80] We take so much pleasure playing with our videos, texting, and Facebooking that we simply do not resist the surveillance. We let our guard down. We care less—we don't read the terms of service, we don't clean out our cookies, we don't sign out of Google. We just *want,* we just *need* to be online, to download that app, to have access to our email, to take a selfie.

The technologies that end up facilitating surveillance are the very technologies that we crave. We desire those digital spaces, those virtual experiences, all those electronic gadgets—and we have become, slowly but surely, enslaved to them. To them, and to our desires, desires for passionate love, for politics, for friends. The uniformity of Winston's blue overalls, the smell of boiled cabbage, the blunt razors, the rough soap, the houses that are rotting away, the disgusting canteen with its "low-ceilinged, crowded room, with walls grimy from the contact of innumerable bodies; battered metal tables, and chairs ... bent spoons, dented trays, coarse white mugs ... and a sourish, composite smell of bad gin and bad coffee and metallic stew and dirty clothes" (59)— these are the things we have avoided today, and by avoiding them we have made the world so much more palatable. Palatable like the colors of the iPhone, the liveliness of IM, the seductiveness of the new Apple Watch, and the messaging that surrounds us daily.

No, we do not live in a drab Orwellian world. We live in a beautiful, colorful, stimulating, digital world that is online, plugged in, wired, and Wi-Fi enabled. A rich, bright, vibrant world full of passion and *jouissance*—and by means of which we reveal ourselves and make ourselves virtually transparent to surveillance. In the end, Orwell's novel is indeed prescient in many ways, but jarringly off on this one key point. So much so, in fact, that one could almost say that whoever it is who has conspired to create our digital world today—if one were so naive as to believe there was such a conspiracy—has surely learned the hard lesson of Orwell's brilliant dystopia. We live in a world today that has rectified Big Brother's error. And, sadly, we no longer even have the illusory hope that Winston once had—hope in the proles.

TWO

THE SURVEILLANCE STATE

ANOTHER DOMINANT WAY to analyze our new political condition today is through the metaphor of the "surveillance state." Even before the Snowden revelations, some astute observers such as the legal scholar Jack Balkin identified an emerging "National Surveillance State"—in Balkin's words in 2008, "The question is not whether we will have a surveillance state in the years to come, but what sort of surveillance state we will have"—and many others rallied around the term in the wake of the NSA leaks.[1] Bill Keller, former *New York Times* editor, picked up on the trend in an influential editorial titled "Living with the Surveillance State."[2] The term has today become omnipresent across the political spectrum. Some commentators, such as Julian Sanchez at the *Daily Beast,* argue vehemently that we need to "dismantle the surveillance state."[3] Others, including the American Civil Liberties Union (ACLU), maintain that we need to "rein in the surveillance state."[4] Still others, like Thomas Friedman at the *New York Times,* echoing Keller, take the position that we simply need to adjust "to life in a surveillance state."[5] Regardless of the political position, though, there seems to be a consensus now that we are indeed *facing* the surveillance state, as Conor Friedersdorf remarks in the *Atlantic.*[6] Even Glenn Greenwald, in the subtitle of his 2014 book on Edward Snowden, deploys the term: *No Place to Hide: Edward Snowden, the NSA, and the U.S. Surveillance State.*

The metaphor has reached into the legislative and constitutional debate as well, especially in the controversy surrounding the constitutionality of the NSA's bulk telephony metadata program. In the related judicial opinions and commission reports, the new "surveillance state" is being variously described as a great protector and selfless warrior by one federal judge in New York, as Big Brother by another federal judge in Washington, D.C., and as a New Deal–like administrative savior by President Barack Obama's advisers.[7]

At one end, then, the American surveillance state is portrayed as a selfless warrior who, in response to 9/11, threw "the Government's counter-punch" to al-Qaeda's "bold jujitsu."[8] This is the metaphor of the Leviathan state—a knight in armor, faithful and selfless, honest and effective. It is the protector state, a view espoused by federal judge William H. Pauley III. Pauley, a 1998 Clinton appointee sitting on the United States District Court for the Southern District of New York—only a few blocks from the World Trade Center—sees the entire constitutional controversy surrounding the bulk telephony metadata program through the lens of the surveillance state at war against a nimble, seventh-century jujitsu fighter. The very first image in his ruling—in fact, the very first five words of his opinion—invokes "the September 11th terrorist attack."[9] The first emotion is fear ("just how dangerous and interconnected the world is"), and the first image is al-Qaeda's martial arts attack ("While Americans depended on technology for the conveniences of modernity, al-Qaeda plotted in a seventh-century milieu to use that technology against us").[10]

In Pauley's view, surveillance technology is the key weapon of the protector state—and an effective weapon at that. "The effectiveness of bulk telephony metadata collection cannot be seriously disputed," Pauley contends, before describing in some detail three concrete instances where the program contributed to thwarting terrorist attacks.[11] "Armed with all the metadata," Pauley writes, "the NSA can draw connections it might otherwise never be able to find."[12] It is the state's "counter-punch": "connecting fragmented and fleeting communications to re-construct and eliminate al-Qaeda's terror network."[13]

For Pauley, the protector state reaches its apex as warrior. Its "power reaches its zenith when wielded to protect national security," when confronting "a new enemy," when fighting "the war on terror."[14] It is here, in battle, that the surveillance state carries out its most "urgent objective of the highest order."[15] The people, by contrast, appear fickle: they complain about the surveillance state, and yet, naively—or stupidly—they willingly surrender all their information to greedy private entrepreneurs who simply want to make money. "Every day, people voluntarily surrender personal and seemingly-private information to transnational corporations, which exploit that data for profit. Few think twice about it, even though it is far more intrusive than bulk telephony metadata collection."[16] The contrast could not be greater: to protect the fickle population, we desperately need an active surveillance state.

At the other end of the political spectrum, our new surveillance state is portrayed as lying, deceiving, and cheating—a surveillance state that misrepresents, inflates, embellishes, and threatens our individual freedoms. This is the view of federal judge Richard Leon, nominated to the United States District Court for the District of Columbia one day before 9/11, on September 10, 2001, by President George W. Bush. His is the more skeptical, conservative libertarian view: the surveillance state encroaches on our liberties, constantly developing new ways to get its hands into our most intimate relations. This is Big Brother, the spiderweb state. In fact, Leon refers explicitly to "the almost-Orwellian technology" of the government, which he describes as "the stuff of science fiction."[17] Elsewhere, Leon quotes another judge who refers to the "wide-ranging 'DNA dragnets' that raise justifiable citations to George Orwell."[18]

This Orwellian surveillance state deceives. It "engaged in 'systematic noncompliance' with FISC [Foreign Intelligence Surveillance Court]–ordered minimization procedures" and "repeatedly made misrepresentations and inaccurate statements about the program to the FISC judges."[19] This surveillance state has "misrepresented the scope of its targeting of certain internet communications."[20] This state has even, in a "third instance," made a "substantial misrepresentation."[21] This is the "Government [that] wants it both ways."[22] This surveillance state cannot be believed. It is not credible. It tells the people things, even under oath, that cannot be trusted. It is the state whose "candor" or lack thereof "defies common sense and does not exactly inspire confidence!"[23] "I cannot believe the Government," Richard Leon writes in the margin of his opinion.[24]

The contrast could not be sharper. To Pauley, the surveillance state does not misrepresent, it does not lie, it does not deceive. It is our selfless guardian. At most, at the extreme, it may exhibit some honest human error, but in good faith. Pauley refers to "issues," "incidents," "inconsistencies," and "errors," but never to misdeeds.[25] "While there have been unintentional violations of guidelines, those appear to stem from human error and the incredibly complex computer programs that support this vital tool. And once detected, those violations were self-reported and stopped," Pauley emphasizes.[26] The protector surveillance state follows its own rules and protects the people's anonymity. "The Government," Pauley underscores, "does not know who *any* of the telephone numbers belong to," and it does not seek to know.[27] There are layers of protection and rules that protect the people from a malevolent state. On Pauley's view, the surveillance state is, in effect, rule bound.

And then, diagonal to both Pauley and Leon, there is a third image of the surveillance state as a New Deal–like regulatory state, the administrative state. This is the view projected by President Obama's advisers, a committee composed of an eclectic range of former officials and academics, including a former counterterrorism czar under both President George W. Bush and President Clinton, Obama's former deputy director of the CIA, a First Amendment scholar, the former head of the Office of Information and Regulatory Affairs, and a former Clinton administrator who is an expert on privacy law. In their report, they project the image of the surveillance state as FDR's dream of a big government with a large bureaucracy that solves problems by creating new offices and boards, new agencies, new rules, and new regulations. This is the regulatory and self-regulating surveillance state—the administrative, cost-benefit, technocratic surveillance state.

This third image of the surveillance state is focused on "risk management" for risks of every type: "risks to national security," but also "risks to trade and commerce, including international commerce," "risks to our relationships with other nations," "risks to privacy," and "risks to freedom and civil liberties."[28] We live surrounded by risk. And we have to manage those risks by conducting "careful analysis of consequences, including both benefits and costs."[29] It is of great import that the state not rely "on intuitions and anecdotes, but on evidence and data."[30] The key is risk assessment "subject to continuing scrutiny, including retrospective analysis."[31] "We recommend," Obama's advisers write, "the use of cost-benefit analysis and risk-management approaches, both prospective and retrospective, to orient judgments about personnel security and network security measures."[32]

Through risk management with cost-benefit analysis, this surveillance state can achieve great heights. It has tremendous promise. It shows remarkable progress: "The United States has made great progress over time in its protection of 'the Blessings of Liberty'—even in times of crisis. The major restrictions of civil liberties that have blackened our past would be unthinkable today."[33] These words reflect the resounding optimism of this New Deal–like state. Increase the regulatory oversight. Create a "privacy and civil liberties policy official."[34] Add a "civil servant tasked with privacy issues"—a "policy official, who would sit within both the NSS [National Security Staff] and the OMB [Office of Management and Budget], to coordinate US Government policy on privacy, including issues within the Intelligence Community."[35] Reinforce the "Privacy and Civil Liberties Oversight Board."[36] Throw in another official to present privacy interests at the Foreign Intelligence Surveillance Court: "We recommend that Congress should create a Public Interest Advocate."[37] Improve and better "manage an interagency process."[38] More government regulators. Better processes. More review. Enhanced cost-benefit. This is the administrative state. A surveillance state that "publicly disclose(s)," promotes "transparency and accountability," and fosters "public trust."[39] A state governed by civilians, not military men.[40] A more open state, not cloaked in military secrecy. One that engages in "careful deliberation at high levels of government" and "due consideration of and respect for the strong presumption of transparency that is central to democratic governance."[41]

As for President Obama, he waded into the debate carefully, tipping his hat in all three directions. The president borrowed Judge Pauley's rhetoric of effectiveness: "These efforts have prevented multiple attacks and saved innocent lives—not just here in the United States, but around the globe."[42] Like Pauley, Obama imagines a protective surveillance state—in fact, one that dates back to "the dawn of our Republic," when "a small, secret surveillance committee," which included Paul Revere, "would patrol the streets, reporting back any signs that the British were preparing raids against America's early Patriots."[43] That's Obama's first image—not 9/11 but the American Revolutionary War, followed by Union balloons during the Civil War, code breakers in World War II, and the creation of the NSA during the Cold War to better watch the Soviet bloc. Like Pauley's surveillance state, Obama's is one that does not lie. To the contrary, "the men and women of the intelligence community, including the NSA, consistently follow protocols designed to protect the privacy of ordinary people. They're not abusing authorities in order to listen to your private phone calls or read your emails. When mistakes are made—which is inevitable in any large and complicated human enterprise—they correct those mistakes."[44] Obama's surveillance state is one that would be able to protect us against a threat like 9/11. For this, Obama borrows from Pauley's discussion of the interception of Khalid al-Mindhar's phone call from San Diego—which, according to Pauley, could have prevented 9/11 if it had been geolocated.[45]

Nodding in Judge Leon's direction, the president also acknowledges some reservations, in fact some "healthy skepticism," regarding the surveillance state.[46] The president makes repeated references to excesses from times past, including the notorious spying on Martin Luther King Jr. There is a "risk of government overreach"—but of course, that was mostly a problem associated with the prior George W. Bush administration and its use of torture.[47]

Then, borrowing from the New Deal–like vision of the regulatory surveillance state, President Obama tells the American people that he has already "increased oversight and auditing" and implemented "new structures aimed at compliance. Improved rules were proposed by the government and approved by the Foreign Intelligence Surveillance

Court," he claims.[48] "I created an outside Review Group on Intelligence and Communications Technologies to make recommendations for reform. I consulted with the Privacy and Civil Liberties Oversight Board, created by Congress."[49] Going forward, Obama will order the director of national intelligence to "annually review" and "report to me and to Congress," and he will ask Congress to establish "a panel of advocates from outside government to provide an independent voice in significant cases before the Foreign Intelligence Surveillance Court."[50] Obama declares, "I am making some important changes to how our government is organized":

> The State Department will designate a senior officer to coordinate our diplomacy on issues related to technology and signals intelligence. We will appoint a senior official at the White House to implement the new privacy safeguards that I have announced today. I will devote the resources to centralize and improve the process we use to handle foreign requests for legal assistance, keeping our high standards for privacy while helping foreign partners fight crime and terrorism.
>
> I have also asked my counselor, John Podesta, to lead a comprehensive review of big data and privacy. And this group will consist of government officials who, along with the President's Council of Advisors on Science and Technology, will reach out to privacy experts, technologists and business leaders, and look how the challenges inherent in big data are being confronted by both the public and private sectors; whether we can forge international norms on how to manage this data; and how we can continue to promote the free flow of information in ways that are consistent with both privacy and security.[51]

This is the well-regulated surveillance state in all its New Deal glory: new agencies, new officials, more oversight, more expertise. It is by no means the surveillance state of stealth or deception. It is nothing like Leon's view. Recall that for Leon, the government has its hands on the data and is getting information about individuals. The searching is not just algorithmic. It is not automatic. The data are queried, Leon emphasizes, "*manually* or automatically"—not simply anonymously.[52] There is a real risk of overreach and prying in Leon's view. And it is not only well regulated, this New Deal–like surveillance state, it is also effective—again, in direct contrast to Leon's image of the state. Recall that, on the "efficacy prong," Leon reminds us, "the Government does *not* cite a single instance in which analysis of the NSA's bulk metadata collection actually stopped an imminent attack, or otherwise aided the Government in achieving any objective that was time-sensitive in nature."[53] The contrast between President Obama and Judge Leon is stark.

But, of course, no one today is prepared to dismantle the surveillance state or take responsibility for terminating the metadata collection program.[54] Even Judge Leon remains extremely sensitive to the need for a surveillance state—which is why, in the end, he stays the injunction in the case before him. This is security-conscious libertarianism, not radicalism in any sense. Leon stays the injunction; Pauley upholds the program; the Second Circuit reverses Pauley and finds the program illegal, but refuses to issue a preliminary injunction; Congress reauthorizes the program in the USA FREEDOM Act and requires that the telecoms hold the data; and President Obama restructures the Section 215 program accordingly—while he appoints more administrators, delegates more review and analysis, implements new and better procedures, establishes more layers of administration. On one thing, everyone agrees: the country needs a "robust foreign intelligence collection capability" to keep us "safe and secure" and to promote "prosperity, security, and openness," but also, even more importantly, to "increase confidence in the US information technology sector, thus contributing to US economic growth."[55] America needs its "surveillance state."

THE "SURVEILLANCE STATE": guardian, trickster, New Deal–like administrator—or faithful servant? Which version is more accurate? And what exactly shall we do? Shall we retain it, dismantle it, learn to live with it, or just face it? These questions are puzzling and difficult, all the more so because the metaphor itself, although obviously correct in some ways, is also inadequate and inapposite in others. Before being led astray by the metaphor, then, it would be wise to first interrogate closely the very notion of a "surveillance state." Because, it turns out, the supposed "surveillance state" that we face today is neither, really: neither just a state, nor one that employs, at its core or exclusively, surveillance. A little history may shed some light.

During the ancien régime, an eccentric French inventor by the name of Jacques-François Guillauté drew an early blueprint for a perfect surveillance state: divide the city of Paris into twenty-four equal-sized neighborhoods, then subdivide it into districts of twenty houses, each under the supervision of a syndic; minutely alphanumerize and label each building, entry hall, floor, stairwell, and door in order to most efficiently collect and organize the constant flow of information coming from police agents, spies, and informants—*les mouches et les mouchards;* and place each district under the supervision of a police watchman at a console (see Figure 2.1).[56]

Then give each district watchman instantaneous access to the infinite reams of surveillance information by means of a remarkable paper-filing machine with gigantic wheels—twelve feet in diameter, thirty-six feet in circumference—that would rotate the data at a tap of each watchman's foot (see Figure 2.3).

Guillauté estimated that each paper-filing wheel—which he called a *serre-papier,* a "paper-squeeze"—could efficiently organize 102,400 individual pieces of paper.[57] He pitched his invention to Louis XV in a splendid report, published in 1749, with twenty-eight gorgeous drawings *"à la plume"* by Gabriel de Saint-Aubin, titled *Mémoire sur la réformation de la police de France.*

This was a faultless system to perfect omniscient state surveillance. A dream of infinite knowledge, of total awareness, of instantaneous access, of perfect data control. A policeman's fantasy. And at the time, not surprisingly, it was and remained pure fantasy. Louis XV could not be bothered with such fantasies. He had a kingdom to run.

Big Data would change all that. Digital technology would make an eighteenth-century dream come true, with data-mining programs like

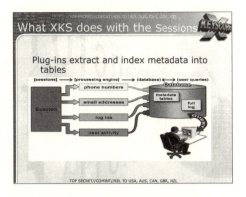

FIGURE 2.1 Jacques-François Guillauté's blueprint for data retrieval (1749). *Source: Mémoire sur la réformation de la police de France, soumis au roi en 1749 par M. Guillauté, illustré de 28 dessins de Gabriel de Saint-Aubin,* ed. Jean Seznec (Paris: Hermann, 1974), 22.

FIGURE 2.2 Top-secret NSA PowerPoint slide for XKeyscore training (2013). *Source:* "XKeyscore: NSA Tool Collects 'Nearly Everything a User Does on the Internet,'" *Guardian,* July 31, 2013.

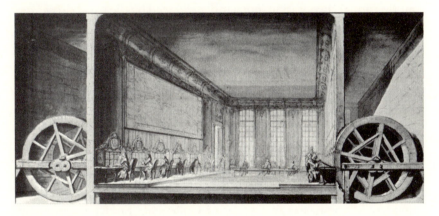

FIGURE 2.3 Jacques-François Guillauté's blueprint for police surveillance (1749). *Source: Mémoire sur la réformation de la police de France, soumis au roi en 1749 par M. Guillauté, illustré de 28 dessins de Gabriel de Saint-Aubin,* ed. Jean Seznec (Paris: Hermann, 1974), 22.

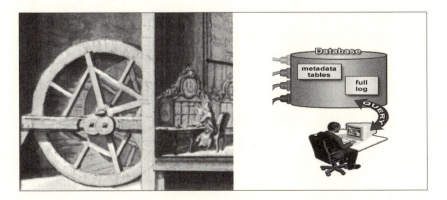

FIGURE 2.4 The watchmen at their consoles (1749 vs. 2013). Composite of Figure 2.2 and Figure 2.3.

XKeyscore, which can instantaneously give an analyst access to "nearly everything a typical user does on the internet."[58] This was illustrated as well, this time in a classified NSA training slide leaked by Edward Snowden in June 2013 (see Figure 2.2).

You may notice an odd, almost eerie similarity. More than 250 years have passed, technology has progressed by leaps and bounds, and yet the little watchmen at their consoles still look awfully alike (see Figure 2.4).

The contemporary resemblance to the "surveillance state" is surprising indeed. But there are certain differences, practically imperceptible at first glance, that are central to understanding our current situation. Because the little watchman on the right, it turns out, may be working for a private consulting company, not the police; querying data provided willingly by enthusiastic users of social media, not the scribbled jottings of a gumshoe; using software manufactured by a multinational corporation, not a tool produced by the royal carpenter; sitting in a location operated by a private telecommunications company, not the *préfecture de police*.

The image on the left, surely, represents *l'État*, but to employ that term under our present circumstances to the watchman on the right seems highly anachronistic—a discourse from an earlier era of modern political theory, with sharply different notions of sovereignty—that may have profoundly misleading effects today. Not just because we have entered an age of neoliberalism that is characterized by privatization, "deregulation," and outsourcing.[59] Nor because, as Pierre Bourdieu would emphasize in a series of Collège de France lectures, the modern "state" is nothing more than a reservoir of symbolic power, a privileged point of view that metes out physical and symbolic violence.[60] Beyond all that, what has changed is the form of surveillance or management or governmentality—or, since the terms themselves are melding together, let's say that the relations of power today have dramatically shifted and circulate in importantly different ways. The form of digital power made possible by our exposure, by data mining, and by modern algorithms undermines the coherence of the notion of the "state," let alone the notion of a "surveillance state."

The fact is, we are no longer simply talking about "the state" when the NSA installs, through its UPSTREAM program, network-tap hardware developed by a Boeing subsidiary, Narus, in a room at the AT&T Folsom Street facility in San Francisco, to splice and make a perfect duplicate of a cable stream of digital communications flowing from the private servers of Facebook, Google, YouTube, Twitter, Myspace, AOL, Hotmail, and so on (most of which are providing data as well under the PRISM program), and has consultants at private firms like Booz Allen monitor the communications using off-the-shelf Silicon Valley or other software, such as the Semantic Traffic Analyzer STA 6400, and then augments the database by means of cooperative agreements, under the LUSTRE program, with France, Germany, the United Kingdom, and other foreign countries.[61]

No, that is not just "the state." When Microsoft collaborates with national intelligence during the fabrication of its new Outlook email software in order to facilitate NSA's decryption abilities; when Microsoft works hand in hand with the FBI to allow access to its new cloud storage service, SkyDrive, used by over 250 million people across the globe; when our network service providers, search engines, and social media monitor our every digital action to recommend products, fuel our consumption, and share our personal data; when employers can access and freely inspect personal emails and the online activities of their employees; when the watchmen at the console can so easily, "out of curiosity," spy on their lovers; when hackers so seamlessly infiltrate the financial systems of large retailers like Target and Neiman Marcus; when so much of Snowden's NSA archive reveals, in Greenwald's words, "what can only be called economic espionage"; when American economic interests are so closely tied to maintaining the country's dominance in communications and the Internet; when the NSA itself is placing undercover agents within private foreign corporations (which themselves may or may not be cooperating with the NSA), then we know we are facing a situation that goes well beyond the modern notion of "the state."[62]

This amalgam of the intelligence community, retailers, Silicon Valley, military interests, social media, the Inner Beltway, multinational corporations, midtown Manhattan, and Wall Street forms an oligarchic concentration that defies any such reductionism. President Dwight Eisenhower cautioned us about the "military-industrial complex" several decades ago.[63] James Bamford warned us of the "surveillance-industrial empire" last decade.[64] But the participants in this new *oligarkhía* are now far too diverse to conjugate into a single neologism. All we can say is that they are *olígos* and appear to rule together, as evidenced by the upward trend in cooperation—reflected in that classified NSA slide disclosed by Snowden—between the NSA and Microsoft (with PRISM collection beginning on September 11, 2007), Yahoo (March 12, 2008), Google (January 14, 2009), Facebook (June 3, 2009), PalTalk (December 7, 2009), YouTube (September 24, 2010), Skype (February 6, 2011), AOL (March 3, 2011), and Apple (October 2012).[65]

. . .

Glenn Greenwald, Laura Poitras, and colleagues at the *Guardian* broke the story in July 2013, a month after the Edward Snowden revelations.[66] Combing through the top-secret NSA documents, Greenwald discovered evidence of close collaboration between Silicon Valley and the intelligence services. The specific documents originated from NSA's elite Special Source Operations (SSO) division, which focuses on domestic communications systems.

The SSO documents reveal that the NSA was particularly worried about its ability to intercept communications on Microsoft's Outlook portal as soon as Microsoft started testing it in July 2012. "Within five months," Greenwald reveals from his review of the documents, "Microsoft and the FBI had come up with a solution that allowed the NSA to circumvent encryption on Outlook.com chats."[67] The NSA document in question, dated December 26, 2012, reads:

> On 31 July, Microsoft (MS) began encrypting web-based chat with the introduction of the new outlook.com service. This new Secure Socket Layer (SSL) encryption effectively cut off collec-

tion of the new service for . . . the Intelligence Community (IC). MS, working with the FBI, developed a surveillance capability to deal with the new SSL. These solutions were successfully tested and went live 12 Dec 2012. . . . The MS legacy collection system will remain in place to collect voice/video and file transfers. . . . An increase in collection volume as a result of this solution has already been noted by CES.[68]

Greenwald, Poitras, and their colleagues additionally report that "another newsletter entry stated that NSA already had pre-encryption access to Outlook email. 'For Prism collection against Hotmail, Live, and Outlook.com emails will be unaffected because Prism collects this data prior to encryption.'"[69]

Another NSA document from April 2013, titled "Microsoft Skydrive Collection Now Part of PRISM Standard Stored Communications Collection," reveals that Microsoft worked hand in hand with the FBI to give the government agency access to the software giant's new cloud storage service, Skydrive. The NSA document states, "Beginning on 7 March 2013, PRISM now collects Microsoft Skydrive data as part of PRISM's standard Stored Communications collection package. . . . This new capability will result in a much more complete and timely collection response. . . . This success is the result of the FBI working for many months with Microsoft to get this tasking and collection solution established."[70] Microsoft's SkyDrive, apparently, is used by more than 250 million people worldwide.[71]

None of this should come entirely as a surprise. There is a long history of cooperation between tech companies—especially in communications and information delivery—and intelligence services in this country and abroad. Episodes of cooperation stretch back for decades. In the 1920s, for instance, the director of U.S. military intelligence, Herbert Yardley, set up what became known as "the Black Chamber," where, thanks to a handshake agreement with Western Union, Yardley got access to all telegrams passing over the company's wires.[72] Before that, during World War I, Western Union was required by law to share all cables with the military, but with the end of wartime censorship, the surveillance of citizens could only be done illegally and in direct violation of the Radio Communications Act of 1912, which guaranteed the privacy of telegraph messages. Faced with that prohibition, Yardley entered into handshake agreements with the presidents of Western Union and All American Cable Company to get complete cooperation and a copy of all their cables.[73] After 1927, when things got dicier, the intelligence unit turned to bribery to obtain communications—which ultimately ended in 1929 with the election of President Herbert Hoover.[74] But the history of cooperation and partnerships would stretch forward from the Black Chamber in 1919 to Operation Shamrock during the post-World War II period, into the mid-1970s, and then to the warrantless eavesdropping post-9/11.[75] And these kinds of cooperation have not only occurred in the United States. In the United Kingdom, a "Secret Department of the Post Office" searched designated "mail of the king's subjects for centuries" before being exposed in 1844.[76]

Cooperation between telecommunication companies and intelligence agencies has been facilitated by a series of revolving doors and lucrative relationships between executives of both sectors. There is, in fact, a long history of board swapping between top intelligence officials, the corporate heads of the telecommunications companies, and the CEOs of related industry leaders. James Bamford's investigative reporting, especially in his book *The Shadow Factory*, provides a long list of collaboration:

- In September 2004, Narus, the Boeing subsidiary that has developed so much of the data-mining software and hardware, appointed William P. Crowell to its board. Crowell was the former deputy director of the NSA.[77]
- Former NSA director Mike McConnell left the NSA in February 1996 and became a vice president at Booz Allen with a $2 million salary.[78] (Booz Allen serves as a consultant to the NSA; in fact, that is where Edward Snowden worked.)

- SPARTA, a private security firm in the surveillance consulting industry, "hired Maureen Baginski, the NSA's powerful signals intelligence director, in October 2006 as president of its National Security Systems Sector." Another firm, Applied Signal Technology, "put John P. Devine, the [NSA's] former deputy director for technology and systems, on its board of directors."[79]
- Former NSA director Adm. William Studeman was hired to manage intelligence programs at TRW Inc., which was then purchased by Grumman.[80] Upon his retirement, Admiral Studeman described his responsibilities as follows: "Retired in 2005, from Northrop Grumman Corporation (NGC) as Vice President and Deputy General Manager of Mission Systems (NGMS) (now consolidated into NGIS), a $5.2B annual revenue Sector (formerly part of TRW) focused on System Integration/System Engineering of large complex systems. In this position, he focused on strategies, programs, business development, marketing related to Intelligence and Information Warfare, as well as corporate cross-Sector integration, and on managing technology partnerships and concepts related to Net Centricity/Cyber matters, ISR, IO/IW and advanced command environments. He served in this position for approximately 9 years, and has continued as an NGC consultant."[81] Admiral Studeman was also elected to a number of boards, including the board of Paracel Inc., "a provider of high-performance genetic and text analysis systems."[82]
- Another major firm, Cylink, which develops encryption products, hired as a vice president former NSA director Mike McConnell's former deputy director, William P. Crowell. As Bamford explains, "Crowell had been through the revolving door before, going from a senior executive position at the NSA to a vice presidency at Atlantic Aerospace Electronics Corporation, an agency contractor, and back to the NSA as chief of staff. Another deputy director of the agency, Charles R. Lord, left the NSA and immediately became a vice president at E Systems, one of the NSA's biggest contractors."[83]

Bamford goes on to note:

> With about forty-four thousand employees each, [defense con-
> tractor] SAIC and NSA are both heavyweights, and they have a
> decidedly incestuous relationship. After first installing the former
> NSA director Bobby Inman on its board, SAIC then hired top
> agency official William B. Black Jr. as a vice president following
> his retirement in 1997. Then [NSA director] Mike Hayden hired
> him back to be the agency's deputy director in 2000. Two years
> later SAIC won the $280 million Trailblazer contract to help de-
> velop the agency's next-generation eavesdropping architecture,
> which Black managed. Another official spinning back and forth
> between the company and the agency was Samuel S. Visner.
> From 1997 to 2001 he was SAIC's vice president for corporate de-
> velopment. He then moved to Fort Meade as the NSA's chief of
> signals intelligence programs and two years later returned as a
> senior vice president and director of SAIC's strategic planning
> and business development within the company's intelligence
> group.[84]

The revolving door has become even more troubling since 2001,
when the NSA began an all-out effort to outsource many of its operations to private contractors.[85] This has resulted in lots of lucrative
consulting contracts that have made board relations even more important and sensitive. The growth of outsourcing since 2001 has been
absolutely exponential: whereas, for instance, in October 2001, the intelligence agency had only 55 contracts with approximately 144 contractors, four years later, by October 2005, the NSA had 7,197 contracts
with 4,388 contractors.[86]

And the cooperation between the private sector and the intelligence services has only continued to blossom post-9/11 to the point
where today, although the NSA itself has about 40,000 employees, it
has contracts with private companies for another 60,000 or so people.[87]
As Greenwald reports, approximately "70 percent of our national intelligence budget is being spent on the private sector."[88]

The relations with the private sector got especially cozy after 9/11. At around the same time that NSA head Michael Hayden began secretly bypassing the FISC—the specially designed intelligence court— to engage in monitoring of all international communications, regardless of whether a domestic caller was involved, Hayden got AT&T to allow the NSA to make copies of all fiber-optic cable communications at key switches like Folsom Street in San Francisco using "optical splitter[s] [that] would create a mirror image of the contents of the entire cable."[89] James Bamford recounts:

> By the late fall of 2001, Hayden succeeded in gaining the secret cooperation of nearly all of the nation's telecommunications giants for his warrantless eavesdropping program. Within a year, engineers were busy installing highly secret, heavily locked rooms in key AT&T switches, among them Bridgeton, New York City, and the company's major West Coast central office in San Francisco. From then on the data—including both address information and content—would flow through the PacketScopes directly to the NSA. And Bridgeton would become the technical command center for the operation. "It was very hush-hush," said one of the AT&T workers there at the time. "We were told there was going to be some government personnel working in that room. We were told, 'Do not try to speak to them. Do not hamper their work. Do not impede anything that they're doing.' "[90]

As Bamford explains in an article published in the *New York Review of Books* on August 15, 2013, there was ongoing cooperation between the telecommunications companies and the NSA at least through 2009 (and possibly continuing since then). Based on a "draft of a top secret NSA inspector general's report leaked by Snowden," a report that apparently dates to 2009, Bamford reports:

> As part of its cable-tapping program, the NSA has secretly installed what amount to computerized filters on the telecommunications infrastructure throughout the country. According to

the leaked inspector general's report, the agency has secret co-operative agreements with the top three telephone companies in the country. Although the report disguises their names, they are likely AT&T, Verizon, and Sprint:

NSA determined that under the Authorization it could gain access to approximately 81% of the international calls into and out of the United States through three corporate partners: Company A had access to 39%, Company B 28%, and Company C 14%.[91]

Bamford goes on to report, "According to a recent slide released by Snowden, the NSA on April 5, 2013, had 117,675 active surveillance targets in the program and was able to access real-time data on live voice, text, e-mail, or Internet chat services, in addition to analyzing stored data."[92]

One NSA document leaked by Snowden, related to the FAIRVIEW program targeting corporate partnerships, refers to an unidentified telecom company as "aggressively" providing access: "FAIRVIEW—Corp partner since 1985 with access to int. cables, routers, switches. The partner operates in the U.S., but has access to information that transits the nation and through its corporate relationships provide unique accesses to other telecoms and ISPs. Aggressively involved in shaping traffic to run signals of interest past our monitors."[93] Another typical NSA dispatch regarding its "corporate portfolio" refers to metadata and content received from "commercial links managed by the NSA Corporate Partner."[94] Under the FAIRVIEW program, the NSA was collecting around 200 million records per day in December 2012, for a monthly total of over 6 billion records.[95]

WHATEVER IT IS that is surveilling us, then, is not simply "the state." It is more an amalgam of various national intelligence services, Google, Microsoft, other Silicon Valley firms, Facebook and other social media corporations, private surveillance industry companies and consultants, IT departments everywhere, and, as we will see, local police departments, friends, hackers, and curious interlopers. It ranges in scale from the NSA and Apple to your neighbor with a "packet sniffer" or free Mac software like Eavesdrop, and it includes friends stalking on Facebook as well as the companies Amazon and Netflix.

We had come to think of ours as a time of "liquid modernity," as Zygmunt Bauman famously wrote—as a new form of modernity, in which institutions and social formations no longer have sufficient time to become hardened structures, to "solidify."[96] We had come to think of ours as a more fluid world. And this oligarchic amalgam certainly bears a family resemblance to more fluid power structures. But at the same time, it is hard to reconcile with liquidity, as the oligarchic power structures grow ever more commanding and formidable.[97]

The amalgam is the product, in part, of shared interests in security among government and tech companies—especially shared interest in security from foreign corporate espionage, cyber hacking, and threats from countries that are viewed as malevolent. It is also largely the product of the rise of neoliberalism over the past four decades, and the associated trend toward "deregulation," outsourcing, and privatization. This is surely what accounts for the proliferation of contracts, consultancies, and private-public partnerships—and the revolving doors between boardrooms, CEO offices, and the upper echelons of intelligence agencies.

To give but one example, the breakup of the Bell telephone system in the 1980s and the advent of mobile phones in the 1990s produced a fragmentation of the phone network and posed problems for government surveillance. A 1997 law allowing telephone customers to retain their telephone number when they switched carriers only made things worse, because it made it harder for the states and local law enforcement to know which carrier had the phone data they were looking for.[98] So the federal government put in place a system—kind of like an air traffic control system—for routing all phone calls and text messages, and then bid the system out to private enterprise.[99] A private Virginia company, Neustar, got the multibillion-dollar contract; in 2013, the company took in about $446 million. But the privatized routing system is, of course, overseen by the FCC, which ensures that the interests of law enforcement and national security are protected. So what we end up with is a cooperative medley of private telecommunications companies and the government, all working together to ensure national security interests. As the *New York Times* reports, "The routing network that was put in place, with Neustar as its administrator, was designed partly to allow the government nearly instant access to the data on where calls were being routed."[100]

When the FCC called for new bids in 2014, as the *New York Times* reported, "officials from the F.B.I., the Drug Enforcement Administration, the Secret Service and the Immigration and Customs Enforcement agency . . . weighed in on the debate, as have senators and House members who supervise American intelligence operations."[101] And while these agencies expressed, to use their own words, "no position" on who should win the contract, they did, according to the *Times,* "want to make sure that their professional needs were adequately addressed and that there would be no disruption in access to call-routing data 'in real time or near real time.' "[102] Specifically, "the agencies expressed particular concern that a contractor with access to the phone system from outside the United States could mean 'unwarranted, and potentially harmful' access to American surveillance methods and targets."[103]

The resulting routing system that is now in place is typical: a lucrative, complex, privately operated but governmentally overseen regulatory network in which everyone's "professional needs" are taken care of—especially "the government's ability to trace reams of phone data used in terrorism and law enforcement investigations."[104] In the process, Neustar hired Michael Chertoff, former U.S. secretary of homeland security, as a consultant, and Chertoff published a forty-five-page report about the national security issues surrounding the contract.[105] The other bidding company, Telcordia, a U.S. subsidiary of the Swedish company Ericsson, hired a retired rear admiral, James Barnett Jr., as outside counsel.[106] As usual, everybody is making out like a bandit.

NONE OF THIS IS TO DENY in any way the breadth, sweep, and depth of the NSA's own surveillance capabilities, which are jaw-dropping. Nor is it to deny that there is, indeed, a surveilling "state" in this amalgam, and that it is truly awesome. We know its contours thanks to the recent disclosures by Edward Snowden and other recent leaks and planted leaks—"pleaks," as David Pozen would say—that have emerged in the press.[107] The NSA and its "Five Eyes" peers—the intelligence agencies of Australia, Canada, New Zealand, and the United Kingdom—have access to and the ability to record and monitor all our emails, Facebook posts, Skype messaging, Yahoo video chat platforms, Twitter tweets, Tumblr photos, Google searches—in sum, all our traffic on social media and on the Internet.

The NSA's program XKeyscore, for instance, can give a government analyst real-time access to practically everything that an Internet user does, including the content that she types in an email, the websites that she visits, and the searches she performs, as well as all of her metadata.[108] In essence, the NSA has access to all the Internet user information it could want, and it retains that information for as long as it wants, sometimes for only twenty-four hours, but other times much longer. The NSA keeps records of most calls made in the United States; records of emails, Facebook posts, and instant messages "for an unknown number of people, via PRISM"; and "massive amounts of raw Internet traffic" in the form of metadata—which alone is extremely powerful.[109]

The extent of the surveillance is practically boundless. Glenn Greenwald reports on the scope of the technology:

> Beyond emails, the XKeyscore system allows analysts to monitor a virtually unlimited array of other internet activities, including those within social media.
>
> An NSA tool called DNI Presenter, used to read the content of stored emails, also enables an analyst using XKeyscore to read the content of Facebook chats or private messages.
>
> ... An analyst can monitor such Facebook chats by entering the Facebook user name and a date range into a simple search screen.
>
> Analysts can search for internet browsing activities using a wide range of information, including search terms entered by the user or the websites viewed.

As one slide indicates, the ability to search HTTP activity by keyword permits the analyst access to what the NSA calls "nearly everything a typical user does on the internet."

The XKeyscore program also allows an analyst to learn the IP addresses of every person who visits any website the analyst specifies.[110]

This is a remarkable amount of access, and a staggering amount of data available to the state. According to Greenwald:

> One NSA report from 2007 estimated that there were 850 billion "call events" collected and stored in the NSA databases, and close to 150 billion internet records. Each day, the document says, 1–2 billion records were added. William Binney, a former NSA mathematician, said last year that the agency had "assembled on the order of 20 trillion transactions about US citizens with other US citizens," an estimate, he said, that "only was involving phone calls and emails." A 2010 *Washington Post* article reported that "every day, collection systems at the [NSA] intercept and store 1.7 billion emails, phone calls and other type of communications."[111]

And what is not turned over by Google, Facebook, and other companies can simply be seized by the state through duplicating technology that makes copies of communications streams. This is the program called UPSTREAM, discussed earlier, which operates at the Folsom Street AT&T facility in San Francisco, a facility through which pass most of the communications in the U.S. Northwest.[112] UPSTREAM captures about 80 percent of the information that the NSA collects; the other 20 percent is retrieved with PRISM.[113] According to William Binney, the former NSA senior official who resigned in protest in 2001, "the NSA has established between ten and twenty of these secret rooms at telecom company switches around the country."[114] Moreover, the technology to access and repair deep-sea fiber-optic cables already exists, and there are some reports of undersea splicing, which adds an entire other dimension to the collection capabilities.[115] Several years ago it was reported that the General Dynamics shipyard in Groton, Connecticut, was hard at work retrofitting the nuclear-powered USS *Jimmy Carter* to equip it with "state-of-the-art technology for undersea fiber-optic taps."[116]

The United States is, of course, a central node for communications, giving the NSA remarkable global access to Internet traffic: one-third of all international telephone calls in the world go through the United States, as does about 99 percent of Internet traffic.[117] That makes it mightily useful to be able to duplicate communications streams right here on American soil.

David Cole reports on a chilling PowerPoint slide, leaked by Edward Snowden, that gives a sense of NSA's capabilities and ambition. The NSA document, which was produced for and shown at a gathering of the "Five Eyes," contends that "the NSA's 'new collection posture' is to 'collect it all,' 'process it all,' 'exploit it all,' 'partner it all,' 'sniff it all,' and, ultimately, 'know it all.' "[118]

What makes this all possible, of course, is the actual size of the NSA. The size and scope of the signal intelligence agency—only one of many intelligence services in the United States—are truly staggering. Back in 2000, before 9/11, the NSA facility in Fort Meade, MD, was a city that had "32 miles of roads ... covering 325 acres," with "more than four dozen buildings containing more than seven million square feet of floor space."[119] At the time, it had upward of 30,000 employees and contractors. "The secret city's police force employed more than 700 uniformed officers and a SWAT team."[120] After 9/11, Michael Hayden, then head of the NSA, began an "enormous building campaign" at the NSA's four locations in Georgia, Texas, Colorado, and Hawaii.[121] In Georgia alone, the listening post had 1,200 people, and after 9/11 the number immediately doubled to 2,400.[122] Today, the agency is made up of multiple little cities.

No, none of this is to deny the breathtaking surveillance capabilities that the "state" has today. It is to suggest, instead, that the state

FIGURE 2.5 United States National Reconnaissance Office NROL-39 satellite logo
Source: Matt Sledge, "America Is Launching a Giant, World-Sucking Octopus into Space,"
Huffington Post, December 5, 2013.

is not alone in surveilling us, that it forms part of a much wider amalgam. We are faced not so much with a "surveillance state" as with a surveillance amalgam or, in the words of John Gilliom and Torin Monahan, a "surveillance society."[123]

More radically, we may need to entirely rethink the very notion of "the state" in this digital age—a notion which was, of course, itself subject to dramatic contestation and transformation over the course of the twentieth century. Those who theorize about the state traditionally begin with Max Weber and his celebrated definition of the state as that which has the monopoly on the legitimate use of physical force.[124] To a certain extent, Judge William Pauley's conception of the guardian surveillance state is Weberian—as is, perhaps, the conception of the New Deal–like administrative state and Judge Leon's image of the deceptor state. The history and theory of the state have featured ebbs and flows of deconstruction and reconstruction—movements to destabilize, followed by waves of trying to "bring the state back in," especially in relation to other institutions and social actors, to better understand this emerging amalgam. The problem with the Weberian ideal type has always been the tendency to reify the state as the single entity that achieves legitimacy in force, when we have known for so long that so many other corporate entities also exercise not only symbolic but legitimate physical force. The outsourcing of the government function, including the use of force, to private enterprise has, for a long time, been too pervasive to ignore. The advent of the digital age is perhaps the final straw.

The proper metaphor, then, is not the government agent at his console, but a large oligopolistic octopus that is enveloping the world—perhaps the none-too-subtle mascot of the United States National Reconnaissance Office's mission USA-247, which was painted on a top-secret surveillance satellite launched into space in December 2013.

This octopus is the amalgam of Google and NSA, of Microsoft and the FBI, of Skype and AT&T, of Netflix and the New York Police Department, of the IT department upstairs and the Facebook stalker next door. And it is nourished, fed, kept alive by our digital exposure and by the technologies themselves, technologies that mysteriously fuse information and data into the surveillance mechanism itself—as if the security industry were designing the technology for its own purposes. We reveal our secrets in texts, emails, photos, videos, and social media posts. We have become our own informants. What we are facing today is not so much a "surveillance state" as an amalgam, an oligarchy, a knot of tenticular statelike actors that see through us and our desire-filled digital lives.

THREE

JEREMY BENTHAM'S *PANOPTICON*

A THIRD DOMINANT WAY of conceptualizing our current condition is through the metaphor of Jeremy Bentham's panopticon as interpreted through the theoretical lens of Michel Foucault's work *Discipline and Punish*. Here too, the use of the metaphor is varied and ranges the political spectrum. Some writers, such as Glenn Greenwald, draw a parallel to the panopticon to warn us of the dangers of NSA surveillance.[1] "In *Discipline and Punish*," Greenwald writes, "Foucault explained that ubiquitous surveillance not only empowers authorities and compels compliance but also induces individuals to internalize their watchers. Those who believe they are watched will instinctively choose to do that which is wanted of them without even realizing that they are being controlled."[2] When this happens, Greenwald cautions us, there is no longer any need for overt government repression, nor any evidence of it. Other commentators, such as William Simon, embrace the "panopticism" of Big Data, and view it as a potentially favorable development. "For democratic accountability," Simon writes, "panopticon-style surveillance has an underappreciated advantage. It may more easily accommodate transparency. Electronic surveillance is governed by fully specified algorithms."[3] In good hands, Simon suggests, panoptic power may be a step in the right direction and serve to police discriminatory practices.

Here too, though, before embracing the metaphor—pro or con—it is even more important to question it.

DURING THE EARLY 1970S, Foucault himself was struggling to properly theorize state repression in the wake of May 1968 and the French government's severe, authoritarian response to the student and worker uprisings.[4] Following the street revolts, the French government of President Georges Pompidou outlawed a number of non-parliamentary leftist political parties, and arrested and imprisoned hundreds of leftist militants. The incarceration of political prisoners, mostly Maoists, led to greater general awareness of prison conditions, and an outbreak of prison riots across France in 1971 and 1972—riots that left the country in a state of turmoil. Not enough turmoil, though, to bring about political reform or a wider social movement. And so Foucault turned all his attention to the questions of state repression, state punitive strategies, and relations of power in what he called our "punitive society."[5] He also turned to the larger question of how, despite increasing awareness of state repression and prison conditions, we so easily and so comfortably continue to tolerate what should be intolerable. "No introduction," Foucault jotted down quickly in his notes for his first lecture of the series, *Penal Theories and Institutions.* "One need only open one's eyes. Those who are loath to do so will surely recognize themselves in what I have to say."[6] Foucault's engagement with state repression would be direct and unwavering.

In Paris, on January 3 1973, Foucault dove into his third lecture series at the Collège de France, titled *The Punitive Society:* an analysis of eighteenth- and nineteenth-century forms of social struggle and repression, conflicts that he would model on civil war. His analysis resonated with current events and directly challenged, *sub silencio,* the dominant Marxist theories of state apparatuses, class struggle, and superstructural power.[7] Foucault's lectures—as reflected in the title—placed the emphasis not just on punitiveness but also, importantly, on society, and in particular on relations of power throughout society.[8] The lectures were a deeply historical engagement with specific acts of repression and strategies of governance—seven of the previous lectures would be almost purely historical narrations of the events surrounding the repression of popular movements in 1639. That same year, in May 1973, Foucault presented a series of five conferences in Rio de Janeiro under the title "Truth and Juridical Form" and elaborated on what he called a new form of "disciplinary power" pulsing throughout society— this time placing special emphasis on the way in which various legal forms or devices, such as the trial or the inquest, produce truth in societies. One of those legal devices that would become increasingly important to Foucault would be penitential exclusion: the birth of the prison.

Both in Paris and in Rio, Foucault directed his audience to a certain Nicolaus Heinrich Julius, a doctor of medicine, prison reformer, professor at the University of Berlin, and colleague of Hegel.[9] "This man," Foucault lectured, "this man named Julius, whom I highly recommend that you read, and who delivered for several years in Berlin a course on prisons, is an extraordinary personality who had, at times, an almost Hegelian tinge [*un souffle presque hégélien*]."[10]

This professor named Julius had discerned, during the 1820s, a remarkable transformation in our ways of seeing and knowing. He identified an architectural mutation reflecting a profound shift in power relations across society. Antiquity, Julius observed, had discovered and exploited the architectural form of the spectacle: the arena, the amphitheater where masses of people watch the few. Modern society, by contrast, had accomplished a fundamental shift from spectacle to surveillance, where one person or the few could watch the many. As Julius elaborated in his "Lectures on Prisons" in 1827, in a passage that would become a cornerstone for Foucault in 1973:

> It is a fact worthy of the highest interest, not only in the history of architecture, but in that of the human mind in general: that in the earliest times, not only in classical antiquity, but even in the Orient, the genius mind conceived and then pleased itself

to decorate—with all the treasures of human magnificence—buildings that were designed to make accessible to a large multitude of men the spectacle and inspection of a small number of objects, such as in temples, theaters, amphitheaters, where they would watch the blood of humans and animals flow. All the while, the human imagination never seems to have applied itself to provide a small number of men, or even a single man, the simultaneous view of a great multitude of men or objects.[11]

Julius then adds, jumping to his own time:

It would be a task reserved to modern times (and I intend to develop this idea later), to a period marked by the growing influence of the state and of its many interventions—deeper, day by day, into every detail and every relation of social life—to ensure and perfect the grand goal of constructing and distributing edifices intended to simultaneously surveil [*surveiller*] at one and the same time a great multitude of men.[12]

It is here that Foucault picks up the thread, declaring in his lecture on January 10, 1973: "This is precisely what happens in the modern era: the reversal of the spectacle into surveillance."[13] Lecturing in Paris, Foucault would elaborate:

We are in the process of inventing, says Julius, not only an architecture, an urbanism, but an entire disposition of the mind in general, such that, from now on, it will be men who will be offered in spectacle to a small number of people, at the limit to only one man destined to surveil them all. The spectacle turned into surveillance, the circle that citizens formed around the spectacle—all that is reversed. We have here a completely different structure where men who are placed next to each other on a flat surface will be surveilled from above by someone who will become a kind of universal eye.[14]

Foucault would return to and develop this idea further in 1975, in his chapter "Panopticism" in *Discipline and Punish,* where, again speaking of Julius, Foucault would say: "A few years after Bentham, Julius wrote the birth certificate of this [disciplinary] society."[15]

Julius had discerned a reversal of the architecture of knowledge and power—in Foucault's words, a "reversal of the spectacle into surveillance." Whereas in ancient societies the directionality of knowledge and power flowed from the gaze of the many to the isolated few in the center of the arena or the well of the amphitheater, in modern times power circulated from the individual in the central tower who could watch and see to the masses placed around the periphery in visible ways. This represented not only an architectural development but, in the words of Foucault, "an event in the 'history of the human mind.'" "In appearance," Foucault emphasized, "it is merely the solution of a technical problem; but through it, a whole type of society emerges."[16]

This reversal of spectacle into surveillance was one that Julius himself would put into effect in his capacity as a penal theorist and prison reformer. Julius had absorbed these new architectural ideas and technologies during his several missions to England, Wales, and Scotland in 1827 and to the United States in 1834–1836, where he became an admirer of the Philadelphia system and the technique of individual isolation—in his words, the "principle of uninterrupted solitude throughout the entire period of imprisonment."[17] Like Gustave de Beaumont and Alexis de Tocqueville—whose famous report on American prisons, *On the Penitentiary System in the United States and Its Possible Application in France,* he would translate into German—Julius would become an ardent proponent of the larger American penitentiary system.[18] Julius also disseminated these new architectural ideas in his theoretical work and practical interventions, becoming a leading figure in the science of prisons—*Gefängniskunde* in German.[19] Julius aided the king of Prussia in constructing prisons based on the "modified Pennsylvania system" of labor and solitary confinement, for which London's Pentonville prison, with its radiating design, served as a model.[20] His plans for prisons in four cities—Berlin, Königsberg, Ratibor, and Münster—would be completed in 1843.[21]

Julius, with his Hegelian tinge, personified a historic reversal in technologies of power in society, and his 1827 "Lectures on Prisons" would constitute one of the main inspirations—Bentham, of course, being another—for Foucault's idea of disciplinary power. Bentham had developed the idea of a perfectly disciplinary space and coined the term *panopticon,* borrowing the idea from his brother, Samuel Bentham, during a visit to Russia in 1786–1787. At the time, Samuel was charged with the supervision of the ports, factories, and workshops of the Potemkin prince, and in that capacity he was trying to implement better ways to supervise and control a poorly disciplined British labor force.[22] In sketching out his reflections on the panoptic principle in a series of letters sent back from White Russia in 1787—a principle that applied as easily to ports, factories, and workshops as to penitentiaries, asylums, hospitals, schools, and barracks—Bentham prefigured and captured the new form of power that circulated throughout society.[23] Foucault also identified the panoptic principle in architectural plans for the circular hospital at the Hôtel-Dieu drawn in 1770, with a design based on the shape of a star.[24] The model at the Hôtel-Dieu allowed for total surveillance—a "huge radiating hospital" that would make possible "a constant and absolute surveillance."[25] Foucault traced the origins of "this isolating visibility" to the dormitories of the Military Academy of Paris in 1751, noting that "all the major projects for the redevelopment of prisons ... would take up the same theme [of the complete visibility of bodies], but this time under the sign, as it was almost always reminded, of Bentham. There were hardly any texts or projects concerning prisons where one did not find that 'trick' of Bentham's—namely, the panopticon."[26]

Foucault would appropriate the metaphor of the panopticon to capture the ethos of disciplinary power in nineteenth-century French society. It is important to emphasize that the panopticon served as a metaphor only and therefore captured some dimensions of disciplinary power better than others. In minute detail, Foucault would articulate the central traits and techniques of the larger category of disciplinary power: a spatial organization that ensures the exact observation of human subjects, so that "the techniques that make it possible to see induce effects of power"; an architecture that permits the isolation and confinement of the individuals who are watched, the omnipresence of the few who watch, and the knowledge of constant surveillance; a perfect control over time that allows the maximum extraction of information and work from those who are under surveillance; a normalizing form of judgment that "compares, differentiates, hierarchizes, homogenizes, excludes"; and a generalized form of truth production, the examination, that constantly evaluates and judges those who are being watched, and which ultimately hides the gaze of the watcher, so that those watched begin to internalize the discipline themselves.[27] On this last point, Foucault emphasized, "*The examination transformed the economy of visibility into the exercise of power.* Traditionally, power was what was seen, what was shown and what was manifested. . . . Disciplinary power, on the other hand, is exercised through its invisibility. . . . In discipline, it is the subjects who have to be seen. Their visibility assures the hold of the power that is exercised over them."[28]

Foucault had a razor-sharp conception of "disciplinary surveillance"—in distinction, for instance, from an alternative notion of "control." Surveillance had a French pedigree and descended from French techniques of internment and confinement.[29] Control, by contrast, drew its heritage from English procedures of moral control. In the British case, the individual was monitored by a group of which he formed a part—whether a guild, a religion, or a social group. Think here of the Quakers or Puritans, or of any of the numerous leagues, federations, or associations that formed around work or even property. The overseeing in England began as a form of self-governance—there was no need for a process of internalization because the individual was already part of the group that did the monitoring. By contrast, in nineteenth-century France, the individual was situated as an outsider in relation to the institution that monitored him. The French case of confinement, Foucault lectured, "involved an exclusion, either temporary, in the name of punishment, or a form of exclusion that built on another, that sanctioned a marginality that had already been acquired (unemployed, vagabonds, beggars)."[30] Gradually, even this would change as the French approach embraced a more productive model of inclusion for purposes of normalization—or what Foucault referred to as "sequestration," as in the sequestration of funds, but this time of people.

This notion of sequestration reflected the productive side of the punitive practices—and especially the physical, corporeal effects of discipline. Surveillance, on Foucault's view, served to render the prisoner, the young laborer, or the schoolchild more docile and to fix him to the carceral, factory, or educational system. As Foucault explained in his lecture on March 14, 1973, "the couple 'surveil-punish' [*surveiller-punir*] establishes itself as the relation of power indispensable to fixing individuals on the production apparatus, indispensable to the constitution of the forces of production, and it characterizes the society that we could call *disciplinary*."[31] Disciplinary mechanisms and their effect on our bodies were as central to the advancement of capitalist production as the accumulation of capital itself: they made possible the accumulation of bodies without which the Industrial Revolution would not have taken place.[32] And they gave birth as well to the penitentiary and the modern prison.

Many of the central features of disciplinary power were refracted in the idea of Bentham's panopticon prison, with its central watchtower and circular ring of transparent cells where the prisoners are constantly visible—especially to themselves. The culmination of disciplinary power is when the few who watch no longer need to look because the masses who are surveilled have internalized the gaze and discipline themselves—when the panopticon induces in the inmate, in Foucault's words, "a state of conscious and permanent visibility that assures the automatic functioning of power."[33] And so Foucault appropriated the metaphor of the panopticon, designating this new form of power and knowledge—one that circulates from the individual at the center to the many on the periphery, that allows for absolute surveillance, that inverts spectacle into surveillance—"panoptic," in homage to Bentham. In an unpublished draft of *Discipline and Punish,* Foucault jotted down his reasoning:

> If we characterize as "disciplinary" the apparatuses of spatial distribution, of extraction and accumulation of time, of individualization and subjection of bodies through a game of watching and recording, then let's honor Bentham and call "panoptic" a society in which the exercise of power is ensured on the model of generalized discipline. And let's say that, at the turn of the eighteenth and nineteenth centuries, we saw clearly emerge "panoptic" societies of which Bentham could be considered, depending on your view, the prophet, the witness, the analyst or the programmer.[34]

"Let's honor Bentham": Foucault named our disciplinary society "panoptic" as a metaphor to capture key features of discipline. "The Panopticon," Foucault emphasized in *Discipline and Punish*, "must be understood as a generalizable model of functioning; a way of defining power relations in terms of the everyday life of men."[35] It captured some features of discipline extremely well: how the panopticon preserves the enclosure of the dungeon but offers "full lighting" and makes possible the trap of visibility; how it produces the internalization of power so that "the perfection of power should tend to render its actual exercise unnecessary"; and the experimental, laboratory nature of the edifice—"a privileged place for experiments on men."[36]

"We are much less Greeks than we believe," Foucault declared. "We are neither in the amphitheater, nor on the stage, but in the panoptic machine, invested by its effects of power, which we bring to ourselves since we are part of its mechanism."[37] Foucault's play on words—"the reversal of the spectacle into surveillance"—was undoubtedly a veiled reference to Guy Debord, theorist of the "society of the spectacle."[38] Debord, notably in his 1967 book but also in his earlier Situationist interventions, had placed the form of the "spectacle" at the very center of his theorization of our times: "The spectacle appears at once as society itself, as a part of society and as a means of unification."[39] For Debord, like Julius before him, the spectacle was not only an architectural form but a means to understand the circulation of power throughout society: "The spectacle is not a collection of images; rather, it is a social relationship between people that is mediated by images."[40] Its key trait was precisely the dimension of social relationship, of relations of power, and its connection to capitalist production. As Debord famously wrote, "The spectacle is ideology *par excellence.*"[41] (Debord's project was still steeped in Marxism, and his concept of the spectacle—or rather, his critique of the "society of the spectacle"—was copious: it incorporated a critique of modes of production and a call for revolutionary class struggle. In other words, it included more than what we might first think of when we imagine "the spectacle.")

By contrast to Debord, Foucault would underscore that what characterizes our modern society—or at least nineteenth-century France, insofar as the book *Discipline and Punish* ends in 1840 with the opening of the Mettray Penal Colony, a juvenile reformatory in the Loire region—was not the spectacle, but rather its eclipse and reversal into surveillance. Or, even more strikingly, the creation of a whole panoptic society, of a "punitive society."

THE CONTRAST BETWEEN SPECTACLE and surveillance may have been drawn too sharply. As W. J. T. Mitchell observes in *The Spectacle Today,* the terms can no longer be understood as mutually exclusive. They have begun to function, in our contemporary setting, in constant tension with each other: it is practically impossible today to conceptualize surveillance in isolation from its spectacular dimensions—if it ever was possible. Rather, spectacle *and* surveillance should be seen, as Mitchell explains, as "dialectical forces in the exertion of power and resistance to power."[42] Or, as Mitchell suggested in *Picture Theory: Essays on Verbal and Visual Representation,* spectacle and surveillance are two sides of the formation of subjects today: "Spectacle is the ideological form of pictorial power; surveillance is its bureaucratic, managerial, and disciplinary forms."[43]

I would go further, however, and suggest that both forms—spectacle and surveillance—are overshadowed today by a third: *exhibition,* or *exposition,* or perhaps more simply *exposure.* In our digital age, we are not so much watching a spectacle or being forced into a cell as we are exhibiting ourselves to others. We are not being *surveilled* today so much as we are *exposing* ourselves knowingly, for many of us with all our love, for others anxiously and hesitantly. There are, to be sure, spectacular dimensions to this exposition today, even if we are no longer in the arena. Although we may not be looking together, at least physically, our computer screens, iPads, and smartphones often create a virtual commons, filled with chats and comments, "likes" and shares. There is undoubtedly a spectacular dimension when the videos and memes go viral. And at times there is also, we know, pure surveillance—bugs, hidden cameras, parabolic listening devices. At times we are watched by some agent at his console, and many are still incarcerated in panoptic prisons, such as at Stateville Prison in Joliet, Illinois.[44] But for the most part, we modern subjects give up our information in love and passion. We are not forced; we *expose ourselves*. Rather than a surveillance apparatus stealthily and invasively forcing information out of us, more often than not we *exhibit ourselves* knowingly to that voyeuristic digital oligarchy—and we put ourselves at its mercy. We are confronted less with surveillance than with an oligarchical voyeur taking advantage of our exhibitionism. As François Ewald suggests, today we should no longer "analyze power relations in the data world according to the patterns of the old state power with its technologies of surveillance, control and domination."[45]

Ours is not so much a "society of the spectacle," as Debord thought, nor so much a disciplinary or "punitive society," as Foucault suggested. Instead, it seems as if we are living in an *expository society*. This is not to minimize in any way the punitive element—so much more clearly connoted by the term *surveillance*. Nor is it to minimize the pervasive forms of repression found not just in liberal democracies, in the behavior of police and in carceral practices, but also in authoritarian countries and territories. The state, narrowly defined, will of course use intimate incriminating information to monitor, surveil, and punish. As we will see, the New York Police Department now has its own social media unit that tracks the Facebook postings of suspected youth gang members.[46] Many other law enforcement agencies are collecting phone data—including content—for purposes of data mining and surveillance.[47] The FBI has rolled out new, "fully operational" facial recognition software, what it calls Next Generation Identification (NGI), with a database that "holds more than 100 million individual records that link fingerprints, iris scans and facial-recognition data with personal information, legal status, home addresses, and other private details" and which will house "52 million facial recognition images by 2015."[48]

But the way that power circulates in advanced capitalist liberal democracies—through expository and voyeuristic forms, rather than through surveillance—has to be understood separately from what the state, traditionally defined, will do with the information. It was, after all, a couple: discipline and punish. The same is true for the couple that is constituted by digital exposure and the collection, mining, profiling, and targeting of our data—more simply, the couple "expose and enmesh," to which I will return in Chapter 9. Georges Didi-Huberman discusses our condition as "exposed peoples" in what he calls the "media age."[49] As a people, he explains, we are both underexposed and overexposed—underexposed insofar as we so often have difficulty seeing injustice, overexposed in our spectacular self-presentations in reality shows and shared profiles.[50] This, it seems, captures far better our digital age.

In the end, the metaphor of panoptic power is somewhat inexact today—for these and other reasons. As Siva Vaidhyanathan argues in *The Googlization of Everything,* many of us are not even aware of the myriad ways in which we are being tracked.[51] "Unlike Bentham's prisoners," Vaidhyanathan emphasizes, "we don't know all the ways in which we are being watched or profiled."[52] Our situation resembles perhaps more a "cryptopticon," in his words—reflecting the Kafkaesque dimension of our times. Recall that for Foucault, knowing visibility was one of the most important features of panoptic power.[53] For the panopticon to function, the inmate had to be aware he was being watched—he had to always "[know] himself to be observed."[54] That too, it seems, is no longer the case. In addition, many of us "don't seem to care" about being watched—which is far different from the condition of those in the panopticon.[55] For the panopticon to function, the watching gaze must be internalized, not ignored.

But overall, the most striking inversion is that today we expose ourselves so freely and passionately and anxiously—at least, so many of us in advanced capitalist liberal democracies. It is an expository form of power we face at the present time, one in which we use digital mediums to tell stories about ourselves and create and reshape our identities. One in which we participate fully—willingly or not—through all of our ordinary, mundane, daily actions: carrying a cell phone, sending emails and texts, taking the subway. No, we are not being forced into this situation, we are not enclosed, it is not just disciplinary. But that, perhaps, should not be entirely surprising.

FOUCAULT'S REFERENCE TO JULIUS having *un souffle presque hégélien* may have been a clue to things to come. "Only when the dusk starts to fall does the owl of Minerva spread its wings and fly."[56] Although Foucault coined the conceptual use of the terms "surveillance" and "panoptic power" as early as 1973—in the lectures published as *The Punitive Society,* in which he described so meticulously that "first example of a civilization of surveillance"—by the mid-1970s he himself had already moved beyond the model of the panopticon.[57] Within a couple of years—perhaps even by 1976, with his discussion of biopower in *The History of Sexuality: Volume 1*—Foucault had already begun to question whether surveillance could serve as a model of power relations in contemporary society.[58] Turning his attention more intensely to the rise of a neoliberal paradigm, Foucault began to articulate a different form—or rather a supplemental form—of power relations, namely, *sécurité.* He would elaborate this model in his lecture series *Security, Territory, Population* in 1978 and *The Birth of Biopolitics* in 1979.

The securitarian form of power is tied to the arts of maximizing and minimizing, to those special competences of neoclassical economists. Security reaches for equilibrium points for populations as a whole; it is not focused on the event in a spectacular way, nor on the individual in a disciplinary manner. In this sense, security differs markedly from surveillance. Discipline is centripetal: it focuses on every instance of minor disorder and seeks to eradicate even the smallest occurrence. Security, by contrast, is centrifugal: it is tolerant of minor deviations and seeks to optimize, to minimize or maximize, rather than to eliminate.[59] And both, of course, differ dramatically from the juridical model of sovereignty, the form of law, which distinguishes in a binary way between the permissible and the prohibited, and then penalizes the latter.[60]

Foucault never provided a clear architectural schematic to visualize the notion of security that he associated with the neoliberal era—in contrast to the spectacular *supplices* or spectacles of torture and the disciplinary panopticon. There was, to be sure, the tripartite series that Foucault presented in his *Security, Territory, Population* lecture on January 11, 1978: the juridical exclusion of lepers in the Middle Ages, the gridlike regulation and disciplinary quarantine of entire cities during the plague in the sixteenth and seventeenth centuries, and the medical campaign of security against smallpox in the eighteenth century.[61] Or, even closer to architecture, the other tripartite series that Foucault discussed in the same lecture: Alexandre Le Maître's discourse on the metropole in the seventeenth century, the construction of artificial cities under Louis XIII and XIV, and the redevelopment of Nantes at the end of the eighteenth century.[62] All the same, we lack a precise schema—akin to the arena for the spectacle or the panopticon for surveillance—to properly visualize securitarian power in a neoliberal age, a necessary first step toward better understanding its relationship to the expository society.

I would propose, as the architectural schema that best captures the notion of *sécurité*, the amusement park or the theme park—or even better, the themed shopping mall or, more simply, the themed space. It is a private commercial space, for profit, that has been underwritten and made possible by state and municipal governments; a mixed private-public space—or perhaps, more accurately, a space of private profit and public expenditures—that practices a form of management and control of large flows of populations in order to maximize the number of visitors, to optimize consumption, to attract more advertising, and to facilitate spending. These are quintessential spaces that seek to optimize the movement of large numbers of people in order to fuel their purchases, while at the same time minimizing labor and other expenditures. This is not just a spectacle, because everything is scattered. The goal is to distribute, not concentrate, the population, to avoid amassing consumers at any one spectacle—so that they spend much more at all the various mini-theaters of consumption. The goal is to optimize and redirect queues, to make them feel shorter than they are, to turn them into consuming opportunities, to learn how to manage the patience of consumers who want to see, but who are never, or almost never, actually in the arena. To be sure, surveillance also occurs, but that is only one small dimension of the endeavor. Yes, there are a large number of CCTV cameras. The watchful eye is everywhere. That is, in fact, what renders a place like Disneyland so popular—the feeling that everything is monitored, that everything is secure, that the children, or rather the parents, have nothing to fear. There is continuous surveillance, yes, but not in a panoptic form. We do not internalize the gaze, we are reassured by it.

Some will recall that amusement parks were the subject of extensive theorizing in the 1980s and 1990s. This included the provocative writings of Jean Baudrillard on Disneyland—the simulacrum that becomes more real than reality itself:

> Disneyland exists in order to hide that it is the "real" country, all of "real" America that is Disneyland (a bit like prisons are there to hide that it is the social in its entirety, in its banal omnipresence, that is carceral). Disneyland is presented as imaginary in order to make us believe that the rest is real, whereas all of Los Angeles and the America that surrounds it are no longer real, but belong to the hyperreal order and to the order of simulation. It is no longer a question of a false representation of reality (ideology) but of concealing the fact that the real is no longer real, and thus of saving the reality principle.[63]

Recall the equally insightful way Umberto Eco analyzed theme parks as "hyperreality," or Louis Marin saw them as a type of "degenerate utopia"—as "ideology changed into the form of a myth."[64] And, conversely, recall the rich theorization of the "new city," of strip malls, shopping malls, and gentrified architecture, of Orange County as the exopolis, of all the "variations on a theme park."[65]

The project here, though, is to visualize securitarian power—in order to better understand its relation to the expository society—and toward that end, rather than rehearse these analyses of simulacra and hyperreality, it may be more useful and punctual to explore the line of thought developed in George Ritzer's texts on the "McDonaldization" of society—and, by extension, the analysis by Alan Bryman of the "McDisneyization" of the social.[66] The four main elements of McDonaldization are present, as Bryman suggests, in the world of Disneyland—and, I would add, in the neoliberal condition that is tied to securitarian logics: first, the effective management of large numbers of people; second, the control of their movements to increase their spending and to police their behaviors; third, the predictability of the experience; and finally, the computability, the calculability, the quantification, the measurability of each queue, each amusement ride, every game.

Amusement parks are, in the words of Steven Miles, "the quintessential physical manifestation of the consumer society," "the ultimate commercial environment," and "an expression of capitalism in perhaps its purest form."[67] They constitute "mini-cities of consumption in their own right," Miles adds.[68] Disney has been marketing experiences and memories for profit for years—with resounding success: "In 1998, overall revenues from the Disney theme parks were $6.1 billion, and . . . above 50 per cent of that was from entrance fees."[69] In this sense, they capture well the guiding spirit of the neoliberal management of consumption.

This is the case today not only with amusement and theme parks, but with all the other themed spaces and venues, such as the Hard Rock Café or Planet Hollywood—or, on an even larger scale, the entire town of Celebration, Florida, built by Disney right next to its

Walt Disney World Resort in Orlando, so marvelously theorized in Keally McBride's book *Collective Dreams*.[70] Theming, says Miles, may be "the most successful commercial architectural strategy of the twentieth century."[71] It gives full rein to our modern neoliberal consumerist desires, imagination, and pleasures—modern precisely in the sense of "modern consumerism" or what Colin Campbell describes as "the distinctively modern faculty, the ability to create an illusion which is known to be false but felt to be true."[72] Walt Disney's objective was to create a new kind of space "free from the dirt and danger of the carnival world of freaks, barkers, and thrill rides."[73] For many, it worked—and still does: "Disney's transformation was a blessing, substituting clean, orderly, and family-oriented fun for the grimy disorder and working-class and minority crowds of America's declining urban amusement parks and seaside resorts."[74] In this sense, the Disney theme parks are the quintessential space of modern consumption—more so, even, than a space of leisure.[75]

Themed consumption: if you live in the "small town" of Celebration, Florida, you will be treated during Christmastime to an hourly snowfall in the downtown area—courtesy of Disney.[76] As McBride writes, "Disney Imagineers are offering us fantasy as real life. We can buy it and move in permanently."[77] The themed ideals of the community now "exist as commodities."[78] And not just kitsch commodities at that—it is an elite form of premier, themed consumption. "The array of architects who helped to plan Celebration is a virtual who's who of contemporary architecture," McBride notes.[79] Philip Johnson designed the town hall, Robert Venturi the town bank, Michael Graves the post office, and Cesar Pelli the movie theater; Robert Stern developed the master plan.[80] It is a star-studded cast, all intended to re-create an imaginary Main Street for us to consume and for Disney to profit from—with a model school and teaching academy (again courtesy of Disney) that doles out not grades, but instead "narrative assessments."[81] As McBride recounts, the principal of the school boasts, "This is a place where nobody fails."[82]

The themed space—or, perhaps, variations on the theme park—as architectural representation of neoliberal *sécurité:* that would be an idea to pursue.[83] Perhaps we could go even further, to the themed space of consumption par excellence: today's oversized shopping malls, in urban towers and in the suburbs—the Water Tower in Chicago, the Shops at Columbus Circle in Manhattan, or the Mall of America outside the Twin Cities.[84] In these gigantic spaces, the theme itself has become consumption—consumption as the amusement, shopping as the entertainment. These are distilled spaces of consumption. Some of these malls, in fact, have physically absorbed the amusement park. The Mall of America, for instance, the most-visited mall in the world, has its own amusement park *inside* its walls. Less monstrous than the panopticon as an image of the punitive society, but just as frightening, and focused exclusively on consumption.

HOW THEN DOES SECURITARIAN POWER in our neoliberal age relate to the expository society? Is it just a question of refreshing the architectural schemata of the themed space and digitizing it? Is it just a question of adding the digital technology? Of turning on Wi-Fi? *Daily Show* host Jon Stewart confessed recently, when asked about BuzzFeed and Vice: "I scroll around, but when I look at the internet, I feel the same as when I'm walking through Coney Island."[85] Is digital exposure something like Coney Island or the Mall of America *with* connectivity? Is it the *wired* space of secure consumption?

It may be tempting to answer in the affirmative, and previously I had developed a notion of "digital security" along these lines.[86] But that no longer seems entirely correct. Our expository society is more than just neoliberal security gone digital, because the central mechanisms of optimization and equilibrium tied to *sécurité* do not characterize digital exposure. Although the expository society is undoubtedly fueled by neoliberalism, power circulates differently there than through a securitarian apparatus. Something else is taking place. To understand this, it is crucial to sever the link that ties neoliberalism to the form of power that Foucault identified as *sécurité*. It is important to pull those two strands—neoliberal logics on the one hand, and securitarian power on the other—apart.

To put it simply, our digital age is surely neoliberal, but not exactly securitarian.

Neoliberalism has unquestionably shaped our current condition. The NSA's ability to see through us is the product of outsourcing the collection of information to social media companies, to retailers, to advertisers—*and to ourselves.* We, each one of us, are doing the work of exposing ourselves and sharing our personal information. The security industry has effectively delegated the work to us, and at a pittance: the PRISM program only costs $20 million per year, which is absolutely nothing for such a massive surveillance program. The job of collecting data is being done by private enterprise and individual users. Plus, the growth of the surveillance-industrial empire is itself an artifact of privatization and outside consultancy. As Edward Snowden reminded us so powerfully when he introduced himself from Hong Kong, "I work for Booz Allen Hamilton as an infrastructure analyst for NSA in Hawaii."[87] James Bamford and Glenn Greenwald document the NSA-corporate relationship in intricate detail and underscore that today, more than two-thirds of intelligence dollars are going to the private sector.[88] The public-private collaboration in the surveillance industry reflects all the classic features of neoliberalism.

Now, neoliberalism must be understood not just in practical terms—namely, in terms of the typical policies of privatization and outsourcing—but also along theoretical lines. At the broader theoretical level, neoliberalism represents an attempt to displace politics with a notion of economic orderliness and naturalness—to neutralize the political clash over irreconcilable normative visions of family, society, and nation by privileging the orderliness of the economic realm. This notion of orderliness corresponds to the idea of the ordered market as *the* model of social interaction—a model of purportedly "voluntary and compensated exchange" that "benefits all parties." Early liberal economists in the eighteenth century set the stage for neoliberalism by introducing the notion of natural order into the sphere of trade, commerce, and agriculture—by inserting the divine notion of orderliness into the economic realm. Building on this original impetus, twentieth-century neoliberals, beginning in the late 1930s and in reaction to the rise of fascism and communism, performed two key theoretical moves that ultimately gave birth to the kind of state interventions (such as "deregulation," privatization, and workfare) that have come to be known as post-1970 neoliberalism. The first critical move was to update and render more technical the notion of natural orderliness (for example, with more technical economic theories that gave rise to concepts such as Pareto optimal outcomes and Kaldor-Hicks efficiency); the second was to extend the earlier liberal notion of orderliness from economics to every other domain—the social, the familial, the political. In other words, to extend the model of natural order beyond economic exchange to crime, divorce, punishment, illicit drugs, adoption, and now social interaction and social media. These two moves demarcate the essence of neoliberalism from its earlier liberal kin in the eighteenth century.[89]

These two moves are refracted throughout our digital lives today, and in this sense the digital age is certainly neoliberal at a theoretical level as well. Facebook is a good illustration. As Phillip Mirowski suggests, Facebook is "neoliberal technology par excellence": a "wildly successful business that teaches its participants how to turn themselves into a flexible entrepreneurial identity."[90] Facebook assumes and promotes the idea of the entrepreneurial self, so closely tied to Chicago School theories of human capital.[91] It is also accompanied by many of the techniques of neoliberal governmentality. It projects a type of market rationality on social interaction. It simultaneously hides the profit motive associated with all the advertising, and highlights the open-market features of sociability. As Mirowski describes, "Even though Facebook sells much of the information posted to it, it stridently maintains that all responsibility for fallout from the Facebook wall devolves entirely to the user. It forces the participant to construct a 'profile' from a limited repertoire of relatively stereotyped materials, challenging the person to somehow attract 'friends' by tweaking their offerings to stand out from the vast run of the mill."[92] Facebook models our social interactions on a market with menus of options.

There is also, naturally, an important confessional dimension to Facebook—one that I will return to in Part II. Many of us use Facebook to avow our foibles and secrets, to share things about ourselves that make us more vulnerable or human. For some of us, there is an automaticity to social media—to Facebook, Twitter, or Instagram, especially among the younger generations—that harks back to older forms of confessional logics. There are elements of truth-telling, of saying truth about oneself, that resemble far earlier forms of avowal, of examination of the self, of penitence even.[93] But even these moments of confession are difficult to dissociate from the instrumental and entrepreneurial presentations of the self that flood Facebook. A "friend" tells us all about her children and family life, only to slip in that she is anxious about being away while she heads to the Sundance Festival for a screening of her film. The avowals turn into self-promotion—again, nothing entirely new, since this too occurred before, but it is a stark reminder of the entrepreneurial possibilities of social media. In short, our expository society is suffused with an entrepreneurial logic and fueled by the consumer spending, advertising, and maximization of profit that are key features of neoliberalism. The digital technology, it seems, goes hand in hand with a new "invested self"—it both facilitates and at the same time is the product of our neoliberal age.

But although our digital lives may well be traversed by neoliberal logics and practices, we are far from the form of securitarian power today that Foucault envisaged in the late 1970s. The transparence that Facebook, Google, and the NSA seek is complete: *total awareness*. The idea is not merely to optimize awareness given cost functions, but to achieve *total* awareness. To know every single device that is hooked to the Internet. We are beyond a model of equilibrium or cost-benefit. And the reason, very simply—as we will see in Part II—is that the data are a raw resource that is being given away for free despite the fact that it is worth its weight in gold. As a result, there is practically no cost barrier: on the contrary, it is free revenue, nothing but positive gain—which is what makes it so cheap for the PRISM program to run. The information, in effect, is worth billions of dollars, and it is being doled out gratis by individuals every minute of every hour of every day across the globe.

You may recall the Total Information Awareness (TIA) program that Adm. John Poindexter tried to pioneer in the wake of 9/11. The TIA program had originally been introduced as part of the Defense Advanced Research Projects Agency (DARPA), a lab at the Pentagon, but it was shelved in 1999, in part because of the taint that followed Poindexter. (Poindexter had been the highest-ranking official in the Reagan administration found guilty during the Iran-contra affair.) Nevertheless, after 9/11 Poindexter got the TIA refunded to the tune of $200 million, with the intent to outsource much of the research to Booz Allen and public and private universities, such as Berkeley and Cornell.[94] The program was exposed in the fall of 2002 and the funding was scrapped—but the idea, the notion of "total information aware-ness," captures perfectly the ambition of digital exposure and the dif-ference with securitarian logics.

The financial equation in our digital age is completely distinct from that of a security apparatus: there is no need to maximize the benefit to national security while minimizing costs. The digital surveillance ap-paratus can take a *totalizing* approach—for small sums of intelligence dollars. Plus, the information is sufficiently granular that there is no need to remain at the level of a population. It—or shall I say *we*—can go after individuals and target them uniquely. We can obtain total in-formation and zero in on one person at a time, one person only. We can stalk each and every one.

. . .

Google launched Gmail in 2004. What made it so attractive to con-sumers was that Google promised Gmail users a large amount of free space for emails: 1 gigabyte of free storage for each user account.[95] In 2004, that was a huge amount of free storage. Even before the platform went online, many of us were clamoring—and some, in fact, were paying—to get priority access to Gmail. In exchange for the free ser-vice and all the storage space, Gmail users agreed, effectively, to give Google access to all of their emails and attachments as well as access to the incoming emails of any nonsubscriber communicating with a Gmail user.

Google automatically scanned all the email traffic and contents in order to provide targeted advertising—with which we are now familiar because of the telltale ad recommendations that pop up at the top, bottom, and sides of our screens. We've become accustomed to that; it's even sometimes amusing, as similar hotels in the same location or goods that we purchase remind us of the searches we have been conducting over the past few days.

But it turns out that Google's commercial surveillance goes far beyond that. As an investigative reporter, Yasha Levine, discovered:

> Google was not simply scanning people's emails for advertising keywords, but had developed underlying technology to compile sophisticated dossiers of everyone who came through its email system. All communication was subject to deep linguistic analysis; conversations were parsed for keywords, meaning and even tone; individuals were matched to real identities using contact information stored in a user's Gmail address book; attached documents were scraped for intel—that info was then cross-referenced with previous email interactions and combined with stuff gleamed from other Google services, as well as third-party sources.[96]

Based on a close analysis of two patents that Google filed prior to launching its Gmail service, we now know that the company uses a wide range of technologies to construct complex digital profiles of its users—to build our digital selves. These technologies include analyzing the concepts and topics discussed in users' emails, as well as in their email attachments; analyzing the content of the websites that users have visited; tracking demographic information about users, including their income, sex, race, and marital status, and linking this to their geographic information; inferring their psychological and "psychographic" information, such as their personality type, values, lifestyle interests, and attitudes; dissecting the previous Internet searches that users have made; collecting information about any documents that the user has viewed and edited; analyzing their Internet browsing activity; and studying their previous purchases.[97]

The goal is to produce profiles of each user so as to better target advertising and facilitate consumption—or, in the lingo, to make the online experience "more enjoyable," to "help people get what they want." This capability has been enhanced with the other products that Google has introduced: Google Calendar, Google Docs, Google Drive, Google Groups, Google Hangouts, Blogger, Orkut, Google Voice, and Google Checkout.[98] And now Google Inbox—which is going to replace Gmail.[99] As Levine suggests, "Google isn't a traditional Internet service company. It isn't even an advertising company. Google is a whole new type of beast: a global advertising-intelligence company that tries to funnel as much user activity in the real and online world through its services in order to track, analyze and profile us: it tracks as much of our daily lives as possible—who we are, what we do, what we like, where we go, who we talk to, what we think about, what we're interested in—all those things are seized, packaged, commodified and sold on the market."[100] And it does all this to better recommend products to us—products that Google knows we will like, even though we do not yet know it.

THE KNOWLEDGE THAT DIGITAL surveillance is seeking is far richer and individualized than the stuff of biopower. It is not merely about "the right to foster life, or disallow it to the point of death," as in biopower; nor, for that matter, is it "the right to take life or to let live," as in sovereign power (although there is some of that, of course, as evidenced by the drone strikes, including lethal strikes on American citizens).[101] It is not primarily about life and death, nor bare life. It is about every little desire, every preference, every want, and all the complexity of the self, social relations, political beliefs and ambitions, psychological well-being. It extends into every crevice and every dimension of everyday living of every single one of us in our individuality. It accompanies us to every website and YouTube video we surf in the darkness of the late hours. It records our slightest emotional reaction, the most minor arrhythmia of our heartbeat. We are today beyond the model of security, in a type of digital transparence and total costless awareness, that thrives on individualities, differentiation, and efficiency—and that shapes us into our digital selves.

PART TWO

The Birth of the Expository Society

FOUR

OUR MIRRORED GLASS PAVILION

IF ONE HAD TO IDENTIFY a single architectural structure to best capture our expository society in the digital age, it would not be a panopticon, nor the Mall of America or another themed space of consumption, but instead a mirrored glass pavilion. Part crystal palace, part high-tech construction, partly aesthetic and partly efficient, these glass and steel constructs allow us to see ourselves and others through mirrored surfaces and virtual reflections. They are spaces in which we play and explore, take selfies and photograph others. At times they resemble a fun house; at other moments they make us anxious. They intrigue and amuse us. They haunt us. And they hide pockets of obscurity.

The mirrored glass structures of the artist Dan Graham are an excellent illustration. Made of reflective glass and steel beams, open-topped, these sculptures invite us in and capture our imagination. Graham's "Hedge Two-Way Mirror Walkabout," the Metropolitan Museum of Art's 2014 Roof Garden Commission, is precisely a space of seeing, mirroring, virtual transparence, and opacity.[1] It is a glass-mirrored space that reflects the surrounding buildings and the people walking around the garden terrace. The associate curator at the Met,

FIGURES 4.1A AND B Dan Graham's "Hedge Two-Way Mirror Walkabout" (2014)
Source: Photographs copyright © Tod Seelie, reproduced by permission.

Ian Alteveer, calls the "Mirror Walkabout" "not quite sculpture, not quite architecture," but rather a "pleasure palace for play or leisure."[2] It thrives on the same pleasures as our digital exposure, as people gaze and make faces, bend and stare to see how the mirrored glass reflects their image and those of others. One is not quite sure, looking through the glass, whether one is inside or outside, whether we are watching or being watched. As Dan Graham explains: "The experience, as you walk around it, is designed as a fun house, and you get changing concave-convex mirroring situations."[3] It functions by means of participation and produces, through the transparencies, reflections, and mirroring, a new virtual space. Notice in Figure 4.1.A how we can see both sides of the glass: the reflections of those walking by, the many people taking photos and selfies, but also the man with a hat on the other side of the glass. Notice how the buildings are there too, perhaps a bit skinnier than they usually stand; many of the figures are elongated, some superimposed, looking in different directions, always paying attention to their reflection in the glass.

Karen Rosenberg of the *New York Times* describes the "Mirror Walkabout" almost as if she were speaking about our digital exposure:

> The pavilion consists of an S-shaped curve of specially treated glass, bookended by two parallel ivy hedgerows. The glass, which divides the structure into two equal compartments, is slightly reflective; viewers moving around it will see faint but visibly distorted mirror images (and will inevitably try to capture them on their cellphone screens). They will also experience a kind of false mirroring, observing people on the opposite side of the glass. As in Mr. Graham's earlier pavilions and installations, looking goes hand in hand with being looked at.[4]

Dan Graham's glass pavilion somehow epitomizes the virtual transparency and exhibition of our digital lives.[5] It is a space in which we play and make ourselves at home, where we can try to orchestrate our identities, digital selves, and traces, where we create a space for our pleasure, entertainment, and productivity, while we render ourselves exposed to the gaze of others and, of course, our own. We embrace digital exposure with a wild cacophony of emotions, ranging from fetishism and exhibitionism for some to discomfort, hesitation, and phobia for others—and including along the way curiosity, experimentation, play, lust, some distance, resistance, uncertainty, and even disgust or loathing. Regardless of our emotions and desires, though, we can be seen. We are exposed.

Tony Bennett described the original Crystal Palace in London as emblematic of what he referred to as the "exhibitionary complex," the set of institutions such as museums, dioramas and panoramas, great exhibitions, and world fairs that served to make bodies and objects public—in contrast to the set of institutions of confinement.[6] These institutions and their associated disciplines, Bennett wrote, are oriented toward the "show and tell."[7] The exhibitionary complex, he maintained, "reversed the orientation of the disciplinary apparatuses" by making it possible for the people "to know rather than be known, to become the subjects rather than the objects of knowledge."[8] It is almost as if today each and every one of us has become our own cabinet of curiosity and, in the process, built our own crystal palaces to play in and to be watched—to show and tell.

For many of us, there is a certain exhibitionist pleasure as we look at ourselves in the mirrored reflections in the glass in an increasingly narcissistic way brought upon us by the digital pleasure, the addiction and stimulation—and the distraction of these desire machines: our iPhone and Kindle, quad-core devices and tablets. We constantly check how many times our Instagram has been "liked," our Snapchat viewed. How many times our blog posts have been shared or reposted on Facebook. How many times our tweet has been retweeted. The craving for stimulated distraction—a kind of self-centered distraction that reminds us that we are living, present, seen, clicked on, liked—makes us increasingly focus on our own digital traces, our new digital self. We take selfies—in fact, we publish entire books of selfies.[9] We look through the glass at others; we look at our reflection in the mirror. We want to see our reflection as we exhibit and expose ourselves to others.

For many of us, we lose track of time in the digital space, as the experience becomes addictively pleasurable. We are so engrossed, we do not even feel the craving; we just slide from one digital platform to another—or stay glued to the screen playing a video game into the early morning hours. It feels almost as if the digital experience taps into a pleasure node in our brains, or that we are experiencing the high of Huxley's magical hallucinogen soma. It is almost as if we are lab subjects of neuropsychological research and a lab technician is stimulating our hypothalamus—triggering the kind of "seeking" or "wanting" behavior that sends us into an altered state and evaporates time and space.[10]

We feed our machines all the time, hoping for good news, for an invitation, for a like or share, constantly hoping for something pleasant. We peek, check email, scroll through Facebook, surf the web—and such behavior feeds on itself. The connectivity draws us in even deeper. Upload a snapshot, and you will want to see who liked it. Post a comment, and you will want to know if people are reading it, tweeting it, sharing it. There is an addictive aspect to all this. The screen is addictive. It is mesmerizing—at least, for many digital subjects today.

For many others, this new digital existence is disconcerting and unsettling, but practically unavoidable. Even when we resist, there are few other ways to publicize an event, to coordinate a meeting, to communicate overseas. Today there is practically no way to talk to a loved one in another city without going through VoIP—that is, voice over Internet protocol technologies. There is no way to see a distant friend without Skype or Yahoo Messenger. And even when we resist and secure our devices, our emails and communications to others are exposed at the other end of the conversation. We can be seen and watched and monitored, as if we were in a glass house. Much of the time, we do not even realize it.

. . .

The British intelligence initiative was code-named Optic Nerve.[11] It is not known whether the Government Communications Headquarters

(GCHQ), the British signals intelligence agency, is still conducting the program today, although there is evidence it was doing so in 2012. What is certain is that during a period of six months in 2008, the British signals intelligence division surreptitiously intercepted screenshots of the webcam video communications of about 1.8 million Internet users using video chat platforms like those provided by Yahoo Messenger.[12]

Unlike some other programs that only capture metadata, Optic Nerve was able to view the video communications—the actual video images streaming in the chat. Apparently it "automatically downloaded the content of video communications—taking a screenshot from the video feed every five minutes."[13] According to a secret report revealed by Edward Snowden, British intelligence aspired to capture more images at a faster clip and hoped to get the full webcam videos at some point, at least for surveillance targets, with the intent to "identify targets using automatic facial recognition software as they stared into their computer's webcams."[14] (GCHQ was assisted in these efforts by the NSA: "Webcam information was fed into NSA's XKeyscore search tool, and NSA research was used to build the tool which identified Yahoo's webcam traffic."[15])

Apparently the operation netted a trove of X-rated images. An intelligence document regarding the program noted, "It would appear that a surprising number of people use webcam conversations to show intimate parts of their body to the other person." According to one informal analysis, somewhere around 7 to 11 percent of the recorded images contained "undesirable nudity."[16] As the Associated Press reported, "The collection of nude photographs also raises questions about potential for blackmail. America's National Security Agency has already acknowledged that some analysts have been caught trawling databases for inappropriate material on partners or love interests. Other leaked documents have revealed how U.S. and British intelligence discussed leaking embarrassing material online to blacken the reputations of their targets."[17]

As reported by Reuters, Optic Nerve "was intended to test automated facial recognition, monitor GCHQ's targets and uncover new ones."[18] The images were used "to aid selection of useful images for 'mugshots' or even for face recognition by assessing the angle of the face," notes the *Guardian,* quoting the leaked document.[19] Facial recognition technology is used to identify users and connect them to their Internet activity. Explaining how the program functions, the paper reports that the bulk data obtained by "GCHQ's huge network of internet cable taps" was then "processed and fed into systems provided by the NSA" through its "XKeyscore search tool."[20]

Webcam surveillance of this type is permitted in the United Kingdom under the Regulation of Investigatory Powers Act (RIPA), a piece of British legislation that was passed in 2000 "with this kind of surveillance in mind," the *Guardian* reports.[21] Optic Nerve makes use of an "external warrant" provided under what is called Section 8 of RIPA.[22] As the *Guardian* details:

> In most Ripa cases, a minister has to be told the name of an individual or firm being targeted before a warrant is granted. But section 8 permits GCHQ to perform more sweeping and indiscriminate trawls of external data if a minister issues a "certificate" along with the warrant. It allows ministers to sanction the collection, storage and analysis of vast amounts of material, using technologies that barely existed when Ripa was introduced.[23]

Furthermore, while "additional legal authorisations are required before analysts can search for the data of individuals likely to be in the British Isles at the time of the search," no such laws prevent GCHQ from doing so to individuals from the other "Five Eyes" partners: the United States, Australia, New Zealand, and Canada.[24] As Amy Davidson notes, "There may be times when the N.S.A. considers it easier to have foreigners spy on Americans and then get the information that is collected, however private it may be."[25] The "Five Eyes" partnership, it seems, largely involves "trading loopholes rather than actionable intelligence: you can't do this but I can, so let's each do the other's forbidden thing and then talk about it."[26]

These latest revelations sparked serious privacy concerns, in part because of the nudity issues.[27] Three U.S. senators—Ron Wyden, Mark

Udall, and Martin Heinrich, all members of the Senate Intelligence Committee—are investigating the role of the NSA in the GCHQ activities. In addition, on February 28, 2014, "the Internet Association—a trade body representing internet giants including Google, Amazon, eBay, Netflix, AOL and Twitter— . . . [issued] a statement expressing alarm at the latest GCHQ revelations, and calling for reform."[28] Yahoo, for its part, denied awareness of the program. Company spokeswoman Suzanne Philion told Reuters, "We were not aware of nor would we condone this reported activity. This [Guardian] report, if true, represents a whole new level of violation of our users' privacy that is completely unacceptable."[29] GCHQ sought to defend the program in ways similar to early defenses of NSA surveillance. The Guardian reported that "Sir Iain Lobban, the [former] director of GCHQ . . . likened the gathering of intelligence to building a haystack and said he was 'very well aware that within that haystack there is going to be plenty of innocent communications from innocent people.' "[30]

Other documents leaked by Edward Snowden reveal that the NSA has been investigating the possibility of intercepting video game console video communication. "The NSA were exploring the video capabilities of game consoles for surveillance purposes," the Guardian reported. "Microsoft, the maker of Xbox, faced a privacy backlash last year when details emerged that the camera bundled with its new console, the Xbox One, would be always-on by default."[31]

WHAT WE ARE FACING today is not so much a "haystack" as a glass pavilion. A space where we expose ourselves to virtually everyone, at every moment, and simultaneously watch others. A space where, at any moment, we can gaze at others, check their Tumblr or Google Scholar page, look up their Facebook friends, check with whom they are LinkedIn, observe, monitor, even stalk the other. A space where we take pleasure in watching them, "following" them, "sharing" their information—even while we are, unwittingly, sharing our every keyboard stroke. A space where we exhibit ourselves and become the voyeur to others, side by side with the social media and retailers who follow us, and the intelligence agencies and security firms too.

This new form of digital exposure is not a radical rupture. There is an element of the spectacle, naturally. Kim Kardashian's selfies are intended to be spectacular—and intended to go viral. And such strategies often succeed. Vice President Joe Biden's selfie with President Obama, which Biden posted on his Instagram account, almost immediately got more than 60,000 "likes": a virtual public that would fill a football stadium.[32] When we expose ourselves, we step into the arena, virtually, or at least we often hope to—we want to be as visible in our glass pavilion as we are in the amphitheater.

Sovereign power has not been left behind, either. The use of metadata to locate and assassinate targets in Pakistan, Yemen, and Somalia and to guide drone missiles amply demonstrates how classic forms of the juridical exercise of power can be brought into the digital age. Similarly, the monitoring of Facebook and Twitter feeds of suspected gang members by the new social media units of large metropolitan police forces, and the resulting arrests, convictions, and punishments, illustrate well how classic forms of juridical enforcement can be updated with big data. In liberal democracies—not to mention in police and military regimes—earlier forms of sovereign power continue to infuse the digital age. Recall General Hayden's words once more: "We kill people based on metadata."[33]

There is also, obviously, a large dose of surveillance. We are watched and monitored through the glass by practically everyone, from the NSA to our nosy neighbor with a packet sniffer. Practically everyone is trying to monitor and record our tweets and texts and posts, the apps we download, our Internet surfing, how we spend, what we read, with whom we speak—in short, all our activities are captured and analyzed to better target us, to better punish us when appropriate. Bentham's ambition is not entirely distinct from the present digital condition: both share the ambition of "total awareness." But the symmetries and asymmetries are different. Foucault reminds us that "Bentham in his first version of the Panopticon had also imagined an acoustic surveillance, through pipes leading from the cells to the central tower."[34] Bentham apparently dropped the idea because the acoustic surveillance was not asymmetrical and allowed those watched to listen into the central tower as well.[35] (Julius, with his Hegelian *souffle*, continued to pursue the project and "tried to develop an asymmetric listening system.")[36] Today, there is no longer as much concern or fear of symmetry. Today we are far more comfortable exposing ourselves *and* watching others (even though asymmetries survive, as evidenced by Optic Nerve).

And there is also an element of securitarian power today, especially when we consider the massive mining of the metadata of entire populations; but here too, things are somewhat different. The cost equation, in particular. We have entered an age of costless publicity and dirt-cheap surveillance. It costs practically nothing now to disseminate vast quantities of our private information; the only cost, in effect, is the loss of secrecy. Not only is it practically costless, but the information itself is so highly valuable that it has become one of the leading primary resources in contemporary Western society. Yet we practically give it away for free. More often than not, we do it unthinkingly, we have no choice, we don't even know or realize we are giving it away. We have become our own administrators and publicists, routinely disseminating our private information. We spend our time sharing our vitals, inputting our personal data online, downloading or printing or emailing our financial information, becoming our own travel agents and bank clerks. We enter our private information on websites like Orbitz and Travelocity, we share our calendars with Hotwire and Hotels.com, we give away our identifying data to online stores like Zappos and Amazon, we input our social security number on IRS.gov .us. The digital age has birthed a whole new form of value, in which the clients and customers themselves participate in the work process and contribute to the bottom line while simultaneously producing their own surveillance—a kind of surplus surplus-value unimaginable in earlier times.[37]

No, ours is not a radical departure. There are elements of spectacle, surveillance, and *sécurité*. But to these we must add the element of exhibition, of exposition, of exposure. To make visible, to be seen, to be filmed—to post something on Facebook that will hopefully go viral, to tweet something that might get favorited—surely this was not an integral element for the inmate in the panoptic cell. Nor was the fact that we also watch, look, observe, examine, contemplate those who are willing to expose themselves. We are fine today with exposing ourselves *and* watching others. In fact, we make it our business to share data about others. Dating apps aggregate our observations about intimate relations to create and share ratings of potential dates. One dating app, Lulu, "lets female reviewers anonymously select hashtags that describe male associates, from #DudeCanCook to #SexualPanther, which it then translates (via secret algorithm) into numerical ratings."[38] We love watching others and stalking their digital traces. "People love spying on one another," Eben Moglen reminds us. "And Facebook allows them to spy on their friends in return for giving everything up to the boy with the peephole in the middle."[39] Then we all meet again in front of the screen, spectators of the surveillance and the exhibition, viewers monitoring our performances and expositions, watching and watched by means of our texts, our mobile apps, our photos, our posts.

The reciprocity is different from that of the panopticon, the cost equations are different from those of the securitarian apparatus. We live in a peculiar glass pavilion. What we need, in the end, is to better understand the power of virtual transparence, to better grasp the exposure in digital power, to begin to see how our own glass house is built. Neither spectacle nor surveillance nor *sécurité* is fitted entirely properly to the present. Ours is an expository society.[40]

THE AGE OF SPECTACLE provided evidence of the high cost of publicity in ancient Greece and Rome. To render something public was expensive, and so the ancients would gather together, amass themselves to watch, to share, to partake in a public act of entertainment. There was no replay button, nor were there any video feeds and no mechanical arts of reproduction. The modern era of surveillance, on the other hand, gave proof of the cost of security. To render secure was expensive, and so the moderns discovered ways to surveil more efficiently, to see everyone from a single gaze, to turn the arena inside out, to imagine the panopticon. In the digital age today, publicity has become virtually costless and surveillance practically free of charge. A new form of digital power circulates through society. Here, then, are some of its central features.

1. *Virtual transparence.* There is a new kind of transparence, neither literal nor entirely phenomenal, that characterizes our digital age: *virtual* transparence. It is not predominantly literal in the sense that we are not facing a perfectly clear or see-through glass surface or a translucent object.[41] We are not faced with transparency as "a physical fact."[42] The digital medium distorts the presentation of ourselves and of others, allowing us to emphasize certain traits or desires, to see some things better than others. We can create new profiles and change them, experiment, twist them, disfigure them to a certain degree. Not entirely, of course, because all of our clicks and keystrokes are collected, meaning that all our habits and impulses, even the least thought through, form part of our digital selves. But there is sufficient room for distortion to believe that we are not seeing through the self.

In this sense, Philip Johnson's glass house is almost too modern a symbol for our postmodern times—as is Mies van der Rohe's Farnsworth House, the "Miesian glass box."[43] Though an intriguing metaphor—useful in many ways, as we will see, more ways than we might first imagine—Johnson's glass house is almost too literally transparent. Completed in 1949, it reveals more the modern ambition of genuine transparence rather than our current digital condition, where we are able to see more reflections and overlapping spaces, behind as well as in front of us, but where these figures are more often distorted, elongated, somewhat manipulated. "Modernity has been haunted," Anthony Vidler writes, "by a myth of transparency: transparency of the self to nature, of the self to the other, of all selves to society, and all this represented, if not constructed, from Jeremy Bentham to Le Corbusier, by a universal transparency of building materials, spatial penetration, and the ubiquitous flow of air, light, and physical movement."[44] But the modern ideal is "notoriously difficult to attain" and "quickly turns into obscurity (its apparent opposite) and reflectivity (its re-versal)."[45] This is surely the case today.

Our virtual transparence is somewhat more phenomenal, insofar as it reflects an organization, design, or structure that is intended to com-municate the notion of a transparent space.[46] It produces a perceptual effect from the superposition of overlapping planes. But it also does more than that.[47] In part, as we will see, it incorporates phenomenal opacity, a term that Vidler proposes.[48] Or what Laura Kurgan calls "the opacity of transparency."[49] For now, though, it playfully engages with distortions in order to seduce us to look, to take selfies, to expose ourselves—these are the mirroring effects of the fun house that draw us all in. We experiment with new engineered glass products that create certain odd reflections, new materials that affect what we see and how. We add colors, we insert new technology within the sheets of glass, we double and curve the mirror.

The public sculpture by the artist Sarah Braman at the northern tip of Dante Park on Broadway and 64th Street in New York City does just that.[50] Titled "Another Time Machine, 2014," it is a multicolored glass cube that stands eight feet tall.[51] Despite the fact that it is an en-closed cube that one cannot enter, the quality of the glass makes it feel as if one is inside the structure when one looks at it. Through the com-bination of colors, transparence, and reflections, the structure distorts the surrounding environment, placing people and buildings inside it, being transparent at the same time, bringing into it all of the city, the taxicabs, the skyscrapers, Lincoln Center, the passing tourists and opera singers, the selfies and those photographing, ourselves—everyone, in short, becomes part of the cube. It mirrors and reflects and allows us to see through, creating in the process a virtual reality.

The ambition of virtual transparence magnifies the disciplinary ambition of visibility within enclosed structures. Recall that there was an important gradual evolution from rendering visible to transparency during the disciplinary turn. "The old simple schema of confinement and enclosure," Foucault wrote, "began to be replaced by the calcula-tion of openings, of filled and empty spaces, passages and transparen-cies."[52] Rendering visible would develop into internal transparency, to the point that the panopticon itself would "become a transparent building in which the exercise of power may be supervised by society as a whole."[53] Foucault refers to Bentham's panopticon as his "celebrated, *transparent,* circular cage," and places the element of transparency at the center of the panoptic principle: it is what made "architecture transparent to the administration of power."[54] The element of transparency played an important role in the internal structure of the disciplinary edifices. "The perfect disciplinary apparatus would make it possible for a single gaze to see everything constantly," Foucault emphasized.[55]

But in our digital age, we have moved beyond the internality of transparence. Our ambition is to see through brick walls and physical barriers, to turn internal structures inside out, to break down entirely the internal-external differentiation, in order to see into devices and to decipher the invisible. The mirrored glass structure allows us to do that by using reflections to open up spaces and break down walls. The digital technology allows us to do that by transcending the physical obstacles and barriers. It is not by accident that Admiral Poindexter named his program Total Information Awareness. Neither is it an accident that the NSA's "Treasure Map" seeks to know every single device connected to the web.[56] Throughout the commercial and intelligence-security sectors, there is an unparalleled exhaustivity— internal *and* external—to the drive to know.

Google's mission is "to organize the world's information and make it universally accessible and useful" no matter where it is hidden.[57] What that means is making a mass of chaotic, illegible information seeable, readable, usable. And so Google "filters and focuses our queries and explorations through the world of digitized information. It ranks and links so quickly and succinctly, reducing the boiling tempest of human expression into such a clean and navigable list."[58] It sees through all the opaque structures of data to sort and rank and feed us information.

Meanwhile, the stated ambition of the NSA, in its own words, is to "sniff it all," "know it all," "collect it all," "process it all," and "exploit it all."[59] This is from a top-secret NSA presentation at an annual conference with its "Five Eyes" partners in 2011. The goal with that alliance is to "partner it all."[60] And the mission of those allies, at least for the GCHQ, is similarly to "collect it all"—which is precisely what is happening.[61] In a single month in 2013, a single unit of the NSA, the Global Access Operations unit, "collected data on more than 3 billion telephone calls and emails that had passed through the US telecommunications system."[62] And digital technological developments make this possible—just like the mirrored glass allows us to see in practically every direction.

With digital capabilities, the actual content of communications can easily be recorded, stored, matched, cued up, dumped onto computers, and analyzed—which is precisely what the NSA excels at doing. An analyst at the agency, Sgt. Adrienne J. Kinne, describes the process as she sits in front of her computer, with all its listening and digital technology, on which "the phone numbers of the parties as well as their names would appear on her digital screen": "In our computer system, it would have the priority, the telephone number, the target's name ... And you could actually triangulate the location of the phone if you wanted to. We could ask our analysts to figure out the exact location of the phone."[63] This information can then be fed into automated data-mining programs that can conduct sophisticated analyses of the content. As James Bamford explained, back in 2008 before any of the Snowden revelations, "to find hidden links between [intercepted calls], [the NSA uses] a technique known as 'call chaining analysis.' ... To do the call chaining, the analysts use a program known as PatternTracer, made by i2 Inc., a company in McLean, Virginia. 'When suspected terrorists go to great lengths to disguise their call activity, Pattern-Tracer can be used to find hidden connections in volumes of call record data,' says the company."[64] Bamford goes on:

> In addition to PatternTracer, the analysts at NSA Georgia have an alphabet soup of data mining, traffic analysis, and social network analysis tools—secret and unclassified—within their computers. They include Agility, AMHS, Anchory, ArcView, Fastscope, Hightide, Hombase, Intelink, Octave, Document

Management Center, Dishfire, CREST, Pinwale, COASTLINE, SNACKS, Cadence, Gamut, Mainway, Marina, Osis, Puzzlecube, Surrey, Tuningfork, Xkeyscore, and Unified Tasking Tool. The NSA also maintains large databases containing the address information, known as externals, on millions of intercepted messages, such as the Externals Data Global Exploitation (EDGE) database, the Communication External Notation list, and the Communications Externals Data Repository and Integration Center.[65]

The result is a combination of mission, desire, and capacity that is truly unmatched—an ambition for virtual transparence. As Glenn Greenwald suggests, the explicit aim of the NSA is "to collect, store, monitor, and analyze all electronic communication by all people around the globe. The agency is devoted to one overarching mission: to prevent the slightest piece of electronic communication from evading its systematic grip."[66] Virtual transparence is precisely this combination of fun house entertainment, which brings us all in, with the desire and technology of total awareness.

2. *Virtual seduction.* One of the most powerful ways to achieve this total awareness is through our pleasure and lust: to seduce us into buying the most recent smartphone, downloading an irresistible application, clicking on a tantalizing image, giving free rein to our curiosity, addictions, fetishes, and ambitions. To recommend things to us we did not even know we wanted—but do, it turns out. To make us take selfies in the glass mirror and share them with everyone. To want to expose ourselves.

Apple introduced the Apple Watch in a ten-minute film that is enough to lure in the most hardened ascetics. The videography is astoundingly seductive. The voice-over, by Jony Ive, senior VP of design, is perfectly accented with a baritone Commonwealth touch.[67] The background music glides and pulses while the images roll over the screen. Against the hard, shining, polished steel and the slick black lines, there rotates the "digital crown" that "fluidly zooms into apps," "enables nimble, precise adjustment," and gently changes the face of the watch, which is "laminated to a machined and polished single crystal of sapphire, that's the second-hardest transparent material after diamond."[68] The images and the quality of the visuals are stupefyingly beautiful. The videography achieves perfection—and makes even the monastic among us salivate.

"You know, it's driven Apple from the beginning: this compulsion to take incredibly powerful technology and make it accessible, relevant and ultimately personal." The watch is the perfect combination of total individuality and high-tech precision, of the unique individual and technical exceptionalism. It is "a completely singular product" with "an unparalleled level of technical innovation, combined with a design that connects with the wearer at an intimate level *to both embrace individuality and inspire desire.*"

The video is all about inspiring desire, about wrapping this seductive object around yourself—and, in the process, disclosing your innermost being. You touch it, you tap, you press, you push. It senses the difference. With its "taptic engine" and the "four sapphire lenses on back, LEDs and photo sensors," the watch can detect your pulse, your activity, and your movement, when you are standing and when you are running—it gives you a "comprehensive picture of your daily activity." It records it all and rewards you for "fitness milestones." It captures every aspect of your personal life: all your apps, all your mail, your contacts, your messages, your locator information, your friends—any of which you can contact in just seconds—all your family photos, even your heartbeat.

"You can send a quick sketch, or you can even share something as personal as your own heartbeat," Jony Ive tells us in his seductive, sophisticated voice. "These are subtle ways to communicate, that technology often inhibits rather than enables"—while images of red heartbeats pulse on the laminated screen. The technology is all about self-expression and individuality. There is a range of watch faces, which can be personalized both for appearance and for capability, and a wide selection of straps, including sport bands, loops in soft quilted leather, solid metal clasps, stainless steel, and the ever seductive "Milanese loops" that magnetically clasp together. "Creating beautiful objects that are as simple and pure as they are functional, well, that's always been our goal at Apple," Ive tells us in his baritone whisper. "We designed Apple Watch as a whole range of products enabling millions of unique designs, unparalleled personalization both in appearance and capability. I think we are now at a compelling beginning, actually designing technology to be worn, to be truly personal."

Yes, and it is to be worn on your wrist, capturing all your data, transmitting everything you do, from your heartbeat and your pictures to every email and PDF, every app you use . . . clasped to your wrist and transmitted to the cloud. We strap on the monitoring device voluntarily, proudly. We show it off. There is no longer any need for the state to force an ankle bracelet on us when we so lustfully clasp this pulsing, slick, hard object on our own wrist—the citizen's second body lusciously wrapping itself around the first. Virtual transparence functions through seduction.

3. *Phenomenal opacity.* While so many expose themselves on social media, many of the most voyeuristic among us try to dissimulate our gaze and shield our own information. The truth is, expository power functions best when those who are seen are not entirely conscious of it, or do not always remember. The marketing works best when the targets do not know they are being watched. Information is more accessible when the subject forgets that she is being stalked.[69]

The retail giant Target, which excels at data mining and targeting customers, has this down to an art form and has developed "best practices" to make sure that the targeted consumers are not aware they are being targeted. As a marketing analyst at Target explains: "As long as a pregnant woman thinks she hasn't been spied on, she'll use the coupons. She just assumes that everyone else on her block got the same mailer for diapers and cribs. As long as we don't spook her, it works."[70] And so Target will mix in ads for lawn mowers or wineglasses next to diapers and infant clothing. And as long as it looks or feels like the items are chosen by chance, the targeting pays dividends. It depends on creating pockets of opacity.

Frank Pasquale explores this in *The Black Box Society,* where he ably demonstrates how "firms seek out intimate details of potential customers' and employees' lives, but give regulators as little information as they possibly can about their own statistics and procedures. Internet companies collect more and more data on their users but fight regulations that would let those same users exercise some control over the resulting digital dossiers."[71] This is the central recurring problem, in Pasquale's words, of "transparent citizens vs. an opaque government/corporate apparatus."[72]

Over time, these pockets grow and shrink. They vary fluidly over time. 9/11 produced a significant shift in the security apparatus toward greater transparency of intelligence information throughout the defense establishment; the Manning and Snowden leaks will undoubtedly lead to less sharing of intelligence and more pockets of opacity. The same vagaries are seen in the corporate field. Tim Wu documents the transformations at Apple in his book *The Master Switch*. He shows how the company, originally dedicated to openness, gradually evolved into a closed environment, reflecting the historical tendency of information sectors to become closed empires—while nevertheless rendering the rest of us transparent.[73]

The artist Andrew Norman Wilson documents pockets of obscurity at Google, where digital laborers known as the "ScanOps" or "Yellow Badged" workers at the "3.1459 Building," predominantly people of color, are hidden from view, kept in deep secret, and denied the benefits of Google meals, Google bikes, Google shuttles; they cannot set foot anywhere else on Google's campus.[74] They are kept in complete obscurity. They are called, disparagingly, the "digital janitors."

This is, surprisingly, where the modernity of Philip Johnson's glass house, the literal transparence of the structure, offers unexpected insight. You may recall that at its innermost core the glass house contains a closed opaque cylindrical shape made of solid brick—dark brown, rock-solid blocks from floor to ceiling, even protruding out the top of the structure. It turns out that at the heart of literal transparency, there needs to be—there must be—a closed cell. (In Johnson's glass house, it serves as the fireplace/chimney and the bathroom). In addition to the distortions of the mirrored glass, it turns out, we also revert to the shuttered space—the locked trunk, the safe, the closet, the iron cage—and reproduce its epistemology.[75] No one can see in, and from within, no one can see out—or at least, we think.

These pockets of opacity represent, in effect, the opposite of the panopticon: there is no ambition here for ordinary citizens to *internalize* the surveillance, since that would render them less legible and would reduce their consumption. This is certainly true for advertisers. As Siva Vaidhyanathan suggests in *The Googlization of Everything*, "ChoicePoint, Facebook, Google, and Amazon want us to relax and be ourselves."[76] This is also true for the NSA, which has entire units devoted to surreptitiously planting surveillance devices in sealed, packaged routers, servers, and other computer network devices— technologies of surveillance that only work through deception and opacity.[77] Just like the Stasi used vapor to unseal and reseal envelopes so that the recipients never knew their mail had been opened, the NSA magically undoes factory seals and carefully reseals them so that the buyers do not know that backdoor surveillance has been inserted into the machinery. A top-secret NSA document, dated June 2010, leaked by Snowden, states the following:

> Here's how it works: shipments of computer network devices (servers, routers, etc.) being delivered to our targets throughout the world are *intercepted*. Next, they are *redirected to a secret location* where Tailored Access Operations/Access Operations (AO-S326) employees, with the support of the Remote Operations Center (S321), enable the *installation of beacon implants* directly into our targets' electronic devices. These devices are then re-packaged and *placed back into transit* to the original destination. All of this happens with the support of Intelligence Community partners and the technical wizards in TAO.[78]

It is too facile to suggest that the intelligence apparatus wants everyone to feel that they are watched in order to instill generalized fear and control the population better. To be sure, the NSA wants the world, especially other nations and potential enemies, to know just how powerful the United States is and how much computing technology it has.[79] But it also depends and thrives on people forgetting how much they are surveilled. The watching works best not when it is internalized, but when it is absentmindedly forgotten.

4. *Virtual authenticity.* What virtual transparence claims is an even deeper penetration into the "authentic" self. The ambition is to excavate a genuine self, a self that is not just the artifice of advertising and consumerism, that is not just molded by the digital devices—or so we tell ourselves. We want these devices to mine our soul, to excavate deep into the biological, to peel away the psychological.

The sports edition of the Apple Watch is a perfect illustration. Again, the seductive advertisement video—called, this time, the "Health and Fitness Film"—documents how the device serves to better extract deeper personal information, in "one device you can wear all the time": "It can track a wider variety of activities because it's able to collect more types of data. It uses an accelerometer to measure your total body movement, it has a custom sensor that can measure intensity by tracking your heart rate, and it uses the GPS and Wi-Fi in your iPhone to track how far you have moved."[80]

This device will know everything about our body and our selves. This is precisely the promise of big data: such deep and accurate knowledge of our innermost being. The analysis of our heart rhythms will detect and warn us of early symptoms of a heart attack hours before it actually occurs, hours before ordinary medicine would detect anything. There is an urgency to this technology, one that requires the deepest intimacy:

> We wanted to give you the most complete picture of your all-day activity, and not just highlight the quantity of movement, but the quality and frequency as well. So the activity app on Apple Watch measures three separate aspects of movement, with the goal of helping you sit less, move more, and get some exercise.
>
> The movement measures the calories you've burned. That gives you the best overview of how active you are. The movement is customized to you and you close it when you hit your personal calorie goal for the day. The exercise ring captures the brisk

activity that you have done. . . . The stand ring shows how often you've stood up to take a break from sitting.[81]

The Apple Watch integrates the "quantified self" seamlessly. It feeds and nourishes the desire to quantify, to know, to share our true selves. "Over time," the film continues, "Apple Watch actually gets to know you the way a good personal trainer would."[82] It makes the subject feel unique and uniquely satisfied. It makes the subject flourish in his or her individuality. It is catered to satisfy the individual. The uniqueness of the wearable to the individual is key: although everyone will want to buy this object, each one will be individualized, tailored to your color preferences and your favorite texture—whether it's the soft leather band, the cold metallic mesh, or the solid plastic sports band. All that in the service of strapping a device on, voluntarily, to capture all your intimate information.

5. *Digital narratives.* Our digital self is a narrative self, one that we construct through our presentation of self and telling of stories. With apps like Storify, which allows us to connect together our bits of self (those 140-character tweets) into continuous, apparently coherent accounts, we share our timelines and our histories, our innermost thoughts, our secrets. For adolescents and those who have grown up with these mediums, communication begins to take, predominantly, the form of storytelling. This is, according to the sociologist Shan-yang Zhao, a key feature of our digital selves. It is, in fact, in the very "process of narrating to others who we are and what we do, [that] the digital self begins to take shape."[83]

Drawing on the work of John Thompson, Zhao sketches out how we, as digital selves, construct our self-identities online by telling others—others who, importantly, are not "corporally copresent" but only "telecopresent"—our own narratives of self.[84] Zhao quotes Thompson: "To recount to ourselves or others who we are is to retell the narratives—which are continuously modified in the process of retelling—of how we got to where we are and of where we are going from here. We are all the unofficial biographers of ourselves, for it is only by constructing a story, however loosely strung together, that we are able to form a sense of who we are and of what our future may be."[85]

This dimension reflects the *expository* nature of our exposed society—the narrative aspect of our surveillance structures. The visibility and data mining practices piece together stories about our lovers and passions, about our political contributions and disobedience, about our fantasies and desires. They glue together all the little pieces of all different colors to make a mosaic of ourselves.[86] And we piece together little mosaics of others as well. We not only expose ourselves, we surveil others and build narratives around them too. There is no clean division between those who expose and those who surveil; surveillance of others has become commonplace today, with nude pictures of celebrities circulating as "trading fodder" on the more popular anonymous online message boards, users stalking other users, and videos constantly being posted about other people's mistakes, accidents, rants, foibles, and prejudices.[87] We tell stories about ourselves and others. We expose ourselves. We watch others.

6. *Digital avowals.* Finally, we constantly emit little puzzle pieces through our increasingly confessional digital presence: our selfies, our "quantified" data, our reality videos. These acts of self-revelation betray our desire for attention and publicity. The urge may not be new, but the medium changes it, creating a potential audience that could never have been imagined before.

The confessional dimensions of these digital times are marked, first, by a more public, exposed confession. These are no longer purely internal—like the stoic examination of conscience at nighttime—nor limited to a lover or minister. They are logged for others to see and hear. Second, they have an element of permanence. They will be cached somewhere, preserved forever. Even if we erase them or delete them, someone will be able to find them in an unknown part of our drive or the cloud. They are not fleeting or defined by their phenomenal presence. They are burned into the digital in the same way that a mark of penitence tattooed on us might last forever. Third, they are lighter and more malleable than the face-to-face confession: there is no risk of blushing, no body language, no visual cues to absorb. The relationship to authenticity and fiction is looser, more supple. In the digital age, we are not forced to avow, we are not required to perform penance at regular intervals—there are no rules, no cold showers.[88] We embrace avowal more entrepreneurially, something made possible and magnified by the publicity and reach of the new medium.

During the protest marches in New York City in December 2014 following two grand juries' failure to indict police officers in Ferguson, Missouri, and Staten Island, New York, protesters in the streets took selfies and shared them on social media. According to the *New York Times,* thousands of such images were posted "on every social-media platform available, captioned with a condensed, hashtagged version of the sort of political chatter that usually populates timelines and newsfeeds (#ICantBreathe, #NoJusticeNoPeace, etc.)."[89] The selfies documented our presence, proved our concern, told truth about ourselves. They represented our digital expression of outrage—and at the same time curated presentations of the self. In the words of the *Times,* somewhat tendentiously but in part accurately, "The protests in New York City were, in large part, a staging ground for people to take and upload personal images, and that, more than the chanting or the marching, seemed to be the point."[90] Perhaps, more fairly, *a* point.

We avow our different selves differently in different contexts— whether on web-based platforms where we can be identified, like Facebook, or in anonymous chat rooms. Neither is more nor less authentic than the other, but they differ. Mediums affect how we present ourselves and how we confess.[91] Our Facebook selves, for instance, appear, to researchers at least, to approximate more closely the more socially desirable identities that we aspire to have offline but haven't been able to achieve. Some believe that we present less filtered or constructed selves in anonymous environments such as online chat rooms—and that, by contrast, the physical selves we present in localized face-to-face interactions are more highly sensitive to visual cues and physical interpersonal dynamics, to the myriad interactions that Goffman and other symbolic interactionists so carefully identified.

Our curated identities and presentations of self vary based on the technology. In contrast to anonymous chat rooms, for instance, "public proclamations of non-mainstream or gay sexual orientations seemed to be rare on Facebook"—or at least more infrequent.[92] More common are mainstream, heterosexual displays of affection like this one, by a female student in a Facebook study: "I am currently married to a man named xxx [real name was provided originally but removed here to protect privacy]. He is the reason I wake up ever[y] morning with a smile on my face & the reason why I look forward to living another day. He is my lover & best friend."[93] Apparently, only recently are people beginning to post on social media about getting divorced. It is still relatively rare to see "the documentation of strife, anxiety, discord or discontent."[94]

In sum, we might call these six different dimensions of our digital age "expository power." Our new technological capabilities make it possible for us to believe that we can achieve total awareness and simultaneously forget. Through lust, anxiety, and distraction, we are made to want these new digital gadgets, and though we marvel at the risks to our autonomy, the very next moment we strap them on ourselves. Virtual transparence gives us the kind of confidence that makes possible the NSA's "Treasure Maps" and Google's Street View—the gall to believe that these wild dreams are within our reach—and simultaneously the diversion that distracts us into forgetting that it is each one of us that is being mapped and viewed.

THE RAPIDITY WITH which we have built our mirrored glass pavilion is remarkable. The explosion of data and computing capacity is a phenomenon that dates back just a few decades. Facebook was only founded in 2004. YouTube only started in 2005. Twitter was only launched in 2006. Email only got going in the mid-1990s, and mobile phones only became popular around 2000. The rich digital life that we live today only really began in the third millennium. We are dealing with a radically new and young turn-of-the-century phenomenon: we have built our glass pavilion with lightning speed. Almost in a frenzy. To get a sense of it, one need only look at the way in which the digital has so quickly overcome the analog—how "big data" (such a multifaceted term) has skyrocketed past analog technology.[95] It is all radically new.

"Big data" refers to data sets that are so copious and complex that they have to be processed by computing power to be useful. The data themselves include a wide variety of objects, including mobile communications, social media, emails, videos, chats, vital statistics, government census data, genomics, satellites, and sensors.[96] They include all the data produced by emailing, texting, tweeting, and videotaping— in other words, by using computers, mobiles, and any digital device. We are often not even aware of these objects. Hospital and home sensors used to monitor key biochemical markers, for instance, are increasingly important and transmit, collect, and analyze massive real-time flows of information in the health care industry.[97] The space where the data are stored can be Google's servers, hospital computers, AT&T's records, or IRS archives of tax returns. Doug Laney, in 2001, famously pointed out the three variables that make such a data set "big": volume, velocity, and variety. These vast sets of data can then be analyzed, mined, and probed across multiple dimensions and for multiple ends.

They are growing at exponential speed. IDC, a premier IT research and consulting firm, provided a basic measure of the size of "big data" in 2013.[98] The total amount of information stored in what IDC refers to as the "digital universe" is roughly 4.4 zettabytes (a zettabyte is equivalent to $1.18059162 \times 10^{21}$ bytes). By 2020, this number is expected to increase to 44 zettabytes.[99] The *Harvard Business Review* reports, "As of 2012, about 2.5 exabytes of data are created each day, and that number is doubling every 40 months or so."[100]

The most scientific, rigorous measurement of the world's technological capacity to store, communicate, and compute data was collected and presented in a research article in *Science* magazine in April 2011.[101] That article is the source of most of the descriptions today regarding the gargantuan quantity of data and those folkloric metaphors of stacking CD-ROMs of data to the moon. The authors, Martin Hilbert and Priscila López, measure and compare the technological

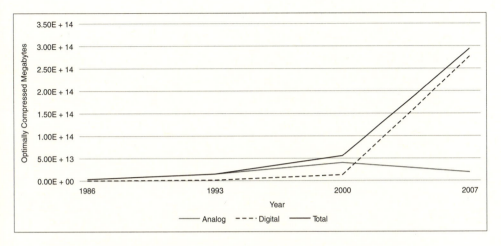

FIGURE 4.2 World's technological capacity to store information, in optimally compressed megabytes (MB) per year, for 1986, 1993, 2000, and 2007. *Graph by* Bernard E. Harcourt. *Data source:* Martin Hilbert and Priscila López, "Supporting Materials for 'The World's Technological Capacity to Store, Communicate, and Compute Information,'" *Science* 332, no. 6025 (2011): 6–7, table S A-1.

capacity to store, communicate, and compute, using analog versus digital technologies, over the period 1986 to 2007.[102] Their findings are remarkable.

The growth in the capacity to store data has been exponential since the turn of the twenty-first century. While overall capacity was originally growing steadily because of analog developments, storage capacity exploded beginning in 2000, as illustrated in Figure 4.2.

Looking at the growth in storage capacity alone, one could almost say that although the foundations were laid in the 1990s, our mirrored glass pavilion started to be built only in 2000. The exponential trend line is equally arresting when it comes to computation. Now, computation has always been digital, so there is no direct point of comparison in this regard, but there is evidence of an exponential increase since 2000 as well, as shown in Figure 4.3.

Third, although a lot of the capacity in broadcast and telecommunication remains analog, there has been a dramatic shift in velocity since 2000 in the digital area. As a result, the overall percentage of these technological capacities that have been taken over by digital media

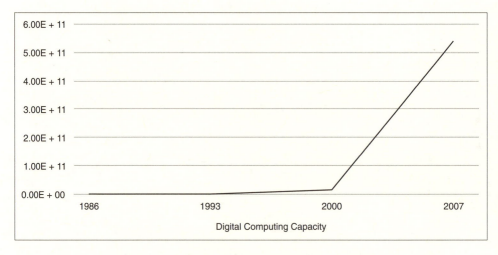

FIGURE 4.3 World's technological capacity to compute information on general-purpose computers (gross usage), in MIPS (million instructions per second), for 1986, 1993, 2000, and 2007. *Graph by* Bernard E. Harcourt. *Data source:* Martin Hilbert and Priscila López, "Supporting Materials for 'The World's Technological Capacity to Store, Communicate, and Compute Information,'" *Science* 332, no. 6025 (2011): 6–7, table S A-1.

increased at a remarkable rate beginning in 2000, as demonstrated in Figure 4.4.

These graphs powerfully illustrate the historical trajectory of our digital universe. Although the cornerstone was set in the 1990s, the construction of the edifice essentially began in 2000. And the overall growth rates are simply staggering. If we pull together data on compound annual growth in all three areas, there is evidence of a 23 percent annual growth in digital storage between 1986 and 2007 and a 58 percent annual growth in computation capacity during the same period.[103] According to the consulting firm McKinsey, the worldwide data totals are expected to increase at a rate of 40 percent per year, with both data storage and data computation capacity continuing to rise.[104] Storage limitation, in fact, is practically the only thing that is slowing down our virtual transparence—including in the signals intelligence area. NSA programs like XKeyscore are collecting such vast quantities of data that a large amount of it can only be retained for short periods of time. So, for instance, in 2012, that program was cap-

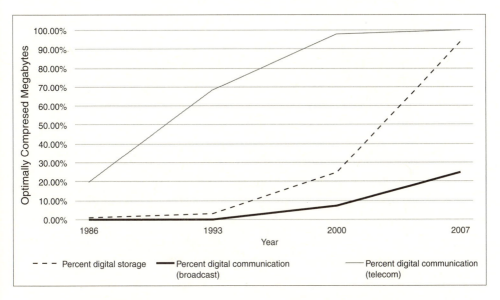

FIGURE 4.4 Evolution of the world's capacity to store and communicate information, percentage in digital format (1986–2007). *Graph by* Bernard E. Harcourt. *Data source:* Martin Hilbert and Priscila López, "The World's Technological Capacity to Store, Communicate, and Compute Information." *Science* 332, no. 6025 (2011): 63.

turing at some sites more than 20 terabytes of data, but these could only be stored for about twenty-four hours because of the vastness of the collected troves. Other documents reveal that certain data were stored for three to five days.[105]

To take another perspective, the IDC measures the prevalence of "digital things" within the total universe of things that manage the physical world—trying to map onto the physical world old analog things, like cars and turbines, alongside the new "digital things." It defines the "Internet of Things" as consisting of "adding computerization, software, and intelligence to things as varied as cars, toys, airplanes, dishwashers, turbines, and dog collars." In 2013, "connected 'things' were 7% of the total" number of connectable things in the world; however, "by 2020, that number will grow to 15%." The Internet of Things is thus exploding in the first decades of the twenty-first century, as "analog functions managing the physical world migrate to digital functions."[106] The Internet of Things is a primary driver of big data accumulation. This is especially the case for mobile devices—these include "devices such as RFID tags, GPS devices, smart cards, cars, toys, and even dog collars"—which in 2014 was expected to "generate 18% of the digital universe" and grow to 27 percent in 2020.[107] In 2011, McKinsey reported some amazing statistics about what it called this "growing torrent" of data and digital things:

- 5 billion mobile phones in use in 2010
- 30 billion pieces of content shared on Facebook every month
- 235 terabytes of data collected by the US Library of Congress as of April 2011
- 15 out of 17 sectors in the United States have more data stored per company than the US Library of Congress[108]

It is estimated that Walmart alone "collects more than 2.5 petabytes of data every hour from its customer transactions."[109] On YouTube, there is nearly an hour of video being uploaded each second.[110] According to the *Harvard Business Review,* "More data cross the Internet every second than were stored in the entire Internet just 20 years ago."[111]

Our mirrored glass pavilion only dates back to the turn of the twenty-first century. The World Wide Web was launched on Christmas Day 1990.[112] Facebook is just over a decade old. As Mark Zuckerberg emphasized in his 2014 Thanksgiving greeting: "In just 10 years, you've built a community with more than 1.35 billion people. You've shared your happy moments and your sad ones, you've kept your friends and families closer together, and you've made the world feel a little smaller and a little warmer.... I hope you have a great Thanksgiving, and thank you for helping to make the world more open and connected."[113] Indeed, it is a world that is more connected and at the same time more open—virtually transparent. And this world has been built practically overnight.

THE ARCHITECTURAL DESIGN—THE ELEMENTS, the structure, the texture of the glass pavilion—is unique, and raises many questions about the actual makeup of this virtually transparent space, what it consists of, and how it can be subdivided. The answers, it turns out, are somewhat murky and depend to a certain extent on the angle or perspective that we take. Big data is described and categorized in a number of different ways—by form, by source, by type, by purpose. Here are a few ways to describe it and to try to understand it better.

One way is in terms of the *sources of data.* The IDC reported in 2014 that roughly two-thirds of the information in the digital universe, or about 2.9 zetabytes (ZB), is "generated by consumers"; the rest, or about 1.5 ZB, is generated by enterprises. Business ends up touching about 85 percent, or 2.3 ZB, of the data.[114] There are certain sectors of the economy that dominate the digital sphere. So, for instance, in 2009, according to McKinsey, the three top sectors relative to the quantity of information stored by sector were (1) discrete manufacturing (966 petabytes), (2) government (848 petabytes), and (3) communications and media (715 petabytes).[115] Government figures second on this list, which reflects the fact that a significant portion of data is still collected and stored by local, state, and federal governments. In the United States, "the amount of information the federal government captures has increased exponentially" since 2000.[116] To give a sense of this, in 2009 alone "the U.S. Government produced 848 petabytes of data and U.S. healthcare data alone reached 150 exabytes. Five exabytes (10^{18} gigabytes) of data would contain all words ever spoken by human beings on earth. At this rate, Big Data for U.S. healthcare will soon reach zetabyte (10^{21} gigabytes) scale and soon yottabytes (10^{24} gigabytes)."[117] The other sectors of the economy that store most of the country's data include banking and investment services, health care providers, retail, education, insurance, and transportation.[118]

Another dimension is the *composition* of the data itself. In this regard, there are a number of ways to analyze and categorize the data:

1. *Smart, identity, and people.* Higinio Maycotte, in an article in *Wired,* suggests that there are three particularly important and prevalent types of big data that should be distinguished: "smart data," "identity data," and "people data."[119] "Smart data" is the term for digital data that has been "siloed, segmented and then visualized" for a particular

need.[120] It has been transformed from its original state as a large collection of binary numbers into something legible that no longer requires an expert to analyze it. "Identity data" is possibly the most important type of data going forward. It is "the force behind predictive modeling and machine learning," and it seeks to "tell the story of who you are in the digital age, including what you like, what you buy, your lifestyle choices and at what time or intervals all of this occurs."[121] It includes social media activity, purchases, behavior analytics, and more. As Maycotte notes, when Target's customer data was hacked in 2013, "it was the loss of identity data . . . that became the biggest issue"; in this case, the identity data stolen included "credit card numbers associated with names and physical addresses, as well as email addresses."[122] Lastly, "people data" are created by "aggregating social data over time."[123] This is collected by looking at the data exhaust of a large number of users: "who your audience likes and follows on social media, what links they click, how long they stay on the site that they clicked over to and how many converted versus bounced."[124]

2. *Social, machine, and transactional.* This first tripartite division overlaps, in part, with another way of dissecting data into "social data," "machine data," and "transactional data."[125] "Social data" refers largely to the "people data" described previously by Maycotte. It is generated mostly by consumer behavior on the Internet. It includes the massive amount of social media data, "with 230 million tweets posted on Twitter per day, 2.7 billion Likes and comments added to Facebook every day, and 60 hours of video uploaded to YouTube every minute."[126] Machine data "consists of information generated from industrial equipment, real-time data from sensors that track parts and monitor machinery (often also called the Internet of Things), and even web logs that track user behavior online."[127] An example of this is the Large Hadron Collider (LHC) at CERN beneath the Franco-Swiss border, the largest research center on particle physics on the globe, which generates approximately 40 terabytes of data per second when experiments are ongoing. Lastly, transactional data is generated by aggregated, recorded transactions from a company's daily business, including the items sold,

"product IDs, prices, payment information, manufacturer and distributor data."[128] Think here of Amazon.com, eBay, and Domino's Pizza—the last of which serves about a million customers per day—which produce huge quantities of big data on a daily basis.

3. *Individuals, public sector, private sector.* The World Economic Forum (WEF), by contrast, tends to think of data as generated from three different sectors. The first is individuals, who provide a particular type of data: "crowdsourced" information, social data from mobile phones, and other media that tend to be referred to as "data exhaust." The WEF emphasizes, in this regard, that "the data emanating from mobile phones holds particular promise, in part because for many low-income people it is their only form of interactive technology, but it is also easier to link mobile-generated data to individuals."[129] And there is tons of this data being produced: "Online or mobile financial transactions, social media traffic, and GPS coordinates now generate over 2.5 quintillion bytes of so-called 'big data' every day."[130] The second involves the public and development sectors. This includes governmental census data, health and vital statistics and indicators, tax and expenditure data, and other "facility data." The third sector is the private sector, and it includes transaction data and information on purchases, spending, consumption, and use.[131]

4. *Structured and unstructured.* Finally, another way to understand the architecture has to do with the texture of the data, whether it is structured or unstructured. The *Economist* reported in 2010 that "only 5% of the information that is created is 'structured,' meaning it comes in a standard format of words or numbers that can be read by computers. The rest are things like photos and phone calls which are less easily retrievable and usable. But this is changing as content on the web is increasingly 'tagged,' and facial-recognition and voice-recognition software can identify people and words in digital files."[132] By 2013, the TechAmerica Foundation reported that about 15 percent of today's data is structured, consisting of rows and columns of data; about 85 percent consists of unstructured information that is humanly generated.[133] This means a colossal amount of unstructured data that is

difficult to process.[134] But new technology is rapidly changing that. A good illustration is face recognition technology, which is rendering video data increasingly useful to intelligence services.

IN THE NINETEENTH CENTURY, it was the government that generated data: statistics, vital records, census reports, military figures. It was costly. It required surveys and civil servants. There was a premium associated with publicity. But now we have all become our own publicists. The production of data has become democratized. And we digital subjects have allowed ourselves to get caught in a tangled web of commercial, governmental, and security projects. Walter Benjamin remarked, "To live in a glass house is a revolutionary virtue par excellence. It is also an intoxication, a moral exhibitionism, that we badly need. Discretion concerning one's own existence, once an aristocratic virtue, has become more and more an affair of petit-bourgeois parvenus."[135] Indeed, it seems that discretion is a thing of the past. At breakneck speed, we have built ourselves a mirrored glass house— translucent to different degrees, reflective in part, distorting as well, with pockets of obscurity. Just like Dan Graham's "Two-Way Mirror Walkabout," we can see ourselves in it and through it, all too well.

FIVE

A GENEALOGY OF
THE NEW *DOPPELGÄNGER* LOGIC

THE CRIMINAL TRIAL OF Arnauld du Tilh in 1560 in the French
Midi-Pyrénées has fascinated generations since the mid-sixteenth
century.[1] It was a remarkable case—truly incredible—of imposture
and identity theft, which carried the risk of convictions for rape, adul-
tery, sacrilege, enslavement, robbery, and other capital crimes. The
question presented to the diocese at Rieux was whether the man who
for more than three years had claimed to be the long-lost Martin Guerre,
who slept with Martin Guerre's wife, gave her two daughters, acted as
a father to Martin Guerre's son, and laid claim to Martin Guerre's estate, was in fact Martin Guerre—or, instead, the man named Arnauld
du Tilh, known by most as "Pansette" (belly), who was from the
nearby village of Sajas, a good day's ride from Martin Guerre's home in
Artigat.[2]

The real Martin Guerre had left his young wife, Bertrande de Rols,
and their son, Sanxi, some eleven years earlier—abandoning them
without leaving a trace. There had been rumors that he had joined the
Spanish army. It had even been said that he had been injured in

battle and had a wooden leg.[3] Eight years after his disappearance, though, during the summer of 1556, a man claiming to be Martin Guerre returned to Artigat and, received by his mother, four sisters, and neighbors as the true Martin Guerre, resumed his conjugal life with Bertrande—eventually living with her "like true married people, eating, drinking, and sleeping as is customary."[4] The man claiming to be Martin Guerre gave Bertrande, in the words of Jean de Coras, the reporter of the case, "several private and personal proofs" that were "much easier to understand than it is proper to speak of or write."[5]

The true Martin Guerre had left some traces, but few: his appearance in the eyes of loved ones and neighbors, personal memories of events and places, perhaps distinct physical marks on his body, the way he touched his wife. But few if any of these could be documented or tested in peasant life during the Middle Ages.

One can imagine, with Natalie Zemon Davis, that Bertrande may have had an inkling that the man claiming to be Martin Guerre was an impostor and that, for a certain period of time at least, she actively conspired with him to conceal the fraud.[6] We may never know the truth of the matter, but it does seem telling that even after Martin Guerre's uncle, mother, and four sisters—in other words, practically every close family member—and many neighbors had turned against the man who claimed to be Martin Guerre, after he had spent time in jail accused of arson, after a fellow soldier had told everyone in the village that the real Martin Guerre had lost his leg serving in battle for the king of Spain, Bertrande nevertheless stayed true to the man. Bertrande "received him [back from jail], and caressed him as her husband, and as soon as he arrived, gave him a white shirt and even washed his feet, and afterwards they went to bed together."[7] Bertrande also physically threw herself to his defense when her uncle and his sons-in-law physically attacked the man, striking him with hard blows. They likely would have killed him had it not been for Bertrande "stretch[ing] herself upon him to receive the blows."[8]

There was, indeed, little by which to test his identity. And so at trial in Rieux, the witnesses compared his appearance to what they remembered, recounted tales of what they knew, and vouched for or against the man. One hundred and fifty witnesses were heard. About thirty or forty testified that they recognized his features or marks or scars on his body. A slightly larger number claimed to recognize another man, Arnauld du Tilh, the man called "Pansette." Sixty or more other witnesses would not commit. And on appeal to the high court of Toulouse, the judges, after hearing from Bertrande—who at this point had finally turned against the man claiming to be Martin Guerre—from the uncle, and from the man himself, concluded that the court in Rieux had erred. The appellate court decided to conduct its own inquiry. Of the thirty-five or forty witnesses heard by the appellate court in Toulouse, nine or ten vouched for the man who claimed to be Martin Guerre, seven or eight identified him as Arnauld du Tilh, and the rest would not commit one way or the other.[9] There were other, special witnesses—six types, according to Jean Coras. Some had particular knowledge or relations that would make them interested parties. For instance, one witness, Carbon Barrau, who was Arnauld du Tilh's true uncle, broke down in tears, "immediately started to cry, and groaned bitterly" when confronted with the man claiming to be Martin Guerre. Others had been witnesses present at contracts, or apparently had firsthand knowledge of the man's fraud.

The visual identification was practically evenly split, and unreliable. As Davis suggests, "the Guerres had no painted portraits by which to recall his features."[10] As for other biological or biographic features that could have served as identification, record keeping was too inexact. "The greater part of the witnesses reported marks and made invincible presumptions—to wit, that Martin Guerre had two broken teeth in his lower jaw, a scar on his forehead, an ingrown nail on a forefinger, three warts on the right hand and one on the little finger, and a drop of blood in the left eye, which marks have all been found on the prisoner."[11] But none of those traits had been documented or recorded—and none of them could be certified.

There was only one piece of evidence of a slightly different character, one data point of a different kind: the size of Martin Guerre's foot. "The cobbler who made shoes for Martin Guerre deposed that this Martin was fitted at twelve points, whereas the prisoner is fitted at only nine," the reporter in Toulouse noted.[12] The cobbler gets little play in the ensuing narrative. Natalie Zemon Davis notes in passing— in a lengthy discussion of the other witnesses—that "the shoemaker told his story about the different shoe sizes, Martin's larger than the prisoner's."[13] Perhaps, as Davis speculates, the cobbler had a motive to testify against the man who claimed to be Martin Guerre. Perhaps the latter had a good response to the cobbler. "What he said, we can only imagine," Davis writes; "to the shoemaker: 'This man is a drinking companion of Pierre Guerre. Let him show his records about the size of my feet. And who else can support his lies?' "[14] Regardless of the man's imagined response, the cobbler, it turns out, had some data: the shoe size of his customers. And he could use the data to check the man's identity. It was a relatively small data set, naturally, subject to error and surely subject to human foibles. It was not digital, of course, and it was by no means "big," like today. But it was some data.

The cobbler's data, in the end, were not determinative and did not weigh heavily, though they did confirm the judge's ruling in Rieux: Arnauld du Tilh was convicted of imposture and sentenced to be beheaded and quartered. The man who claimed to be Martin Guerre was executed a few days later on the threshold of Martin Guerre's house in Artigat. Reflecting on the full episode, Coras writes prosaically about the indeterminacy of physical features, observing that "it is no new thing that two people should resemble each other, not only in features and characteristics of the face, but also in some specific bodily marks."[15]

Today, by contrast, we could so easily identify Martin Guerre and track his movements, follow his ATM and bank withdrawals, locate him with GPS, catch him on CCTV. We could use the FBI's facial recognition technology, Next Generation Identification, to find him at the train station, pub, or ballpark. We could test his fingerprints using the FBI's Integrated Automated Fingerprint Identification System, IAFIS, which contains more than 100 million prints and returns results in less than twenty-seven minutes. (IAFIS processed in excess of 61 million ten-print electronic fingerprint submissions in 2010.)[16] We could use the FBI's electronic Combined DNA Index System (CODIS). We could turn to computerized cryptanalysis. Today there are myriad ways to digitally differentiate and verify an identity. Our identities are etched in the cloud in ways that even the cobbler could not have dreamed of. Computer algorithms can analyze our syntactic style and mannerisms, track our eye movements, study our habits and quirks, collect our discarded fragments, note our consumption, trace our friendships, and read right through us—at the same time as they shape the very identities they predict. It would be unimaginable for most husbands to disappear with such ease, without a concerted, deliberate, and costly effort at covering their tracks.[17]

TODAY, WE HAVE MASTERED identification and differentiation— and, in the process, we have learned to perfectly exploit resemblance. It is indeed nothing new, as Jean de Coras observed in 1561, that "two people should resemble each other." But we have elevated that to an art form. Our digital condition is marked precisely by a new form of rationality that centers on matching each and every one of us, individually, to our twin—to find, for each of us, our other half. Not just to identify—we mastered and surpassed that in the twentieth century—but *to match*. It is what I would call a *doppelgänger* logic: Find the person who fits our consumption patterns and digital behavior, and use that person to anticipate our every action, in a back-and-forth manner, constantly chasing the trace of the last digital click. Record what the other person is searching on the web, which items he is buying, which books he is reading, which sites he is surfing—in sum, know what he wants, in order to predict what the other *will want*. This new rationality suffuses the digital age.

The idea is to find our *doppelgänger* and then model that person's behavior to know ours, and vice versa, in the process shaping each other's desires reciprocally. It is not just feedback but *feed forward*: to look forward by looking backward, back and forth, so that someone is always in front of the other, behaving in a way that the other will want next. You never double back; you are always at the same time leading and following, abuzz with action, shaping and being shaped, influencing and being influenced. Netflix will tell you which film to watch next based on the one I just watched because I watched a film that you had just seen—like a Möbius strip that circulates round and round between us.

This new logic—this *doppelgänger* logic—differs from the forms of rationality that governed earlier, but it grew out of them. The previous models did not involve finding the perfect match. In the nineteenth and early twentieth centuries, the main idea was to categorize people into groups and predict group behavior, intersecting groups where possible to increase accuracy. This started soon after the emergence of state statistics and probability, with a nascent logic of prediction based on actuarial methods and a rationality focused on group membership, on categorical reasoning. With the advent of mainframe computers and multivariate computing abilities, there emerged in the mid-twentieth century a "variables paradigm" that transmuted categorical thought into more linear relationships; with enhanced data storage and analysis, with the power to engage in large-n statistical regressions, a new logic of data control surfaced that sought to identify the exact equations that would properly relate different variables to each other. The actuarial approach gave way to a more algebraic relationship that sought to relate in a multivariate equation the different factors that produced an outcome—looking for causal connections, and trying to capture them errorlessly. Now, the dominant logic is one that tries to match at an individual level, to know exactly, for each one of us, regardless of the causal mechanism or reason. And with all the data we produce, mine, and analyze, it is becoming increasingly possible to do just that.

Today, the capabilities and logic have evolved toward the ideal of the perfect match. Yet there is one thing that runs through these different logics: data control. In the digital age, this means the control of all our intimate information, wants, and desires. This may be novel, but each step of the way, there was a machine, a desire-producing machine, a way to automate and perfect, to mechanize, to model, to take it out of human hands and place it in the object realm: the actuarial instrument, the multivariate regression, today the recommender algorithm. And each step of the way, one leading into the other, there was a progression in exactitude, from the group to the relationship to the individual—or, now, the perfect match.

THERE ARE MANY PLACES one could start—the Renaissance, the classical age, the birth of statistics—but one truly stands out: the birth of actuarial instruments and logics at the turn of the twentieth century. I have written at length and traced the history of the actuarial turn in a previous book, *Against Prediction,* and will not rehearse that history here.[18] Let me be brief with the chronology, then, in order to elaborate somewhat on the context and the form of rationality that arose and helped pave the way to our *doppelgänger* logic today.

The actuarial logic that emerged so robustly at the turn of the twentieth century reoriented thought—or what we might call dispositions of the mind—toward the *group* and *classification,* toward categorical ways of seeing. The use of actuarial tables extended rapidly from the narrow realm of the insurance industry and nascent workmen's compensation systems to many other domains, ranging from criminology and delinquency studies to the prediction of divorce outcomes, employment, and so on. These developments were founded on a logic of group identification. The prediction instruments invented in the 1920s and early 1930s rested on the comparison of group base rates. They depended on a subgroup's relative position vis-à-vis the overall group's behavior: for instance, how did the recidivism rate of inmates with fathers of German ancestry, paroled from Stateville Penitentiary in Joliet, Illinois, compare to the average recidivism rate for all parolees from Stateville? How did the group of "ne'er-do-wells" released from Pontiac prison compare to the average of all those paroled from Pontiac? In the actuarial approach, each point of comparison required dividing the group into subgroups, in order to analyze their comparative rates of offending or other behavior. This was group-based prediction, grounded on membership in categories: the individual could only be understood—analyzed, studied, predicted—in relation to the groups that he or she belonged to.

The turn to a group-based, actuarial logic was made possible by the accumulation of statistics, developments in empirical methods, and a perceived increase in the ability to predict human behavior. It went hand in hand with a probabilistic, risk-based way of thinking that had significant effects on the practice of social control. As Ian Hacking persuasively demonstrates in *The Taming of Chance,* the laws of probability had largely displaced the laws of necessity in much of Western discourse by the late nineteenth century, and the turn from natural law to probabilistic reasoning, paradoxically, facilitated efforts to control human behavior.[19] It was behind one of the first major drives toward social engineering in the early twentieth century. It served to discipline uncertainty, as Nikolas Rose suggests, "by bringing uncertainty under control, making it orderly and docile."[20] "Risk thinking tames chance, fate, and uncertainty by a paradoxical move," Rose explains. "By recognizing the impossibility of certainty about the future, it simultaneously makes this lack of certainty quantifiable in terms of probability. And once one has quantified the probability of a future event's occurring, decisions can be made and justified about what to do in the present, informed by what now seems to be secure, if probabilistic, knowledge about the future."[21]

Predictive risk analysis based on groups—using the laws of chance and the normal curve—would allow for greater control over behavior.[22] "The cardinal concept of the psychology of the Enlightenment had been, simply, human nature," Hacking explains.[23] "By the end of the nineteenth century, it was being replaced by something different: normal people."[24] Normal distributions, the Gaussian bell curve, and the laws of chance would allow us to categorize people into the classifications of "normal" and "abnormal," or "delinquent," or "dangerous."[25] Most of this predictive analysis was first done in the area of delinquency—crime, suicide, madness, and prostitution. Adolphe Quetelet, the Belgian statistician, would write as early as 1832 of the statistical regularities concerning deviancy.[26] He described the phenomenon as a "kind of budget for the scaffold, the galleys and the prisons, achieved by the French nation with greater regularity, without doubt, than the financial budget."[27]

By the 1920s, there was a real thirst for group-based actuarial prediction on this side of the Atlantic—a strong desire to place the study
of social and legal behavior on scientific footing in order to better correct and discipline the delinquent. Doctoral students and researchers
were practically mesmerized by the possibilities. George Vold, a doctoral student in sociology at the University of Chicago, captured well,
in reasoned and deliberate cadence, the sentiment of the times: "The
principle of predicting future human conduct . . . seems sound and
worthwhile. Insurance companies have built up large business enterprises on this principle of predicting the future from the past. The
present study [on prediction methods and parole] . . . seems to establish the validity of that general principle for other phases of human
conduct as well."[28] One of the leading minds behind the turn to group-
based actuarial prediction, University of Chicago sociology professor
Ernest W. Burgess, expressed the sensibility and the aspirations of the
times. "Predictability is feasible," he declared in 1928.[29] "There can
be no doubt of the feasibility of determining the factors governing the
success or the failure of the man on parole."[30]

What emerged were group-based prediction tables for prisoners—
an entire actuarial logic that would categorize and place people with
certain risk factors in designated groups in order to determine their
likely behaviors. It was a radical transformation that inaugurated a
generalized turn to actuaries and actuarial logics. In 1933, by way of
illustration, Ferris F. Laune, Ph.D., was hired as a "sociologist and ac-
tuary" at the Illinois State Penitentiary at Joliet to conduct actuarial
analyses of all potential parolees.[31] Laune would be the first to offi-
cially implement the actuarial method of parole prediction in the
country and to produce the first *prognasio,* as he called it—a report,
based on group-trait offending rates, that evaluated an individual prisoner's probability of violating parole if released. By 1935, group-based
actuarial methods were being deployed in penitentiaries throughout
Illinois.

The actuarial turn reflected three important dimensions, the first
of which propelled the actuarial turn itself, but also the subsequent
stages of technological development.

1. *The desire to know.* There was a palpable desire to know, scientifically, these groups of delinquent men and to predict their behavior. There was a drive to operationalize and model group behavior in the most parsimonious way, a quest for the most efficient instruments to better anticipate future actions. It was the same drive that had inspired the development of statistics in the eighteenth and nineteenth centuries—and which would eventually push the actuarial toward even greater statistical power with multivariate regression analysis.

The desire to count, to predict, to know the group—the desire, in Hacking's words, to tame chance—reflects precisely this desire to control the future. Hacking describes well how this turn to probabilities, originating in the natural sciences, fueled ideas of social control:

> The more the indeterminism, the more the control. This is obvious in the physical sciences. Quantum physics take for granted that nature is at bottom irreducibly stochastic. Precisely that discovery has immeasurably enhanced our ability to interfere with and alter the course of nature. A moment's reflection shows that a similar statement may be attempted in connection with people. The parallel was noticed quite early. Wilhelm Wundt, one of the founding fathers of quantitative psychology, wrote as early as 1862: "It is statistics that first demonstrated that love follows psychological laws."[32]

This is precisely what is reflected so resoundingly in the triumphant declarations of researchers at the time: "Prediction is the aim of the social sciences as it is of the physical sciences."[33] Group prediction is not only feasible, it is *necessary.* "Prediction as to probable success on parole is," Ferris Laune declared in 1936, "a self-evident necessity."[34]

2. *The desire to categorize.* This desire to know went hand in hand with an urge to categorize—to slot people into the right box, to fit them into the right rubric. This was the logic of groups. It formed a categorical way of interpreting reality—a group logic or sensibility that was at one and the same time embraced and promoted, but often resisted as well. There was, simultaneously, a critique of the subjective

nature of prior categories and an embrace of new (but surprisingly subjective) ones to replace them. Younger scholars would criticize older ones, but then ultimately propose similar group labels. The "ne'er-do-well" category, for instance, would be recycled alongside the "irresponsible youngblood," the "weak character," and the "country bully," replacing the "hobo," "drifter," and "tough guy."[35] Here, for instance, are the group labels developed by a younger scholar in 1931:

1. Responsible and substantial citizen
2. Respected workman
3. Irresponsible youngblood
4. Recent immigrant
5. Weak character
6. Ne'er-do-well
7. Transient worker
8. Small town or country "bully"
9. City tough[36]

Despite all the introspection and criticism, scholars would recycle these labels in their research, grasping on to the logic of group identity. So Daniel Glaser, in his 1954 study, would redefine the "social types" into "seven general life patterns toward which the subjects seemed to be developing prior to their offense," but nevertheless retain the "ne'er-do-well" category, as well as the "floater" and the "dissipated."[37] The urge to categorize in groups was simply overwhelming.

3. *The desire to insure.* Third, the group logic was inflected with a desire to insure—to spread risk, to diffuse responsibility, to eliminate blame. François Ewald, in his book *L'état providence* (The welfare state), traces this subtle shift in ways of thinking—in rationality—to the late nineteenth century. It was a transformation from an earlier time, marked by a stronger sense of individual responsibility and accountability, to a logic of risk, categorical harms, and group compensation. With regard to accidents, for instance, the model in the earlier nineteenth century was civil liability, with the associated allocation of fault; it relied on an individualized fault-based regime. The twentieth

century, by contrast, saw the emergence of workmen's compensation regimes: accidents became perceived through the lens of risk and the predictable probability of harm. Employers were no longer seen as responsible for factory injuries; those accidents became part of the process of production; and compensation would become regularized by means of categorical tables of injuries.[38] Ewald traces this moment to an important legal development in France in 1898 that inaugurated a workmen's compensation regime for accidents, a system of automatic indemnities for workers injured in industrial accidents. The law of 1898 launched, Ewald would write, "the philosophy of risk."[39] It was a philosophy based on categories of harm and groups of injured workers. It rested on group-based logics.

These notions of risk assessment and insurability would play an important role in the early twentieth century—one that served to spread responsibility and to reinforce group-based determinations. There was a strong desire to mimic in other domains the supremacy of the actuarial as seen in the insurance industry. In his very first study in 1928, Ernest Burgess had emphasized that "the practical value of an expectancy rate should be as useful in parole administration as similar rates have proved to be in insurance and in other fields where forecasting the future is necessary."[40] Practically all of the subsequent literature would return to this leitmotif.

The comparison to the insurance industry served, in part, to normalize the research and to mainstream its application. The comparison was meant to reassure parole board members. After all, the insurance industry had been using these group-based methods successfully for many years; parole boards did not need to fear them. There was, in addition, an effort to relieve the decision makers of responsibility—to turn parole decisions into automatic or automated processes about *groups,* not individuals. The goal of analogizing to insurance, at least in part, was to assuage those who had to make difficult decisions, to make sure that they could sleep at night.[41] Over and over again, the insurance analogy and the model of group prediction were deployed to placate concern. The focus on the group was key to assuaging the anxieties of the time. The insurance model would calm nerves. So, for instance, Walter Argow would write in 1935, speaking about parole prediction: "This idea is not totally new or peculiar to our field, for insurance companies have been using a similar device to compute the 'probable life-range' of an individual on the basis of data regarding others in similar circumstances."[42] By using a rationality that focused on groups, the idea was ultimately to create a machine—a mechanistic way—to solve problems.

THE MID-TWENTIETH CENTURY would experience a genetic mutation from this logic of groups to a more individualized, causal, linear rationality represented by multivariate statistical analysis. In large part, enhanced computing power caused the transfiguration, though it was also driven by a similar dream of machine knowledge and fantasy of social control.[43] In the process, a different rationality would emerge, an entirely different logic based not on groups but on relations between individuals.

At the height of the Cold War in the late 1950s and early 1960s, a distinctive type of reasoning called systems analysis was perfected and began to be applied broadly to matters ranging from nuclear defense strategy to domestic crime policy.[44] The brainchild of the RAND Corporation, systems analysis drew on the innovations of "operations research" and the knowledge and techniques developed in the U.S. Army Air Force's Office of Statistical Control during World War II.[45] The idea of statistical control itself came from Harvard Business School and the management sciences, which had looked increasingly to the model of data control—and to the very notion of "control"—to improve manufacturing, production, and budgeting processes.

The Army Air Force had turned to Harvard Business School in 1942 to set up a training camp for their new officers of statistical control, and the Air Force College began operating at Harvard that same year. Officers would learn there everything they needed to know about data collection and quantitative analysis to effectuate studies of bombing accuracy and every other kind of efficiency analysis regarding refueling, repairs, manpower, stocking, and so on. From there, the "stats control" officers would be scattered across the globe to different air divisions to engage in data collection, data mining, research, and statistical analyses. The officers would then report back to their superior commanding officers with what they had discovered: the optimal altitude, for instance, or the optimal flying formation for a bombing mission in order to maximize the accuracy of the operation and minimize casualties and equipment losses. But for that, of course, it was necessary to collect more and more data—about sorties, bombing accuracy, losses, weather, cloud formations, wind conditions, and flying formations. More data were needed, and better machines to store the data, retrieve it, and analyze it. Everything needed to be machine accessible. And, of course, better mathematical and statistical models were needed to figure out the relations in the data. To mine the data. To discover a linear relationship between the individual observations.

Statistical control would mushroom after the war. Data-driven systems research would expand dramatically as the research wing of the air force peeled off and become a major research institution, the RAND Corporation, in 1948. (The air force itself had become a separate branch of the military in 1947.) Statistical control would migrate to the private sector as former "stats control" officers returned to civilian life—several of them, notably, moving to Detroit, Michigan, to head research and management at the Ford Motor Corporation. And statistical control would make its way back into government and the core functions of governmentality when, for instance, Robert S. McNamara took over the Pentagon in 1961 as secretary of defense and, a few years later, President Lyndon B. Johnson imposed the methods of systems analysis across the board to all federal agencies under the rubric of "planning-programming-budgeting systems" analysis. It would reach down into more localized forms of government as locally elected politicians began to deploy the same technologies. John Lindsay would be elected mayor of New York City in 1966 and take office promising to reform city government—specifically, by using planning-programming-budgeting systems analysis "to improve budgeting and operations."[46]

As the private sector, federal government, local officials, higher ed-ucation, and research all turned to methods of statistical control, there was increasing pressure to collect and store more data, to develop com-puterized algorithms, to mine and analyze the data, and to build larger and better machine tools to read, store, manipulate, and organize the massive quantities of data being accumulated. But it was not simply a question of collecting more data and analyzing it. There was also a cer-tain logic, a certain type of rationality, that emerged with statistical control and would expand, extending into every sphere and colonizing every domain. It was a type of rationality, a way of making decisions, that focused on the *relationships between individual observations in a large data set* and that let those relationships drive the decision making. That demanded more and more data in order to let the data *speak*. To let the data do the work.

The logic was simple: all that was needed was a lot of data about dif-ferent possible policies and a clear, singular objective in order not just to evaluate the relationships individually (itself a data-intense project) but to compare the evaluations of different programs. The data anal-ysis and computations could determine the correct ordinal ranking of the policy alternatives along any criterion. This would also eliminate the need for political wrangling, for value judgments. Data-driven systems analysis, in effect, would put aside partisan politics, personal preferences, subjective values, and overinflated expectations. As a RAND analyst and future secretary of defense, James R. Schlesinger, would explain: "[Systems analysis] eliminates the purely subjective ap-proach on the part of devotees of a program and forces them to change their lines of argument. They must talk about reality rather than mo-rality."[47] And all that was necessary—that is, necessary to avoid talking about morality—was a lot of information and good statistical anal-yses. If we could just collect more and more data about all of these dif-ferent policies and grind them through our machine models, it might be possible to optimize any objective.

Data-driven, statistical systems analysis: a machine that compares relationships. This was precisely the rationality that McNamara would impose on the military upon taking office as secretary of defense. Mc-Namara's objective was to refine and perfect military decision making and, beyond that, to gain control of the military establishment. As a civilian with no combat experience, McNamara needed to find a way to assert and maintain legitimacy with, and authority over, military generals. And that is precisely what he was trying to achieve by imposing at the Pentagon a new form of reasoning—an entirely new form of economic discourse that the military generals had no access to and no proficiency in.

The drive for data and computational power during the Cold War would have profound transformative effects on ways of thinking, on forms of rationality. It would induce a foundational shift from a group-based, actuarial logic to an individualized, "fitted" way of thinking, associated with multivariate regression analysis. It would move us from the actuarial table to the regression equation.

The drive for more data and greater computing power would eventually make possible the kinds of complex statistical analyses unimaginable in an earlier day—from ordinary least-squares regression analyses to hierarchical linear or fixed-effects models, logit and probit models, and Poisson regression models. All of these could now be performed in nanoseconds on mainframe computers, but would have taken days or weeks or months to accomplish only a few decades earlier. This computational power would transform our way of thinking, reorienting us from those earlier two-dimensional group-based tables to the statistical equation with its coefficients, error terms, and confidence intervals. The statistical analyses could tell us the function, the curve, the contributing factors. They would offer a more individually tailored, more fine-grained probability for each individual observation, for each individual case, and highlight the potential magnitude of error. They would shift the logic from the group to the fitted individual.

THE SHIFT FROM THE STATISTICAL LOGIC of the late twentieth century to today's digital era would represent another quantum leap, a leap from individualized statistical prediction to the "match"—the logic of the *doppelgänger*. The object of the algorithmic data-mining quest of the digital age is to find our perfect double, our hidden twin. It deploys a new rationality of similitude, of matching, without regard for the causal link. The goal, the aspiration, the object is to find that second identical person, almost practically perfectly individualized, but not so individualized that she cannot be matched: not the unique individual, perhaps, but rather the matched *duodividual*.[48]

Expository power represents a particular form of power that deals less with total populations, groups, or individuals than with the twins, the matches, the doubles. NSA surveillance and Google recommendations are not about groups or populations but about the *doppelgänger:* they are about matching, not about populating. It is about finding our twin in order to get us to download more apps and to do more on our computers—and to expose ourselves even more to the digital itself. It is about identifying our digital self by matching us to our digital double—what Kevin Haggerty and Richard Ericson might refer to as our "data double."[49] It requires making sure that everything we do goes through the digital sphere, so that everything gets counted, all the clicks, all the watches, all the likes, all the attention, and also so that everything we do is seeable, transparent, surveillable. It works by accumulation and multiplicity. More is better. More information about us is needed so that we can be better matched to our twin. So that we can know more accurately, recommend more precisely.

A top engineer at Netflix, which has become a leader in the field, explains the shift in logics from group prediction to matching: "Testing has shown that the predicted ratings aren't actually super-useful, while what you're actually playing is. We're going from focusing exclusively on ratings and rating predictions to depending on a more complex ecosystem of algorithms."[50]

Netflix employs about 800 engineers at its headquarters in Silicon Valley to do just this. The goal is to refine recommend technology to the utmost. And they seem to be succeeding. In March 2013, Netflix sent out its 4 billionth DVD, and in the first quarter of 2013 alone, Netflix streamed more than 4 billion hours of film. Of this, about

75 percent was driven by recommend technology.[51] The key, according to Carlos Gomez Uribe, vice president of product innovation and personalization algorithms, is just to look "at the metadata."[52]

The algorithms themselves are extremely complicated and derive from different kinds of technologies, depending on whether they are memory-based or model-based and whether they take a content-based or collaborative filtering systems approach. For instance, memory-based algorithms can take a collaborative filtering approach and try "to find users that are similar to the active user (i.e. the users we want to make predictions for), and use their preferences to predict ratings for the active user."[53] The competition to improve these recommend technologies is intense. In 2006, Netflix offered a million-dollar prize to anyone who could improve its recommend algorithm, Cinematch, by 10 percent.[54] In 2014, Twitter opened up the source code of one of its algorithms, Dimension Independent Matrix Square using Map-Reduce (DIMSUM), in order to get other computer techies to improve on it.[55]

Although the science behind the recommend algorithms may be extremely complex and require hundreds of engineers—and is too intricate to detail here—the logic of matching is quite simple. As Xavier Amatriain, engineering director at Netflix, explains:

> We know what you played, searched for, or rated, as well as the time, date, and device. We even track user interactions such as browsing or scrolling behavior. All that data is fed into several algorithms, each optimized for a different purpose. In a broad sense, most of our algorithms are based on the assumption that similar viewing patterns represent similar user tastes. We can use the behavior of similar users to infer your preferences.[56]

Recommend technology is key to Netflix's success, as it is at Amazon.com and many other digital retailers. As Amatriain of Netflix explains, "Almost everything we do is a recommendation.... Recommendation is huge, and our search feature is what people do when we're not able to show them what to watch."[57]

The next frontier is to refine these matching algorithms by exact hour and location—to completely contextualize the recommend technology so we know exactly what changes depending on the hour of the day or where you are. As Amatriain explains, Netflix has "data that suggests there is different viewing behavior depending on the day of the week, the time of day, the device, and sometimes even the location."[58] Netflix does not yet know how to capitalize on this data, but will soon. "We are currently working on [it]. We hope to be using it in the near future."[59]

All these matching algorithms serve to nudge our desires and curiosities in an infinitely looping forward-and-backward motion, as we each glom on to new and different "duodividuals." And the algorithmic logic coaxes us to believe in its truth. As Tarleton Gillespie writes, "Google and its algorithm help assert and normalize this knowledge logic as 'right,' as right as its results appear to be."[60]

. . .

Finding each of our *doppelgängers* takes work. Cookies help—or helped. So did "people-based marketing," which can follow us on our mobile phones when cookies fail. But now there is even more: "onboarding."[61] Onboarding is the practice of merging all the digital knowledge we have about an individual, culled from cookies, web tracking, and data mining, with the physical or material information we have about that person as well.

The link happens when someone sheds some of his digital information in person, at a store, for instance. This allows the trackers to attach a real identity to the digital address. For example, when a shopper at the mall gives his email address to the store at the checkout counter, a data broker can take all of the physical information about the buyer (credit card, name, purchases, and any other identifying information) and link it, through the buyer's email address, to an IP address. And with the IP address, there is full access to all the buyer's web activities—surfing, buying online, email contacts, social media, and so forth.

The tracking power here is truly awesome. In the words of Scott Howe, chief executive of one of the largest data brokers in the United States, Acxiom, "The marriage of online and offline is the ad targeting of the last 10 years on steroids."[62]

Many social media are now using and offering these digital advertising steroids. Twitter and Facebook, for example, offer their advertisers "onboarding."[63] Twitter reports to its advertisers how it works. This is directly from the Twitter blog, straight from Kevin Weil (@kevinweil), who is Twitter's "Senior Director of Product, Revenue":

> Let's say a local florist wants to advertise a Valentine's Day special on Twitter. They'd prefer to show their ad to flower enthusiasts who frequent their website or subscribe to their newsletter. To get the special offer to those people who are also on Twitter, the shop may share with us a scrambled, unreadable email address (a hash) or browser-related information (a browser cookie ID). We can then match that information to accounts in order to show them a Promoted Tweet with the Valentine's Day deal.[64]

In case you are wondering what a "hash" is, Facebook explains it nicely here: "Hashing summarizes text into a short fingerprint that can't be decrypted."[65] Apparently Facebook uses this hashing technology a lot in the context of trying to match email addresses and other information such as cell phone numbers to users. As the company explains on its website, "Facebook calculates the hash of the email addresses and phone numbers that people have given us and stores these hashes with the corresponding person."[66]

What this kind of onboarding technology permits social media, data brokers, and advertisers to do is to identify with much greater precision each of our *doppelgängers*. Facebook refers to those twins of ours as folks who are in our "Lookalike Audiences."[67] That's a good synonym—"lookalike"—and it reflects well the underlying rationality: to try to match an individual to another. Facebook explains its "Lookalike Audiences" feature in relation to its other advertising tool, "Custom Audiences," which are compiled "using email addresses, phone numbers, Facebook user IDs or mobile advertiser IDs."[68] Here's how Facebook explains it all, in lay terms:

> What are lookalike audiences?
>
> Lookalike audiences let you reach new people who are likely to be interested in your business because they're similar to a customer list you care about.
>
> Lookalike audiences help you reach people who are similar to your current customers for fan acquisition, site registration, off-Facebook purchases, coupon claims and brand awareness.[69]

And once you have a "Lookalike Audience," you can then "optimize" the matches using other tools (again, straight from the Facebook page):

> What does it mean to optimize for "Similarity" or "Greater Reach" when creating a lookalike audience?
>
> Lookalike audiences help you find the people who are most similar to the people in your Custom Audience.
>
> · When you optimize for Similarity, your lookalike audience will include the people in your selected country who are most similar (down to the top 1%) to your custom audience. The reach of this new audience will be smaller, but the match will be more precise. This option is best when you want a very specific match or have a limited budget.
> · When you optimize for Reach, your lookalike audience will include up to the top 10% of people in your selected country that are similar to your custom audience, but with a less precise match. This option is best if you want to get your message in front of more people who are likely to be interested in your business.[70]

THE LOGIC HAS SHIFTED toward the lookalike, the *doppelgänger.* The transformation from data to digits has changed the way we think.

There are, of course, some continuities and commonalities. All three rationalities share a common desire to know and a thirst for prediction—for forecasting deviance and consumption, for example. Each values objectivity, the scientific method, and efficient technologies. In a very real sense, each one of these logics aspires to a form of positivism that avoids morality or anything approximating a normative judgment.

But at the same time, they are deeply different, these logics. The new matching technologies are potentially more accurate—at least with respect to our digital selves: they reflect progress from a rudimentary focus on group traits (actuarial) to more individualized determinations (regression equations) to the perfect match (algorithmic data mining). They may be more powerful computationally: they evince a continuously growing technicity and exponentially increased computational abilities. They present as more objective, if possible: they are devoid of the moralizing touch. Consider that some of the early actuarians would refer to "white marks" and "black marks"; for example, Clark Tibbitts wrote in 1931, "A record of no work . . . which shows a violation rate of 38.5 percent would be an unfavorable sign or what we have chosen to call 'a black mark,' while a good work record with only 5.6 percent failure would be favorable or 'a white mark.' "[71] Notice how seamlessly the science slipped into morality back then. Not so with statistical control, and certainly not with data mining in the digital age. We have entered the age of amoral machine logics.

And just as there is movement away from morality, there is also today a gradual shift away from causality. The algorithmic approach does not care anymore about causation, as long as it has a match. Kenneth Neil Cukier and Viktor Mayer-Schoenberger develop this insight in their book *Big Data: A Revolution That Will Transform How We Live, Work, and Think.* They run through the ever increasing number of examples where correlations alone—at Amazon, at Walmart, at Target, at UPS, at the Centers for Disease Control and Prevention—result in nontheorized solutions. Where just the collection and analysis of millions of data points produce connections that solve problems. "Causality won't be discarded," they predict, "but it is being knocked off its

TABLE 5.1

The birth of a doppelgänger *logic*

1920s	1960s	2010s
Actuarial Logics	Statistical Control	*Doppelgänger* Logics
Group-based predictions	Relationships between individuals	Matching technology
Actuarial prediction	Statistical regression	Data mining
Actuarial tables	Regression equations	Recommend algorithms
Ernest Burgess	The RAND Corp.	Netflix/Amazon/NSA
Mechanical calculator	Mainframe computer	Supercomputer
Moralistic	Amoral	Acausal

Table by Bernard E. Harcourt.

pedestal as the primary foundation of meaning. Big data turbocharges non-causal analyses, often replacing causal investigations."[72]

What we have, then, is a genealogy of logics that have fed into each other, culminating in the present *doppelgänger* rationality of the expository society (see Table 5.1).

This is not to suggest that in our new digital age, the prior logics no longer hold sway. To the contrary, they continue to dominate in particular fields—actuarial logics in criminal justice, for instance, the variables paradigm in political science. Logics have long shadows—or bad hangovers. Technological innovation takes a long time to colonize a field, but when it does, it lasts and lingers for decades. Burgess introduced actuarialism in the parole decision context in the late 1920s, but it is only today that the approach has colonized the field. It was only when the federal government began to use an actuarial instrument, the Salient Factor Score, in the early 1970s that many other states would gravitate toward the actuarial method. Actuarial logics took about fifty years to take hold of the criminal justice field, but now they have a lock on the business. Similarly, the discipline of political science is today overrun by regression analyses and rational choice models. It may have taken a number of years and much conflict in the discipline, but the variables paradigm ultimately prevailed—and is hanging on, despite new methodological developments, the advent of experimentalism, the emergence of data mining, and so on. These logics come with a heavy investment in methodological training, a buildup of know-how, and steep professional costs. They do not cede easily. They often overstay their welcome.

Hegemony takes time. In a number of domains, the *doppelgänger* logic has not yet displaced these others. But it is all around us. We experience it daily in our routine activities, in our social relations, in the commercial sphere—and, for many, in the carceral field as well. The new logic of the digital match is here, and growing.

It is a new logic, a new rationality—but it evinces a much older drive to know. The drive to quantify everything, the ambition for total awareness has been with us for centuries. Recall the "paper-squeeze" of Jacques-François Guillauté. Or think back further to William Petty, who in his *Political Arithmetick,* published in 1691, proposed to measure and to count everything. "Instead of using only comparative and superlative Words, and intellectual Arguments, I have taken the course (as a Specimen of the Political Arithmetick I have long aimed at) to express my self in Terms of *Number, Weight, or Measure,*" Petty declared.[73] Petty even calculated the value, in pounds sterling, of "each Head of Man, Woman, and Child" in England, using a complex or convoluted method of data extraction and inference:

> I shall here digress to set down the way of computing the value of every Head one with another, and that by the instance of People in *England, viz.* Suppose the People of *England* be Six Millions in number, that their expence at 7 *l. per* Head be forty two Millions: suppose also that the Rent of the Lands be eight Millions, and the profit of all the Personal Estate be Eight Millions more; it must needs follow, that the Labour of the People must have supplyed the remaining Twenty Six Millions, the which multiplied by Twenty (the Mass of Mankind being worth Twenty Years purchase as well as Land) makes Five Hundred and Twenty Millions, as the value of the whole People: which number divided by

Six Millions, makes above 80 *l.* Sterling, to be valued of each Head of Man, Woman, and Child, and of adult Persons twice as much; from whence we may learn to compute the loss we have sustained by the Plague, by the Slaughter of Men in War, and by the sending them abroad into the service of Foreign Princes.[74]

These were, perhaps, wild ruminations—a quixotic drive for exact empirical measurement, quantification, and knowledge—but Petty was not alone. Richard Cantillon, in the eighteenth century, had the same ambition. In his *Essai sur la nature du commerce en general,* published in 1755 (though written and distributed to specialists as early as 1728), Cantillon included an entry for "coût de la vie" (cost of life).[75] There he calculated, with exactitude, the amount of land necessary to sustain a peasant and, separately, a bourgeois: "le produit d'un arpent et demi de terre de moyenne bonté, qui rapporte six fois la semence, et qui se repose tous les trois ans" for the peasant, and "le produit de quatre à cinq arpents de terre de moyenne bonté" for the bourgeois.[76] These ruminations proved to be fertile. Marx quoted Petty extensively in his *Capital,* specifically regarding Petty's theory of surplus value, as did Adam Smith.[77] And the development of political arithmetic was important to the larger scientific development in the seventeenth century, exemplified especially by Petty's mentor, Thomas Hobbes, for whom "reality expresses itself by means of a quantitative structure."[78] The drive to quantify, to mine data, and to achieve total awareness was not born yesterday. The *doppelgänger* logic, though, is new—and has significant and unique effects on our subjectivity today.

SIX

THE ECLIPSE OF HUMANISM

THE BIRTH OF THE expository society has gone hand in hand with a gradual erosion of the analog values we once prized—privacy, autonomy, some anonymity, secrecy, dignity, a room of one's own, the right to be let alone. Commenting on a survey of British consumers—which found that most respondents "were happy to have companies use their personal data, on the condition that they receive something in return"—Kevin Haggerty and Richard Ericson note that "privacy is now less a line in the sand beyond which transgression is not permitted, than a shifting space of negotiation where privacy is traded for products, better services or special deals."[1] Facebook founder Mark Zuckerberg infamously remarked in 2010 that privacy is no longer really the social norm.[2] Truth is, most Americans have reacted little to the Snowden revelations and the stories of social media and corporate surveillance. Only three-fourths of American Internet users report that they are aware of the Snowden affair, and of those, only about a third report taking steps "to protect their online privacy and security as a result of his revelations."[3] According to a Pew survey from June 2013, "A majority of Americans—56%—say the National Security Agency's (NSA) program tracking the telephone records of millions of Americans is an acceptable way for the government to investigate terrorism."[4] Overall, there is a lack of concern about ongoing infringements of our privacy, whether by corporate giants like Google, Microsoft, and Facebook or by security agencies like the NSA, Britain's GCHQ, or France's DGSE. The most significant popular reform to date, the USA FREEDOM Act, barely scratches at the surface of the NSA's myriad boundless collection programs.

Many of us have become complacent today—that is, when we are not actively craving publicity, or learning to love what we can't do without. We have gotten accustomed to the commodification of privacy, of autonomy, of anonymity—to the commodification of what D. A. Miller refers to as that "integral, autonomous, 'secret' self."[5] Whereas we once viewed privacy and dignity as necessary ingredients for a fulfilling life, as basic human needs, as the psychic equivalent of air and water, today we tend to view them as market goods, as commodities that are to be bought or sold in a marketplace. This has coincided with a larger shift toward a neoliberal worldview, in which market rationalities dominate every sphere of life, including the social and personal. We have begun to think of ourselves, more and more, as calculating, rational actors pursuing our self-interest by means of cost-benefit analyses that convert practically every good into commoditized form. This way of thinking and behaving has had tangible effects, not least on our valuation of privacy, autonomy, and anonymity.

In an earlier period, in the 1950s and 1960s, privacy was thought of along more humanistic lines—as something that we needed, like air or water, for our very existence. Autonomy and anonymity were viewed as integral parts of our environment and ecology, as essential ingredients necessary for human beings to thrive. We needed a room of our own, space to think and experiment, a place to be left alone.[6] We had a more spiritual or poetic appreciation of these values. With the increased commodification of things today, though, privacy has itself been transformed into a type of good that can be traded, bought, or sold. Rather than a *human* property, privacy protections have become *private* property. Privacy has been *privatized*.

Although not logically necessary, this gradual transition from a humanist mind-set to a rational choice framework has coincided with a shift in our conception and valuation of autonomy and anonymity—from an earlier period marked by the idea of a penumbral protection necessary for human existence to the more contemporary view of privacy as a commodity that has to be purchased. And even though there is no necessary relation between the commodification of privacy and its evisceration, there nevertheless has been a historical correlation between the dominance of market rationalities and the eclipse of privacy and anonymity. Ideas have effects. They may not always be necessary effects, in the sense that they could well have produced other consequences, but often they have concrete effects, actual impacts. And in this case, there is a natural affinity that has tied the commodification of privacy to its lower valuation. The point is, privacy fared a lot better when it was seen as an integral part of our human essence than when it is placed on a scale to be measured, monetized, and sold on the free market.

As privacy has become commodified, many of us have begun to question its value. Many no longer believe that there is anything fundamentally essential about a more private or anonymous human existence. Many have lost faith in the autonomous realm of the self—in that mysterious physical space that once surrounded and protected us, in that cocoon of self and privacy. Many have embraced a more instrumental view of life, where everything, including privacy and anonymity, is up for grabs. This is tied, in part, to a broader tendency toward the commodity form and neoliberalism during the past several decades. With the fall of the Berlin Wall and the collapse of the Soviet Union, there has come a renewed faith in market efficiency and notions of natural order—in the Mandevillian fable of the bees, the principle of "private vice, public virtues."[7] In the idea, in short, that each and every one of us should simply focus on our personal desires and self-interests. This is particularly true in the digital sphere, where entrepreneurial dot-coms and start-ups rise and fall, where the free market and free enterprise reign. Siva Vaidhyanathan describes this tellingly in *The Googlization of Everything:*

In the 1990s—heady days of global prosperity, burgeoning freedom, and relative peace—I saw in digital networks the means to solve some of the problems we faced as a species. Back then I took seriously the notion that the world had stepped beyond the stalemate of the Cold War and had settled on a rough consensus on competitive open markets, basic human rights, and liberal democracy—even if the road to those goals was still long and rocky in much of the world. I assumed digitization would level the commercial playing field in wealthy economies and invite new competition into markets that had always had high barriers to entry. I imagined a rapid spread of education and critical thinking once we surmounted the millennium-old problems of information scarcity and maldistribution.[8]

Many of us internalized the basic tenets of economic neoliberalism and applied them to the digital sphere—this idea that government is incompetent in economic matters and should not be involved in regulating or planning the digital realm. Many of us internalized the neoliberal worldview, captured so succinctly by Google's CEO, Eric Schmidt: "There are models and there are countries where in fact the government does try to do that, and I think the American model works better."[9] And our growing faith in free enterprise, especially in the digital sphere, has gone hand in hand with a loss of humanist faith that has gradually marginalized traditional values like privacy.

None of this is to suggest that a humanist outlook is, in itself, any better or more accurate or more truthful than an economistic worldview. None of this is to suggest that there really is such a thing as "human essence" or even "humanity." No, humanism and the belief in human nature may well be just as much of an illusion as the free market, and surely they do not guarantee good outcomes. But in this case, the move away from a more spiritual humanism has had effects. We have witnessed a discursive shift from the dominance of a humanist sentiment in the postwar period to the reign of an economistic attitude in the twenty-first century, and this shift has made it easier to commodify privacy and, gradually, to eviscerate it. It is worth tracing that transformation in some detail.

THE POST–WORLD WAR II period was marked by a humanist orientation that, sad to say, was in large part a response to the atrocities of the Holocaust. In the wake of a historical period that had witnessed so much destruction and slaughter, there was a resurgence of interest in the capacities and nature of the human being and in the value of the human condition. This was certainly the case in Europe, in literature and philosophy. It was reflected by the dominance of thinkers such as the young Jean-Paul Sartre, who espoused a form of existential humanism succinctly articulated in his popular tract *Existentialism Is a Humanism*.[10] In Europe, humanist sentiment was reflected not only in existentialist thought but also in more philosophically oriented Marxism, in Christian humanism, and in liberal philosophies. Key thinkers associated with these European trends included, in addition to Sartre, François Mauriac, Henri Lefebvre, and Karl Jaspers.

Existentialist humanism sought to understand "man" in relation to his exercise of "authentic freedom." For many, as evidenced by the work of Karl Jaspers, it was a specific reaction to the horrors of German fascism.[11] Jaspers argued for liberal democratic institutions as an appropriate vehicle through which human beings could authentically realize themselves.[12] Marxist humanism, meanwhile, sought to emphasize continuity between the early philosophical writings of the young Marx and his more mature economic writings, in order to show that his later economic theories did not abandon humanist principles but rather enriched them—for instance, by means of Marx's analysis of how alienation under capitalism affected human relations. This was best captured by the French Marxist thinker Henri Lefebvre, who studied the alienation of man in urban settings and talked of the "right to the city" as a way for human beings to reclaim that social space from the forces of capital.[13] Meanwhile, French thinker François Mauriac articulated a humanist vision from a very different perspective, namely,

Catholicism. Continuing his prewar political commitments, Mauriac tirelessly advocated for human perseverance in the face of the horrors of World War II and his own nation's racist policies in its colonies.[14] His form of activism, though scorned by Sartre as being little better than "bourgeois humanism," gave birth to other, diverse strands of humanist discourse.[15]

And it was not Europe alone that was ensconced in humanism. In the United States, a parallel trend toward humanism took shape in the postwar period. Its importance cannot be underestimated. In the words of the historian Carl J. Richards:

> Immediately following World War II, the United States witnessed a resurgence of humanism. As a result of the nation's victory over depression and the Axis powers and its uncontested command of the global economy . . . the United States . . . largely reverted to humanism, which had . . . strong roots in traditional American optimism. . . . [P]ostwar prosperity formed the natural foundation for a resurgence of humanism. Furthermore, the totalitarian threats posed first by the Axis powers, and then by the Soviet bloc, reinforced the American people's sense of their liberal humanist mission to spread toleration ("human rights"), democracy, and capitalism.[16]

There were several currents to humanist thought in postwar America, as its roots here were somewhat different than in Europe and its expressions diverse—more Protestant and Jewish.[17] Émigré intellectuals from Europe played a critical role in creating a distinctive American humanist ethic—one focused on the importance of a distinctive realm for the self. Hannah Arendt gave American humanism a distinctly political dimension, famously writing of *vita activa,* of active life, of truly human life as central to human existence.[18] Another major influence on American humanism came from Paul Tillich, whose 1952 book *The Courage to Be* proved popular in the burgeoning field of humanistic psychology. Although American postwar humanism developed out of the same contextual impetus as its European counterpart, they parted company when it came to policy recommendations, with the American version privileging liberal democratic themes of human betterment, as opposed to the political radicalism espoused by certain postwar European humanists such as Sartre and Lefebvre. American humanism arose from a position of strength, bearing confidence in the role of the United States as a harbinger of democratic values in the face of tyranny. Like its European counterpart, it was guided by a kind of secular spiritualism that would value the more intangible and ethereal realms of the self.

At the same time, American discourse was absolutely saturated with discussions of privacy, especially during the Cold War. As Deborah Nelson underscores in *Pursuing Privacy in Cold War America,* "During the period from 1959 to 1965 the perceived need for privacy was utterly transformed by its sudden visibility as a dying feature of modern and cold war American society."[19] Nelson demonstrates this in "an astonishing variety of locations—journalistic exposés, television programs, law review articles, mass-market magazines, films, Supreme Court decisions, poems, novels, autobiographies, corporate hiring manuals, scientific protocols, government studies, and congressional hearings—and in response to an extraordinary range of stimuli—satellites, surveillance equipment such as 'spike mikes' and telephoto lenses, job testing, psychological surveys, consumer polls, educational records, databases and computers in general, psychoanalysis, suburbs, television, celebrity profiles, news reporting, and more."[20]

The combination of humanist belief and privacy discourse would have tangible effects in the second half of the twentieth century. The liberal democratic tilt of American humanism—as opposed to the greater radicalism of European humanism—translated well into legal discourse. And as a result, this prevalent postwar humanism made its way into judicial opinions on constitutional issues of privacy and security. At the height of the Cold War, in the mid-1960s, the Supreme Court would create a constitutional right to privacy and peg it to the idea of "reasonable expectations of privacy." This idea was intimately tied to a humanist conception of subjectivity, and there are many ringing examples of humanist discourse in the Supreme Court deci-

sions of the 1950s and 1960s. Justice William O. Douglas's opinion in
Papachristou v. Jacksonville in 1972, a famous case striking down laws
against loitering on the grounds of vagueness, reflects this humanist
tendency well.[21] Douglas's opinion in *Papachristou* reads like an ode or
a poem to the spiritual self—the self of reflection, of meandering, of
the lackadaisical life. It is an ovation to the realm of privacy that surrounds
the self and nourishes it. In elaborating on the value of being free to
wander by oneself, to think while strolling, to be left alone, to have a
realm of privacy, Justice Douglas would describe these activities as
"historically part of the amenities of life as we have known them. They
are not mentioned in the Constitution or in the Bill of Rights. These
unwritten amenities have been in part responsible for giving our
people the feeling of independence and self-confidence, the feeling of
creativity. These amenities have dignified the right of dissent and have
honored the right to be nonconformists and the right to defy submis-
siveness. They have encouraged lives of high spirits rather than hushed,
suffocating silence."[22]

In defending the lackadaisical life, Justice Douglas would refer to
poets and authors for guidance and inspiration. His muses are Walt
Whitman and Vachel Lindsay: these amenities of life, he would note,
"are embedded in Walt Whitman's writings, especially in his 'Song of
the Open Road.' They are reflected, too, in the spirit of Vachel Lindsay's 'I Want to Go Wandering,' and by Henry D. Thoreau."[23] Justice
Douglas's heroes include Luis Munoz-Marin, a former governor of
Puerto Rico, who "commented once that 'loafing' was a national virtue
in his Commonwealth and that it should be encouraged."[24]

Justice Douglas's defense of loitering and strolling transformed the
private realm of the self into a prized and valuable good, "responsible
for giving our people the feeling of independence and self-confidence,
the feeling of creativity."[25] Douglas drops a note in the margin to this
evocative passage from Thoreau's *Excursions*:

> I have met with but one or two persons in the course of my life
> who understood the art of Walking, that is, of taking walks,—
> who had a genius, so to speak, for sauntering: which word is

beautifully derived "from idle people who roved about the country, in the Middle Ages, and asked charity, under pretence of going a la Sainte Terre," to the Holy Land, till the children exclaimed, "There goes a Sainte Terrer," a Saunterer, a Holy-Lander. They who never go to the Holy Land in their walks, as they pretend, are indeed mere idlers and vagabonds; but they who do go there are saunterers in the good sense, such as I mean. Some, however, would derive the word from sans terre, without land or a home, which, therefore, in the good sense, will mean, having no particular home, but equally at home everywhere. For this is the secret of successful sauntering. He who sits still in a house all the time may be the greatest vagrant of all; but the saunterer, in the good sense, is no more vagrant than the meandering river, which is all the while sedulously seeking the shortest course to the sea. But I prefer the first, which, indeed, is the most probable derivation. For every walk is a sort of crusade, preached by some Peter the Hermit in us, to go forth and reconquer this Holy Land from the hands of the Infidels.[26]

To be sure, Douglas was an idiosyncratic justice and his opinion in *Papachristou* was perhaps equally so. But he illustrated a larger tendency toward a humanist conception of the self and privacy. Many of the plurality opinions of the Supreme Court resonated with this kind of soft humanism. The very idea of a right to privacy emanating from the penumbra of various amendments—such as in the famous case of *Griswold v. Connecticut*—was itself infused with a certain form of humanism.[27] The same is true for the importance assigned to the right to privacy in the well-known case of *Katz v. United States,* which extended privacy broadly in the Fourth Amendment context involving the right to be free of unreasonable search and seizure.[28] Similarly for the decision in *Chimel v. California,* where Justice Potter Stewart, writing for the Court, upheld the sanctity and privacy of the home against the power of the state to search incident to arrest.[29] The impact of cases such as *Stanley v. Georgia* (1969), *Eisenstadt v. Baird* (1972), and *Roe v.*

Wade (1972) expanded the scope and reach of the right to privacy. The cases rang of a humanist faith in the essential human need for autonomy. This was reflected even in ordinary national security cases, such as *United States v. United States District Court,* where in 1972 the Supreme Court declared that surveillance without a warrant, even in the name of national security, was not constitutionally sanctioned.[30]

IT WOULD BE HARD to imagine a sitting justice of the Supreme Court today extolling the humanist virtues of privacy in the way that Justice Douglas did in *Papachristou.* Hard to imagine a Supreme Court justice writing today in such prosaic terms about the intrinsic value of being left alone, of not being followed, watched, tracked, stalked—even by colossal social media corporations, giant retailers, and over-sized government intelligence services. Hard to imagine an opinion extolling the virtues of having such a robust personal sphere of privacy and autonomy—of the importance to human flourishing of having control over the dissemination of one's thoughts, wishes, and desires, over the publicity of intimate photos, or over the distribution of private affection. It sounds, today, so poetic, so analog, so passé.

In the digital age, we are far more likely to hear about the cost of privacy, not its virtues or even its value. We are far more likely to hear about trade-offs and opportunity costs. We are more likely to hear about necessary compromise and balancing of interests.[31] We are, in effect, more likely to hear the discourse of economists or economistic judges like Richard Posner, who writes that privacy "is not an interest of private individuals alone": "The Foreign Intelligence Surveillance Act is an attempt to strike a balance between the interest in full openness of legal proceedings and the interest in national security, which requires a degree of secrecy concerning the government's efforts to protect the nation. Terrorism is not a chimera. . . . Conventional adversary procedure thus has to be compromised in recognition of valid social interests that compete with the social interest in openness. And 'compromise' is the word in this case."[32]

"Compromise," "balancing," "the interest in national security"—these terms resonate much better with our current times, a period that is characterized by a dominant economistic rationality and cost-benefit calculations. In *The Illusion of Free Markets,* I trace this rise of neoliberal economic thought during the past forty years, especially at the University of Chicago. I will not review that history here, but rather pick up the thread. Neoclassical economic thought rose in prominence during the 1960s as a lens to decipher various aspects of social life, and it became somewhat dominant during the 1980s and 1990s. Gary Becker's approach of extending economic analysis to facets of life as diverse as marriage, divorce, crime, punishment, and discrimination brought about a generalization of rational choice theory throughout the social realm and the academic disciplines. As a result, postwar humanism got displaced by a far more rational, calculating view of the human subject. The logic of measurable self-interest, of transaction costs, of costs and benefits, and of the commodification and privatization of goods displaced the more nebulous notion of human essence.[33] Quantifiable material interests took precedence over spirituality, lessening the hold of the more ethereal concepts, such as privacy, human self-development, autonomy and anonymity.

This shift in our self-conception of human subjectivity has gone hand in hand with an important change in the extent and scope of privacy. Combined with a general political shift of the Supreme Court toward a more law-and-order framework—which we usually describe as the shift from the Warren Court to the Burger, Rehnquist, and now Roberts Courts—this transformation in conceptions of human subjectivities has facilitated the encroachment of digital surveillance in our lives, unprotected by notions of human essence or of the fragility of our humanity, resulting in an erosion of privacy and autonomy interests. Although it is not logically necessary, the transition from humanist discourse to a rational choice framework is entirely consonant with the shift in conceptions of privacy. Whereas privacy was previously framed in humanistic terms, it is now far more likely to be thought of as a type of property, something that can be bought or sold in a market. Privacy has become a form of private property.

It would be naive to think that the humanist tradition that Justice Douglas drew upon was inherently or necessarily more protective of individual privacy than the current neoliberal framework. Humanism is not necessarily or always more liberal than an economistic worldview—it all depends, naturally, on how one defines human essence, or what costs and benefits are thrown into our economic equations. There are surely many free market economists today who are more liberal, in the sense of respecting individual liberty, than many humanists of the twentieth century. Milton Friedman's position on the legalization of illicit drugs, for instance, was far more liberal than that even of many contemporary humanists. So as a logical matter, there is no necessity to the opposition between Justice Douglas and Judge Posner.

But as a historical matter, turning privacy from an essential human attribute to a market commodity has gone hand in hand with an eclipse of the very values that Justice Douglas extolled in his writings. Those who have deployed most persuasively and powerfully the language of economics in the past forty to fifty years have enlisted the discursive framework of trade-offs, costs and benefits, efficiency, and rational choice—in sum, market logics—in furtherance of a securitarian agenda. They have deployed economic analysis to reinforce a politics of national security, of intelligence gathering, of militarism.

One explanation for the historical conjuncture has to do with material and political interests, and others have written admirably about this, especially Naomi Klein, David Harvey, and Noam Chomsky.[34] The top echelon in the administration of President George W. Bush had significant financial, material, political, and ideological investments in a conception of national security that was tied to the petroleum market and the military. President Bush's close ties to the oil industry, Vice President Dick Cheney's revolving-door relationship with Halliburton, and the ideological commitments of their chief aides, including Donald Rumsfeld, Paul Wolfowitz, and John Bolton, among others, attest to the material and political interests that fueled the war in Iraq. A number of simple facts—for instance, that Vice President Cheney was CEO of Halliburton after having served as secretary of defense during the first war in Iraq; that, under his leadership, Halliburton rose from number seventy-three to number eighteen on the list of top Pentagon contractors, and earned $2.3 billion in government contracts; that Cheney received a retirement package worth more than $33.7 million and sold stock holdings worth more than $30 million when he became vice president; and that Halliburton profited handsomely under his vice presidency—alone are enough to settle any question about motivations and material interests.[35] Moreover, the ideological commitments of many neoconservative thinkers—including many of the top advisers to Republican administrations—to a neoliberal paradigm highly influenced by particular readings of Friedrich Hayek, Milton Friedman, Leo Strauss, and more generally the Chicago School (whether or not faithful to those writings) also linked the neoliberal cluster to a strong national security agenda.[36] The concrete financial ties and elite power relations between the intelligence community, telecommunications companies, multinational corporations, and military leadership—in this country and abroad—reinforce the historical conjuncture as well.

All this is undoubtedly true, and explains in part the turn to national security during the rise of neoliberalism. In part, but not entirely—given that during the Cold War there was also a deeply embedded relationship between the military, intelligence, political, and corporate elite, and yet there were, at the highest echelons of political leadership, strong advocates of the right to privacy. There were leaders, like Justice Douglas and other members of the Warren Court, who defended a humanistic conception of autonomy and control over one's information and surroundings. Today, practically everyone in a position of political leadership seems far more likely to claim that he has "nothing to hide" and is not particularly troubled by NSA surveillance. Justice Douglas's exhortations sound old-fashioned today, out of touch with our present securitarian condition.

This is reflected at the Supreme Court. There has been a tangible shift toward more of a rational-choice approach to human subjectivity, in tandem with a free-market conception of privacy. The movement toward rational choice has been noticeable since the 1970s and 1980s. One can feel the turn away from humanism, for instance, in a case like *Vernonia School District v. Acton,* 515 U.S. 646 (1995), where Justice Antonin Scalia, upholding a drug testing program for school athletes, analyzes everything through the lens of calculated choice. Athletes, essentially, give up their rights to privacy by deciding to participate in school sports, by making that choice. And Justice Scalia is not alone. One can detect the same kind of rational calculation in Justice David Souter's opinion in *Georgia v. Randolph,* 547 U.S. 103 (2006), where the Court held that a co-tenant cannot consent to a search of the tenant's property where the other tenant is present and objecting. The point is not the political direction of the decision but the form of rationality, the conceptualization of subjectivity. Souter focuses there on what a rational tenant, as rational actor, would want, rather than on a humanistic rights discourse. *Roper v. Simmons,* 543 U.S. 551 (2005), is another illustrative case. In holding that the execution of juveniles violates the Eighth Amendment, Justice Anthony Kennedy discusses there how the lack of cost-benefit analysis among juveniles is a reason to not impose the death penalty.[37]

Whether upholding or denying a privacy interest, it seems, the humanist discourse has evaporated. Even in *Kyllo v. United States,* for instance, where the Court finds that the use of heat-seeking technology constitutes a search because it infringes on the intimacies of the home, Justice Scalia mocks the humanist conception of privacy and autonomy.[38] The cases that deny privacy protections are also telling, insofar as they so often eviscerate any notion of human essence or human needs. We see this reflected in the *Greenwood* case, permitting unwarranted searches of garbage, for instance, or in the *King* case, involving DNA testing.[39] The limited set of cases where the Supreme Court extends human rights—for instance, by abolishing the death penalty for individuals with mental retardation or for juveniles, or maintaining habeas corpus protections over Guantánamo detainees—are also telling, for these are the most marginalized populations, the extremities of humanity: children, the mentally ill, bare life. These are the actors who are not always viewed as fully rational and who have not managed to calculate properly. If you restrict the notion of humanism to nonrational actors, these are the only fringes that are left. Their calculations cannot be trusted; the courts must then step in.

In sum, there has been a shift, at the level of reasoning and discourse, regarding the conception of human subjectivity shared by different pluralities of the Supreme Court. In the postwar period, during the 1950 and 1960s, the Court tended to embrace a conception of human subjectivity that was marked by a form of postwar humanism and that exhibited a strong secular spiritual belief in the importance of the inner self and the subjectivity of individuals. This helped to produce a robust and original jurisprudence around the right to privacy, the human need to be left alone by the government, and the importance of due process to human self-development. That postwar humanism was reflected most prominently in the writings of certain individual justices, but it also ran through many of the other decisions of the Warren Court.

The predominant lack of concern about privacy in the wake of the Snowden revelations is entirely consonant with this shift from a humanist conception of human subjectivity to one emphasizing rational choice. And given the present constitutional landscape, it is unlikely that constitutional privacy protections today are going to serve to put a halt to the kinds of NSA programs, like PRISM, that allow our government access to all our Internet activities. Again, this is not to suggest that the connection is logically necessary. It would be possible to justify a position toward privacy that was based either in humanist discourse or in rational choice theory. The discourse is not determinative. But somehow, some discourses *fit better:* humanist discourse, it turns out, fits better with notions of essential human values such as privacy; rational choice fits better with the idea of commodified conceptions of privacy.

In the end, the humanism and spirituality that infused earlier conceptions of privacy simply does not hold up well to economic rationality and commodification, to a more calculating conception of the individual.[40] Along with other cultural and technological changes, these shifting conceptions of human subjectivity have had significant effects on the materialism of our current times.[41] The move from humanist discourse to a more economistic view has facilitated a certain complacency in the face of the evisceration of our privacy and autonomy interests.

WE TELL OURSELVES—and some of us sincerely believe—that there is nothing to worry about. That we have nothing to hide. That transparency benefits us more than it harms us. That our information will be lost in the mass of data, like a needle in a haystack. For some, there is an honest lack of knowledge—many digitally literate people do not even know the extent to which they can be tracked and followed on the Internet. For others, the digital world appears innocuous, or even comforting. "Cookies": who wouldn't want cookies? (We didn't call them "crabs," or "lice," or "bacteria," after all.) Who wouldn't want to store all our heavy baggage in the "cloud"? It's so liberating, light, carefree-sounding. "The metaphors we use matter," as Tim Hwang and Karen Levy remind us.[42]

Many others of us are lulled into giving our most sensitive data. No one sees us when we do it. We give the information in the privacy of our office, or behind the closed doors of our home. We do it quickly, discreetly. We slip our social security number to the IRS, our passport number to the airline, believing that it will get lost in the stream of information, in the flood of digits that are gushing over the Internet. It feels as if no one will even notice, no human will know; it is just data on a screen. From times past—from the analog age—we have been conditioned to give our information, knowing that it would get lost in the mounds of paperwork. That was the beauty of bureaucracy. There was no way that all the paperwork could be retrieved or used. The mounds of handwritten, illegible customs forms at the border—there was no possibility that those could be retained, properly filed, retrieved, or transmitted to the fiscal or regulatory authorities. There were too many cabinets overflowing with forms, far too many for humans to manage. Not so with digits and high-powered computing; but we got used to filling out forms and giving our information in analog times.

And after a few moments of doubt, when we flinch at the disclosure, most of us nevertheless proceed, feeling that we have no choice, not knowing how *not* to give our information, whom we would talk to, how to get the task done without the exposure. We feel we have no other option but to disclose. And though it may feel ephemeral, it is permanent. The information we disclose is captured, recorded, etched into the stone of digital memory. It is seared into the silicon. Engraved. It will not disappear. We think it is nothing more than a cloud, but it is tattooed on our digital subjectivity.

We want to believe the information will not be seen or used. That we are only giving the information to machines. But we know, we've read, that this is not really true. We know far too well that humans are behind the machines. We've heard that the young analysts who work for the NSA are sharing the intimate details they come across. It happens "routinely," as Edward Snowden reveals—it's a type of "fringe benefit" associated with working for the intelligence services.[43] "You got young enlisted guys, 18–to–22 years old," Snowden explains, who in the course of analyzing data come across, "for example, an intimate nude photo of someone in a sexually compromising situation, but they're extremely attractive."[44] Snowden recounts:

> So what do they do, they turn around in their chair and show their coworker. And their coworker says, "Oh, hey, that's great. Send that to Bill down the way." And then Bill sends it to George, George sends it to Tom. Sooner or later, this person's whole life has been seen by all of these other people. It's never reported, nobody ever knows about it because the auditing of these systems is very weak. The fact that your private images, records of your private lives, records of your intimate moments have been taken from your private communications stream, from the intended recipient, and given to the government without any specific authorization, without any specific need, is in itself a violation of your rights. Why is that in a government database?[45]

We also know that some intelligence analysts use the database to spy on their loved ones and lovers. In the business this is called LOVEINT, and it is known to happen with some frequency. In September 2013, Reuters reported that "at least a dozen U.S. National Security Agency employees have been caught using secret government surveillance tools to spy on the emails or phone calls of their current or former spouses and lovers."[46]

But we ignore it. We wipe it out of our minds. We let ourselves get distracted. We tell ourselves it won't happen to us. It may well happen to others, but not to us. Plus, we've got "nothing to hide."[47] So we continue to crave the digital life, to consume and be consumed in a stream of emails, scanned documents, digital photos, tweets, text messages, emoticons, and Instagrams. We contribute constantly to the digital stream, producing and reproducing it, feeding it, incessantly, inevitably, in the process opening ourselves up, delivering ourselves to the digital voyeur. We search the web, buy things online, swipe access cards—to the gym, to work, to the garage, into the subway—retrieve money from the ATM, deposit checks by mobile phone, accumulate points at the grocery, scan shampoo at the pharmacy, swipe the credit card at Starbucks (or, rather, tap our cell phones). We hit the tip button in the taxi, on GrubHub, or on the Domino's website—and never even think twice that our privacy might be worth the cost of that tip.

PART THREE

The Perils of Digital Exposure

SEVEN

THE COLLAPSE OF STATE, ECONOMY, AND SOCIETY

THE DIGITAL ECONOMY has torn down the conventional boundaries between governing, commerce, and private life. In our digital age, social media companies engage in surveillance, data brokers sell personal information, tech companies govern our expression of political views, and intelligence agencies free-ride off e-commerce. The customary lines between politics, economics, and society are rapidly vanishing, and the three spheres are melding into one—one gigantic trove of data, one colossal data market, that allows corporations and governments to identify and cajole, to stimulate our consumption and shape our desires, to manipulate us politically, to watch, surveil, detect, predict, and, for some, punish. In the process, the traditional limits placed on the state and on governing are being eviscerated, as we turn more and more into marketized malleable subjects who, willingly or unwittingly, allow ourselves to be nudged, recommended, tracked, diagnosed, and predicted by a blurred amalgam of governmental and commercial initiatives.

The collapse of these different spheres is disempowering to us as individuals. Resisting state excess—whether in the form of J. Edgar

Hoover's FBI and surveillance programs like COINTELPRO, loyalty oaths and McCarthyism, or ordinary civil rights violations—can be harrowing and difficult.[1] But trying to rein in a behemoth that includes the NSA, Google, Facebook, Netflix, Amazon, Samsung, Target, Skype, and Microsoft, to check a tentacular oligarchy that spans government, commerce, surveillance, and the private sphere, can feel even more daunting. Before, in the analog age, there were ways to divide and conquer, or at least to try—to attempt to build bridges with civic institutions to check the powers of the state. But today the interests are so aligned that the task feels practically impossible.

At the root of it all is the fact that the line between governance, commerce, surveillance, and private life is evaporating. What we face today is one unified, marketized space. The famous lines drawn in the nineteenth century—for instance, in Max Weber's seminal work on economy and society—are vanishing.[2] Governing is collapsing into commerce as states such as China, Russia, and the United States increasingly seek to secure what Evgeny Morozov calls their "digital sovereignty" through trade regulation—passing laws that require tech companies to store their citizens' data on servers located within the state's territorial boundaries or placing restrictions on Internet providers for services such as Gmail.[3] Commerce is collapsing into surveillance, right before our eyes, as retailers begin to collect all our data. And commerce is turning into governance as new data markets emerge, allowing businesses, employers, salespeople, bureaucrats, advertisers, the police, and parole officers to track our physical movements, to follow our Internet browsing, to know what we read, what we like, what we wear, whom we communicate with, what we think, how we protest, and where we spend our money.

. . .

Josh Begley is a former graduate student at New York University who has been trying to raise awareness about the United States government's use of drones and the number of civilian casualties resulting from drone strikes.[4] Begley would like drone strikes to be known here at home—to be felt here when they hit in faraway locations, when they accidentally kill children and civilians. Begley wants drone strikes to interrupt our lives, not just the lives of others across the globe. He wants the strikes to ping us when we are playing our video games, to disrupt our Facebook session, to interrupt our web browsing and shopping.

In the summer of 2012, Begley created an application for iPhones called Drones+. His app would provide instant, updated information about each and every drone strike and all of the facts about casualties. As a Vimeo demonstration of the app shows, Drones+ triggers an alert on your iPhone when a drone attack occurs and provides details about the number of people killed, as well as an interactive map of the area where the attack took place that identifies other previous air assaults in the vicinity.[5]

Begley submitted his app to Apple on July 10, 2012, but Apple refused to offer Drones+ through its App Store. According to the *Guardian*, Apple notified Begley that his app was " 'not useful' enough and did not appeal to a 'broad enough audience.' "[6] Begley reapplied in August 2012 and got rejected again because, according to a Mashable article, "the app used Google Maps images without the 'associated Google branding.' "[7] Begley tried to fix that problem and reapplied a third time, on August 27, 2012; this application was also denied. This time, Apple reportedly wrote him, "We found that your app contains content that many audiences would find objectionable, which is not in compliance with the app store review guidelines."[8]

Begley decided to move on and instead create a website called Dron-estre.am that would provide, according to the site itself, "real-time and historical data about every reported United States drone strike."[9] The website has a searchable database and a real-time Twitter account, and it posts news stories about drone strikes.

With that website up and running, Begley tried again a year later to create an app for smartphones. What he did was to change the name of the app to Dronestream—just like his functioning website. Begley submitted that application on September 10, 2013, and then a second time on September 17. But to no avail—his applications got rejected both times.[10]

Curiously, though, Begley then received an email from someone who worked on the Apple Review Team. It was now September 23, 2013, and in the email the Apple employee asked Begley whether they could talk on the phone about the app. The Apple employee called Begley and asked him whether the application he was pitching was about United States government drone strikes. Naturally, Begley answered yes. The Apple employee then reportedly told Begley, "If it's going to be about that specifically, it's not going to be approved. But if you broaden your topic, then we can take another look. You know, there are certain concepts that we decide not to move forward with, and this is one."[11]

So in early 2014 Begley went back to the drawing board—this time with a new idea. What he did was to create an application that had no content at all, in order to test whether it would get approved. He called it Ephemeral and submitted it to Apple on January 17, 2014, to see what would happen. As he explained, "The point of it was to have no content at all." That application, surprisingly, was accepted without any fuss a few days later.

On January 22, Begley then "submitted yet another empty app called 'Metadata+' which promised 'real-time updates on national security.'"[12] The Mashable piece tells us the rest of the story: "As with the previous test app, this one went through after just six days—no objections on usefulness or questionable content this time. At that point, all Begley had to do was fill the app with the historical archive of drone strikes. Now, apart from some design improvements, the app looks exactly as it did originally."[13] The app had finally been approved.

THE JOSH BEGLEY incident illustrates the collapse of the boundaries between commerce and government. Apple was effectively "governing" for us—for profit, it seems—refusing to allow a controversial or political app on the basis of commercial interests. Not much different, perhaps, from the United States government putting a stop to media coverage of the return to the United States of the bodies of soldiers killed in the Iraq War in 1991. As you may recall, for more than eighteen years the United States prohibited news coverage of the return of our soldiers killed in war.[14] The Obama administration lifted the absolute ban in December 2009, allowing families of the deceased soldiers to decide for themselves whether the media could photograph the flag-covered caskets.[15] The ban had originally been initiated by President George H. W. Bush during the Gulf War in 1991 and was upheld by his son President George W. Bush during the war in Iraq and Afghanistan. It was deeply controversial.[16]

Apple, it seems, has taken on that state function of censorship, though its only motive seems to be profit. Apple may not have wanted Begley's app because it was going to upset people, who would then consume less. And to be honest, that was part of Begley's project: to interrupt our ordinary consumption of everyday life with the reality of drone strikes. To ping our complacency when we, through our government, accidentally or not kill others. But notice how Apple's rejections make the exercise of power so much more tolerable. If the visibility of drone strikes is likely to raise our consciousness about these horrifying practices—for a wrenching taste of drone strikes and the weight of innocent casualties, you must take a look at the video that Pitch Interactive put together, called "Out of Sight, Out of Mind," available on the web[17]—then the invisibility that Apple promoted can only serve to mask and hide the intolerable. By keeping the drones out of sight and out of mind, it makes it far easier for our presidents to continue to deploy drones. The same for body bags and the war in Afghanistan.

In another incident, Facebook deleted a post by a prominent Tibetan writer and critic of Chinese policies toward Tibet, Tsering Woeser. Woeser had written on her Facebook page about the self-immolation of a monk in Sichuan province, and provided a link to a video of the incident. In response to inquiries, Facebook said it deleted Woeser's post because it "didn't meet Facebook's community standards."[18] Facebook followed up with this statement:

> Facebook has long been a place where people share things and experiences. Sometimes, those experiences involve violence and graphic videos. We work hard to balance expression and safety.

> However, since some people object to graphic videos, we are
> working to give people additional control over the content they
> see. This may include warning them in advance that the image
> they are about to see contains graphic content. We do not cur-
> rently have these tools available and as a result we have removed
> this content.[19]

That is an astonishing avowal of technical incompetence in this digital age. (Meanwhile, Facebook is making strides in its efforts to conquer the hearts and minds of Chinese citizens. In October 2014, Mark
Zuckerberg impressed us all when, speaking Mandarin—which he is
learning—he participated in a question-and-answer session at Tsinghua
University in Beijing.[20] One wonders whether there is no connection.)

These are precisely the micropractices that render tolerable the intolerable—in large part by eliding unreviewable, hidden commercial
choices with political functions. The boundaries are vanishing, and in
the process, a new form of power is born. A form of power that is hard
to see, precisely because it feels so natural, like a glove—just ordinary,
everyday business as usual. It is almost transparent, unseeable, invisible, illegible. Yet it is right there, right in front of our eyes:

> "'not useful' enough"
> "does not appeal to a 'broad enough audience'"
> "content that many audiences would find objectionable"
> "not in compliance with the app store review guidelines"

True, many of us would find Begley's app "objectionable" because
we would rather not be reminded that each and every one of us partici-
pates in these drone strikes, with our tax dollars or our bodies. We
don't want to be reminded about the collateral damage to innocent
children when we are taking our own darlings to the movies, spoiling
them, taking them to a ballgame, or watching them score a run. We
don't really want to know every time another Pakistani or Arab child
is killed. We need to carry on, and we must continue to consume.
Consumer confidence is good for consumption, and consumption is
good for the economy and our country, we tell ourselves. We must not

let these drones stop us. We should not let them interfere. We must continue to shop—and in the process produce more data about ourselves that can be shared, mined, analyzed, sold, and surveilled.

And just as the boundary between commerce and governing seems to be evaporating, so is the line between exchange and surveillance.

. . .

From 2007 until it was discovered in 2010, Google equipped its Street View cars—the ones that digitally photograph sidewalks and buildings so we can see them on the Street View function of Google Maps—with special technology that would vacuum up all of the unencrypted Wi-Fi traffic in the surrounding neighborhood, and probably the encrypted traffic as well.[21] Google's special technology captured highly sensitive data above and beyond the basic data that Google seized with the ordinary Wi-Fi antennas and software mounted on its vehicles—basic data that included "the network's name (SSID), the unique number assigned to the router transmitting the wireless signal (MAC address), the signal strength, and whether the network was encrypted."[22] With its special technology, Google would also capture and record, in the words of a federal appellate court, "everything transmitted by a device connected to a Wi-Fi network, such as personal emails, usernames, passwords, videos, and documents."[23] In technical jargon, this is called "payload data," and it can include very private information. As the federal court noted in 2013, "Google acknowledged in May 2010 that its Street View vehicles had been collecting fragments of payload data from unencrypted Wi-Fi networks. The company publicly apologized, grounded its vehicles, and rendered inaccessible the personal data that had been acquired. In total, Google's Street View cars collected about 600 gigabytes of data transmitted over Wi-Fi networks in more than 30 countries."[24]

Google has since filed a petition for writ of certiorari to the United States Supreme Court arguing that the collection of such unencrypted data is exempt from the federal Wiretap Act, 18 U.S.C. Sec. 2511, because the Wi-Fi communications are themselves "readily accessible to the general public"—a term of art that would make them an "electronic communication" exempt under the act. In other words, Google's lawyers maintain that the sensitive payload data should be freely accessible. Google took this position in response to a class action suit alleging that it violated the Wiretap Act.[25]

If Google's position ultimately prevails, then it would be entirely legal for Google—or for anyone else, for that matter—to vacuum up all of the encrypted and unencrypted Wi-Fi traffic in neighborhoods across the country, and possibly around the world. Paradoxically, Google's argument rests in part on the claim that we would all be safer and more secure if that were legal: Google contends that shielding Wi-Fi signals behind the Wiretap Act might actually *decrease* overall computer security because it would hinder legitimate security scanning of those Wi-Fi signals. "IT professionals routinely use the same kind of technology as Google's Street View cars did to collect packet data in order to secure company networks," Google wrote to the Supreme Court.[26] "And unlike Google, which never used the payload data it collected, security professionals also parse and analyze the data collected from wired and wireless networks, including networks operated by other persons or entities, to identify vulnerabilities in and potential attacks on the networks they protect."[27]

HERE TOO, THE LINE between commerce and surveillance is evaporating. We know how the retail giant Target mines its customer's consumption data to predict pregnancy and other life-changing events, in order to target coupons and advertising to them.[28] Target has excelled at data mining, creating for each shopper a unique code, referred to as a "Guest ID," and capturing by its means all possible information about each shopper. As a marketing analyst at Target explained to the *New York Times*, "If you use a credit card or a coupon, or fill out a survey, or mail in a refund, or call the customer help line, or open an e-mail we've sent you or visit our Web site, we'll record it and link it to your Guest ID. . . . We want to know everything we can."[29] So Target links all this data for each Guest ID to demographic data, marital status, residence, income level, and so on, and analyzes it all to predict consumption and target marketing. And once that information is in the hands of third parties, it seems to become fair game for governing, exchanging, and securing.[30]

"Almost every major retailer, from grocery chains to investment banks to the U.S. Postal Service, has a 'predictive analytics' department devoted to understanding not just consumers' shopping habits but also their personal habits, so as to more efficiently market to them," Charles Duhigg reports.[31] These surveillance practices are creeping into the other areas of life, such as education and employment—and, through these, into governance. Universities such as Arizona State are increasingly using surveillance technology to monitor student activity during class and at home, as well as trajectories of student performance over semesters or years. Students receive automated messages telling them when they are "off-track" on a particular course of study, and algorithms are used to inform the student of everything from peers they may like to meet to their likelihood of dropping out of or failing a class.[32]

Many companies have begun to use digital tracking systems to monitor their employees and the efficiency of their work habits across a range of industries, from nursing to trucking.[33] Amazon is a leader in the field, especially at its warehouses in the United States, the United Kingdom, and Germany. Their gigantic warehouses are, naturally, fully digitized, so all merchandise that comes in and is unpacked, transported, shelved, picked, repacked, and shipped out is tracked by computer software. The same computerized processes also direct employees to the right shelves—using software that routes them in such a way as to maximize their efficiency—but at the same time monitor their every behavior and movements. At the facility in Rugeley, Staffordshire, a huge warehouse the size of nine soccer fields, employees reported that "several former workers said the handheld computers, which look like clunky scientific calculators with handles and big screens, gave them a real-time indication of whether they were running behind or ahead of their target and by how much. Managers could also send text messages to these devices to tell workers to speed up, they said. 'People were constantly warned about talking to one another by the management, who were keen to eliminate any form of time-wasting,' one former worker added."[34]

According to Simon Head, author of *Mindless: Why Smarter Machines Are Making Dumber Humans,* this tracking technology is now in operation at Amazon warehouses across the globe: "All this information is available to management in real time, and if an employee is behind schedule she will receive a text message pointing this out and telling her to reach her targets or suffer the consequences. At Amazon's depot in Allentown, Pennsylvania, Kate Salasky worked shifts of up to eleven hours a day, mostly spent walking the length and breadth of the warehouse. In March 2011 she received a warning message from her manager, saying that she had been found unproductive during several minutes of her shift, and she was eventually fired."[35]

These new forms of Taylorism and control of employee time bear an uncanny resemblance to the types of disciplinary methods—regarding timetables and spatial control, as well as human capital—that Foucault described so poignantly in *La société punitive.* Here is a chilling description by Simon Head of an Amazon warehouse in Augsburg, Germany:

> Machines measured whether the packers were meeting their targets for output per hour and whether the finished packages met their targets for weight and so had been packed "the one best way." But alongside these digital controls there was a team of Taylor's "functional foremen," overseers in the full nineteenth-century sense of the term, watching the employees every second to ensure that there was no "time theft," in the language of Walmart. On the packing lines there were six such foremen, one known in Amazonspeak as a "coworker" and above him five "leads," whose collective task was to make sure that the line kept moving. Workers would be reprimanded for speaking to one another or for pausing to catch their breath (*Verschnaufpause*) after an especially tough packing job.
>
> The functional foreman would record how often the packers went to the bathroom and, if they had not gone to the bathroom

nearest the line, why not. The student packer also noticed how, in the manner of Jeremy Bentham's nineteenth-century panopticon, the architecture of the depot was geared to make surveillance easier, with a bridge positioned at the end of the workstation where an overseer could stand and look down on his wards. However, the task of the depot managers and supervisors was not simply to fight time theft and keep the line moving but also to find ways of making it move still faster. Sometimes this was done using the classic methods of Scientific Management, but at other times higher targets for output were simply proclaimed by management, in the manner of the Soviet workplace during the Stalin era.[36]

The gesture to Jeremy Bentham is telling—and reminds us of parallels from the past. The desire to control is not new; it is just the capabilities today that are staggering. The panoptic principle was intended to be applied as easily to penitentiaries and asylums as to factories, hospitals, schools, and barracks. In this sense, Bentham foreshadowed the collapse of the different spheres of life, including punishment, employment, and education. All of those spaces—including the Soviet workplace during the Stalin era—were spatial nodes or architectural forms of power relations, no different from punching the clock or the hourly salary that would emerge in nineteenth-century capitalism. No different from the GPS tracking systems today—but for the awesome power of today's technology.

Commentators have begun to describe commercial surveillance practices as attempts to manufacture a "consuming subject"—or as practices of "manufacturing consumers."[37] On this view, the new information economy is part of a broad shift in the relationship of production and consumption, one expressed through "flexible" production models such as "Toyotism" and "just-in-time and lean manufacturing."[38] Consumption and governing interrelate and operate "as a dynamic process where existing surveillance and profiling systems and personal information continuously inform each other with each new interaction between the system and consumers."[39] Part of the ambition of "manufacturing consumers" is to know in advance what they will purchase. Consumer data is used to narrow down categories of consumers, to filter out a population to whom a product or service might most reliably appeal. The goal in the long run is to predict preferences better than consumers themselves can. Indeed, Google CEO Eric Schmidt states that his company wishes to collect user information such that "the algorithms will get better and we will get better at personalization. . . . The goal is to enable Google users to be able to ask the question such as 'What shall I do tomorrow?' and 'What job shall I take?' "[40] Or even "How shall I be governed?"

In sum, the boundaries between commerce, governing, and surveillance have collapsed. The spheres are converging, and today they coincide in the production, exploitation, and shaping of our digital personalities. Practically every commercial advance in digital technology enhances the security apparatus, increases commercial profits, and facilitates more governing—and vice versa. The intelligence sector fuels business with consulting contracts, new R&D, and outlets for developing technology; new commercial products, like the Apple Watch, Google Drive, and Vine, increase potential surveillance.

WHAT IS EMERGING in the place of the separate spheres is a single behemoth of a data market: a colossal marketplace for personal data. In 2012 alone, the data brokerage market reached $156 billion in revenue, which, as Sen. John D. Rockefeller IV noted, is "twice the size of the entire intelligence budget of the United States government—all generated by the effort to detail and sell information about our private lives."[41] There are currently more than 4,000 data broker companies. Some of them are large, publicly traded corporations, household names like Lexis-Nexis and Experian. Others are much smaller and less well known.[42] The companies troll the Internet to collect all available data. In the words of Frank Pasquale, they "vacuum up data from just about any source imaginable: consumer health websites, payday lenders, online surveys, warranty registrations, Internet sweepstakes, loyalty-card data from retailers, charities' donor lists, magazine subscription lists, and information from public records."[43] They then mine, analyze, organize, and link the data, creating valuable data sets for sale, and generating an entire political economy of publicity and surveillance that can only properly be described as neoliberal—privatized, deregulated, and outsourced. The mass of digital subjects give away their data for free, and private enterprise reaps the benefits, skimming the wealth off the top, while the government facilitates and underwrites the profit in exchange for costless but invaluable intelligence.

Senator Rockefeller held hearings in December 2013 to investigate the data brokerage markets and throw some light on what can only be described as appalling practices. The hearings disclosed, for instance, that one data broker in Lake Forest, Illinois, named Medbase200, offered to sell to pharmaceutical companies a list of "rape sufferers" at a cost of $79 for 1,000 names.[44] Medbase200 marketed this list on its website as follows:

> These rape sufferers are family members who have reported, or have been identified as individuals affected by specific illnesses, conditions or ailments relating to rape. Medbase200 is the owner of this list. Select from families affected by over 500 different ailments, and/or who are consumers of over 200 different Rx medications. Lists can be further selected on the basis of lifestyle, ethnicity, geo, gender, and much more. Inquire today for more information.[45]

Medbase200 took the "rape sufferers" database off its website after the revelations, and removed as well "lists of domestic violence victims, HIV/AIDS patients and 'peer pressure sufferers' that it had been offering for sale."[46] But the number and variety of other lists it offers for sale are simply staggering. Table 7.1 is a list of categories starting with the letter *A*, with the data size and price information (in dollars per 1,000 pieces of information) in the right column.

If that's not enough for your taste, take a peek at the *B*'s in Table 7.2.

As Senator Rockefeller exclaimed, "One of the largest data broker companies, Acxiom, recently boasted to its investors that it can provide

TABLE 7.1

Medbase200 data size and price information (letter "A")

Allergy/Immunology Nurses	53423 Total Universe @ 59/M
AARP Members Mailing List	20435556 Total Universe @ 79/M
Abscess Sufferers	> (Inquire) Total Universe @ 79.00/M
Abuse Sufferers	> (Inquire) Total Universe @ 79.00/M
Acetaminophen Users	21092445 Total Universe @ 79/M
Achondroplasia Sufferers	> (Inquire) Total Universe @ 79.00/M
Acid Reflux Disease (GERD) Sufferers	> (Inquire) Total Universe @ 79.00/M
Acid Reflux Disease (GERD) Sufferers at Home	5456709 Total Universe @ 79/M
Acne Sufferers	> (Inquire) Total Universe @ 79.00/M
Addiction Sufferers	> (Inquire) Total Universe @ 79.00/M
Addiction/Substance Abuse (Drug Abuse) Nurses	38009 Total Universe @ 59/M
Addison's Disease Sufferers	> (Inquire) Total Universe @ 79.00/M
Adenoma Sufferers	> (Inquire) Total Universe @ 79.00/M
Adolescent Medicine Nurses	20198 Total Universe @ 59/M
Adult Medicine/Adult Care Nurses	98996 Total Universe @ 59/M
Advanced Practice Nurses	92231 Total Universe @ 59/M
Aestheticians at Home	116545 Total Universe @ 59/M
Agoraphobia Sufferers	> (Inquire) Total Universe @ 79.00/M
AIDS and HIV Infection Sufferers	> (Inquire) Total Universe @ 79.00/M
AIDS/HIV Nurses	300893 Total Universe @ 59/M
Ailments, Diseases & Conditions—Hispanic Sufferers	17234554 Total Universe @ 79/M
Ailments, Diseases & Conditions—Sufferers	227453121 Total Universe @ 79/M
Ailments, Diseases & Conditions—Sufferers (Vol)	227453121 Total Universe @ 39.5/M
Ailments, Diseases & Conditions—Sufferers via Email	173209889 Total Universe @ 129/M
Albinism Sufferers	> (Inquire) Total Universe @ 79.00/M
Alcoholic Hepatitis Sufferers	> (Inquire) Total Universe @ 79.00/M
Alcoholism Sufferers	> (Inquire) Total Universe @ 79.00/M
Allergies Sufferers	> (Inquire) Total Universe @ 79.00/M
Allergy/Immunology Nurses	57886 Total Universe @ 59/M
Allergy Sufferers at Home	25698121 Total Universe @ 79/M
Alli Users	1985452 Total Universe @ 79/M
Alopecia (Thinning Hair/Hair Loss) Sufferers	> (Inquire) Total Universe @ 79.00/M

(TABLE 7.1 CONTINUED)

Altitude Sickness Sufferers	> (Inquire) Total Universe @ 79.00/M
Alzheimer's Disease Sufferers	> (Inquire) Total Universe @ 79.00/M
Amblyopia Sufferers	> (Inquire) Total Universe @ 79.00/M
Ambulatory Care Nurses	72234 Total Universe @ 59/M
Amebiasis Sufferers	> (Inquire) Total Universe @ 79.00/M
Amnesia Sufferers	> (Inquire) Total Universe @ 79.00/M
Amyotrophic Lateral Sclerosis Sufferers	> (Inquire) Total Universe @ 79.00/M
Anemia Sufferers	> (Inquire) Total Universe @ 79.00/M
Anesthesiology Nurses	172339 Total Universe @ 59/M
Aneurdu Sufferers	> (Inquire) Total Universe @ 79.00/M
Aneurysm Sufferers	> (Inquire) Total Universe @ 79.00/M
Angina Sufferers	> (Inquire) Total Universe @ 79.00/M
Animal Bites Sufferers	> (Inquire) Total Universe @ 79.00/M
Anorexia Sufferers	> (Inquire) Total Universe @ 79.00/M
Anosmia Sufferers	> (Inquire) Total Universe @ 79.00/M
Anotia Sufferers	> (Inquire) Total Universe @ 79.00/M
Anthrax Sufferers	> (Inquire) Total Universe @ 79.00/M
Antisocial Personality Disorder Sufferers	> (Inquire) Total Universe @ 79.00/M
Anxiety and Anxiety Disorders Sufferers	> (Inquire) Total Universe @ 79.00/M
Anxiety Disorders Sufferers	> (Inquire) Total Universe @ 79.00/M
Anxiety Sufferers (GAD) Sufferers at Home	3983434 Total Universe @ 79/M
Appendicitis Sufferers	> (Inquire) Total Universe @ 79.00/M
Apraxia Sufferers	> (Inquire) Total Universe @ 79.00/M
Argyria Sufferers	> (Inquire) Total Universe @ 79.00/M
Arthritis Nurses	180371 Total Universe @ 59/M
Arthritis Sufferers	> (Inquire) Total Universe @ 79.00/M
Arthritis, Infectious Sufferers	> (Inquire) Total Universe @ 79.00/M
Ascariasis Sufferers	> (Inquire) Total Universe @ 79.00/M
Aseptic Meningitis Sufferers	> (Inquire) Total Universe @ 79.00/M
Asperger Disorder Sufferers	> (Inquire) Total Universe @ 79.00/M
Asthenia Sufferers	> (Inquire) Total Universe @ 79.00/M
Asthma Sufferers	> (Inquire) Total Universe @ 79.00/M
Astigmatism Sufferers	> (Inquire) Total Universe @ 79.00/M

(continued)

(TABLE 7.1 CONTINUED)

Atherosclerosis Sufferers	> (Inquire) Total Universe @ 79.00/M
Athetosis Sufferers	> (Inquire) Total Universe @ 79.00/M
Athlete's Foot Sufferers	> (Inquire) Total Universe @ 79.00/M
Atrophy Sufferers	> (Inquire) Total Universe @ 79.00/M
Attention Deficit Hyperactivity Disorder (ADHD) Sufferers	> (Inquire) Total Universe @ 79.00/M
Attention Sufferers	> (Inquire) Total Universe @ 79.00/M
Autism Sufferers	> (Inquire) Total Universe @ 79.00/M
Autism Sufferers at Home	2983342 Total Universe @ 79/M
Avandia Users	6898545 Total Universe @ 79/M

Table and data source: This database was originally on the Medbase200 website but was removed after public backlash, according to Elizabeth Dwoskin, "Data Broker Removes Rape-Victims List after Journal Inquiry," *Digits* blog, *Wall Street Journal*, December 19, 2013.

'multi-sourced insight into approximately 700 million consumers worldwide.'"[47] Rockefeller went on: "Data brokers segment Americans into categories based on their incomes, and they sort economically vulnerable consumers into groups" with names like "Rural and Barely Making It," "Tough Start: Young Single Parents," "Rough Retirement: Small Town and Rural Seniors," and "Zero Mobility."[48]

The *New York Times* recently reported that InfoUSA, one of the largest data brokers in the country, "advertised lists of 'Elderly Opportunity Seekers,' 3.3 million older people 'looking for ways to make money,' and 'Suffering Seniors,' 4.7 million people with cancer or Alzheimer's disease. 'Oldies but Goodies' contained 500,000 gamblers over 55 years old, for 8.5 cents apiece. One list said: 'These people are gullible. They want to believe that their luck can change.'"[49] As you can imagine, these types of lists are often sold to people who then prey on those listed.[50]

Acxiom is "the quiet giant of a multibillion-dollar industry known as database marketing."[51] "Few consumers have ever heard of Acxiom," the *Times* noted. "But analysts say it has amassed the world's largest commercial database on consumers—and that it wants to know much,

TABLE 7.2

Medbase200 data size and price information (letter "B")

Babesiosis Sufferers	> (Inquire) Total Universe @ 79.00/M
Back Pain Sufferers	> (Inquire) Total Universe @ 79.00/M
Bacterial Infections Sufferers	> (Inquire) Total Universe @ 79.00/M
Bacterial Meningitis Sufferers	> (Inquire) Total Universe @ 79.00/M
Bedsores (Pressure Sores) Sufferers	> (Inquire) Total Universe @ 79.00/M
Bedwetting (Enuresis) Sufferers	> (Inquire) Total Universe @ 79.00/M
Bell's Palsy Sufferers	> (Inquire) Total Universe @ 79.00/M
Bends Sufferers	> (Inquire) Total Universe @ 79.00/M
Beriberi Sufferers	> (Inquire) Total Universe @ 79.00/M
Binge Eating Disorder Sufferers	> (Inquire) Total Universe @ 79.00/M
Bioterrorism Sufferers	> (Inquire) Total Universe @ 79.00/M
Bipolar Disorder Sufferers	> (Inquire) Total Universe @ 79.00/M
Bipolar Disorder Sufferers at Home	1985233 Total Universe @ 79/M
Birth Defects and Brain Development Sufferers	> (Inquire) Total Universe @ 79.00/M
Bites and Stings Sufferers	> (Inquire) Total Universe @ 79.00/M
Bladder Cancer Sufferers	> (Inquire) Total Universe @ 79.00/M
Blindness Sufferers	> (Inquire) Total Universe @ 79.00/M
Body Dysmorphic Disorder Sufferers	> (Inquire) Total Universe @ 79.00/M
Body Image Sufferers	> (Inquire) Total Universe @ 79.00/M
Bone Densitometry Nurses	27045 Total Universe @ 59/M
Botulism Sufferers	> (Inquire) Total Universe @ 79.00/M
Brain Injuries Sufferers	> (Inquire) Total Universe @ 79.00/M
Brain Tumor Sufferers	> (Inquire) Total Universe @ 79.00/M
Breast Cancer Sufferers	> (Inquire) Total Universe @ 79.00/M
Broken Bones and Fractures Sufferers	> (Inquire) Total Universe @ 79.00/M
Bronchiolitis Sufferers	> (Inquire) Total Universe @ 79.00/M
Bronchitis Sufferers	> (Inquire) Total Universe @ 79.00/M
Bronchitis, Infectious Sufferers	> (Inquire) Total Universe @ 79.00/M
Brucellosis Sufferers	> (Inquire) Total Universe @ 79.00/M
Bulimia Sufferers	> (Inquire) Total Universe @ 79.00/M
Bullying Sufferers	> (Inquire) Total Universe @ 79.00/M

(continued)

(TABLE 7.2 CONTINUED)

Bunions Sufferers	> (Inquire) Total Universe @ 79.00/M
Burn Care Nurses	45814 Total Universe @ 59/M
Burns Sufferers	> (Inquire) Total Universe @ 79.00/M

Table and data source: This database was originally on the Medbase200 website but was removed after public backlash, according to Elizabeth Dwoskin, "Data Broker Removes Rape-Victims List after Journal Inquiry," *Digits* blog, *Wall Street Journal,* December 19, 2013.

much more. Its servers process more than 50 trillion data 'transactions' a year. Company executives have said its database contains information about 500 million active consumers worldwide, with about 1,500 data points per person. That includes a majority of adults in the United States."[52] Here's a flavor of what that data broker knows: "It peers deeper into American life than the F.B.I. or the I.R.S., or those prying digital eyes at Facebook and Google. If you are an American adult, the odds are that it knows things like your age, race, sex, weight, height, marital status, education level, politics, buying habits, household health worries, vacation dreams—and on and on."[53]

The data brokers obtain information from every possible source. Brokers gather information from store loyalty cards, purchase histories, government records, credit reporting agencies, public voting records, and all of our web surfing and activities—including "information you post online, including your screen names, website addresses, interests, hometown and professional history, and how many friends or followers you have."[54] One data broker, Datalogix, "says it has information on more than $1 trillion in consumer spending 'across 1400+ leading brands'" from store loyalty cards.[55] And Acxiom boasts that it has 3,000 points of data on practically every consumer in the United States.[56]

Alongside these brokered data, there is a whole new and emerging collection of consumer rankings and scoring—what the World Privacy Forum calls "consumer scores."[57] A myriad of public and private entities are now engaged in massive ranking and scoring of us all, trying to place numbers on each of us to "describe or predict [our] characteristics, habits, or predilections."[58] Following in the footsteps of the "credit scores" that were developed in the 1950s, we are seeing today the proliferation and extension of this scoring logic to all facets of life. There are today consumer scores including "the medication adherence score, the health risk score, the consumer profitability score, the job security score, collection and recovery scores, frailty scores, energy people meter scores, modeled credit scores, youth delinquency score, fraud scores, casino gaming propensity score, and brand name medicine propensity scores," among others.[59]

Pam Dixon and Robert Gellman document these different scores in a fascinating and haunting report that includes, for instance, the Job Security Score, produced by Scorelogix and described by the company as follows: "Scorelogix is the inventor of the Job Security Score or JSS. The JSS is the industry's first income-risk based credit score and the only score that predicts borrowers' ability to pay by factoring their income stability. The JSS dramatically improves banks' ability to reduce credit losses and marketers' ability to reduce mailing costs."[60] The DonorScore, created and marketed by DonorTrends, allows nonprofits to rank their contributors according to a model that predicts future donations from the nonprofits' internal data. Says Donor-Trends: "Our scientific DonorScores system assigns a value from 0 to 1,000 to each donor in your database. This value predicts the future actions each donor is likely to take. This enables you to target your donors more effectively to increase revenue and decrease cost."[61]

Practically none of these scores are revealed to us, and their accuracy is often haphazard. As Dixon and Gellman suggest, most consumers "do not know about the existence or use of consumers scores" and "cannot have any say in who used the scores, or how."[62] The result is a type of virtual one-way transparency with complete opacity for those who are ranking and scoring us.

Collecting and providing information about individual consumers and their spending habits is now a competitive market.[63] The going rate for information about individual consumers varies depending on its character, competition between data providers, and the "sheer ubiquity of details about hundreds of millions of consumers."[64] In general, however, "the more intimate the information, the more valuable it is." For instance, Emily Steel writes that "basic age, gender and location information sells for as little as $0.0005 per person, or $0.50 per thousand people, according to price details seen by the *Financial Times*." But information about individuals "believed to be 'influential' within their social networks sells for $0.00075, or $0.75 per thousand people." Data about "income details and shopping histories" sell for $.001—so, $1 per thousand.[65] Particular developments in a consumer's life also increase the price of that individual's information: "Certain milestones in a person's life prompt major changes in buying patterns, whether that's becoming a new parent, moving homes, getting engaged, buying a car, or going through a divorce. Marketers are willing to pay more to reach consumers at those major life events. Knowing that a woman is expecting a baby and is in her second trimester of pregnancy, for instance, sends the price to tag for that information about her to $0.11."[66] Medical information is worth much more: "For $0.26 per person, LeadsPlease.com sells the names and mailing addresses of people suffering from ailments such as cancer, diabetes and clinical depression. The information includes specific medications including cancer treatment drug Methotrexate and Paxil, the antidepressant," according to the *Financial Times*.[67]

These data are complemented with surveillance of consumer behaviors outside the Internet, as department stores increasingly turn to "behavioral tracking" of their customers, involving careful analysis of customer behavior via CCTV, "web coupons embedded with bar codes that can identify, and alert retailers to, the search terms you used to find them," and "mobile marketers that can find you near a store clothing rack, and send ads to your cellphone based on your past preferences and behavior."[68]

The investigative reporter Yasha Levine documents how "large employers are turning to for-profit intelligence to mine and monitor the lifestyles and habits of their workers outside the workplace."[69] As the *Wall Street Journal* elaborates, "Your company already knows whether you have been taking your meds, getting your teeth cleaned and going for regular medical checkups. Now some employers or their insurance companies are tracking what staffers eat, where they shop and how much weight they are putting on—and taking action to keep them in line."[70] The idea, essentially, is to track employee consumption, exercise, and living habits in order to reduce insurance costs—which is attractive from both the employer's perspective and that of the insurance industry. Retailers use the data markets to predict future buying behavior; to classify existing behavior into "predetermined categories"; to associate certain behaviors with others, such as when (to borrow an example from a recent article in the *Atlantic*) Amazon makes a recommendation for martini glasses based on the recent purchase of a cocktail shaker; and to form "clusters" of information based on consumer behaviors, as when a group of consumers are separated into their specific hobbies and interests.[71] Consumer surveillance, as David Lyon notes, is "the most rapidly growing sphere of surveillance ... outstripping the surveillance capacities of most nation-states. And even within nation-states, administrative surveillance is guided as much by the canons of consumption as those of citizenship, classically construed."[72]

The data market—our new behemoth—is this agglomerated space of the public and private spheres, of government, economy, and society. There, corporations surveil and govern, governments do commerce, individuals go public. The private and the public sphere enmesh. Employers spy on their workers, tech companies govern our app tastes, and intelligence services vacuum up corporate secrets. And as the world becomes a digital market, this feeds back and reinforces our complacency: with everything around us being marketized, we no longer have any basis to question or object when we too are catalogued and classified based on our desires and tastes, when we too become marketed, even when the technology invades our privacy. We become marketized subjects of this vast digital economy, driven by the same market logics—ways of seeing and being that gradually displace our social and political selves as citizens and private subjects.

THE COLLAPSE OF THE CLASSIC DIVIDE between state and society, between the public and private spheres, is particularly debilitating and disarming. The reason is that the boundaries of the state had always been imagined in order to *limit* them: Liberalism depended on a clear demarcation, without which there never could be a space for the liberal self to pursue its own vision of the good life. For centuries, liberal legal thinkers and policy makers tried to strengthen the hedges between the public and private realms—between state and society— in order to place more robust limits on the state. With written constitutions and bills of rights that placed restraints on state power, with constitutional doctrines requiring "state action," with common law principles requiring "harm to others" or tangible injuries before government could interfere—all these were efforts to police the public sphere and simultaneously to liberate individuals to pursue their own different ways of life in the private realm. The central device of liberal legalism was to demarcate the public sphere in order to allow the private pursuit of individual conceptions of the good.

Critics, on the other hand, relentlessly tried to show that the neat separation between public and private was an illusion, and that the regulation of the public necessarily bled into the private realm, both personal and socioeconomic. The effort was to demonstrate how the regulation of the public sphere actually pervasively shaped the private sphere: how *public* policies ranging from mandatory arrest in domestic abuse cases to heat-of-passion defenses in homicide cases to resistance requirements in rape prosecutions actually affected *personal* relationships in the home. The slogan "the personal is political" captured the critique perfectly: there was no clean border between the two, and we had to be constantly vigilant about the unexamined effects of liberal governance on our private lives. By exposing the interconnectedness of the public and private spheres, the critics hoped to initiate a robust public debate over our shared values and ideals. By rendering the seemingly invisible visible, the critics hoped to stimulate deliberative democratic conversations—debates in which the values of solidarity, they hoped, would ultimately prevail. The goal was to render everything truly political, and then to persuade on the strength of their political values and beliefs. The key critical move assumed and placed as its target a liberal legal imaginary of rights discourse, boundaries, and progress.

But today there is no longer even the pretense of a liberal ambition to seriously cabin the state. There is no longer a genuine effort to limit governing or commerce—or whatever this behemoth is. Most of our liberal guardians of the Constitution have essentially abdicated their role and do not even pretend to delimit constitutional powers. At the very moment that journalists were revealing NSA programs, such as BOUNDLESS INFORMANT, that were literally "collecting, analyzing, and storing billions of telephone calls and emails sent across the American telecommunications infrastructure," President Barack Obama, a Democratic, liberal, rule-of-law president, former lecturer on constitutional law at the University of Chicago, was simultaneously telling the American people, "If you're a U.S. person, then the NSA is not listening to your phone calls and it's not targeting your emails unless it's getting an individualized court order. That's the existing rule."[73] Apparently the liberal legalists themselves are prepared to fudge the facts and mislead the American people into believing that nothing has really changed since the 1960s. Even after the Snowden leaks, even after we learned about PRISM and UPSTREAM, even after we *know* the intelligence agencies will listen to and read content by an American citizen so long as a foreign national is believed to be involved, President Obama would tell us:

> What I can say unequivocally is that if you are a U.S. person, the NSA cannot listen to your telephone calls . . . and have not. They cannot and have not, by law and by rule, and unless they—and usually it wouldn't be "they," it'd be the FBI—go to a court, and obtain a warrant, and seek probable cause, the same way it's always been, the same way when we were growing up and we were

watching movies, you want to go set up a wiretap, you got to go to a judge, show probable cause.[74]

Neither President Obama nor his liberal legal advisers even recognize that things have changed—or they do not seem to care. They appear to have given up on the boundaries delimiting the state.

This is entirely disarming to the critics, who can no longer even hold the liberals to their own aspirations and principles. There are, to be sure, valiant efforts by the ACLU, the Center for Constitutional Rights, and other nongovernmental organizations to keep the American government accountable to liberal ideals. But it is unclear how the traditional legal remedies—neutral magistrates independently deciding whether there is reasonable suspicion—can function any longer when secretive courts, such as the U.S. Foreign Intelligence Surveillance Court, set up under the Foreign Intelligence Surveillance Act (FISA), do not deny a single request by the NSA and when, in the face of that, liberals maintain, in the words of President Obama, that the FISA court "is transparent, that's why we set up the FISA court."[75] Given that the liberal guardians of the Constitution themselves no longer genuinely care about the limits of the state, there is hardly anything left for the critics to critique. Liberalism may have been mistaken in assuming that state action could be cabined, but the ambition was noble. We seem to have lost even that ambition today.

This is all the more troubling because the combinations we face today are so utterly powerful. The combination of the computing power and financial resources of the federal government with those of digital giants like Apple, Hewlett-Packard, Intel, Cisco, Samsung, and Microsoft, with the colossal advertising power of Google, Facebook, and Yahoo!, with the marketing ambitions of gigantic digital advertisers, and with the security interests of the intelligence industry is simply daunting. Apple earned $182 billion in revenue for the fiscal year ending October 2014. It made $39.5 billion in net income and has cash reserves around $137 billion—yes, *cash reserves* of $137 billion. Samsung reported net income of over $30 billion on revenues of $327 billion in 2013. It employs almost half a million people. Hewlett-Packard had revenues over $111 billion in 2014, and Facebook has an equity value over $15 billion.[76]

The idea that individuals could securely encrypt their personal communications in order to maintain some level of privacy in the face of Microsoft working with the NSA and the FBI to allow them better access to its cloud products—in the face, essentially, of the combined computing power of Microsoft and the intelligence agencies—is hard to imagine. Not impossible, but certainly daunting. Eben Moglen beseeches us to not give up hope. "Our struggle to retain our privacy is far from hopeless," he emphasizes. "Hopelessness is merely the condition they want you to catch, not one you have to have."[77] True, but maintaining hope in the face of these giant collaborators requires real force and courage.

CRITICS TRADITIONALLY RESISTED definitional clarity at the boundary of state and society, precisely to politicize the personal and to personalize the political—to get each and every one of us invested in the public debate.

Many critics, in times past, expressly avoided the term "the state," in order to better understand political relations, both in the United States and on the Continent. A number of American political scientists, as early as the 1950s, tried to eliminate the term from their lexicon. David Easton, Gabriel Almond, and others turned to the systems analytic framework discussed earlier—to the broader notion of the "political system" that included political parties and the media—precisely because the state was nothing more than an ideological construct, or, in their words, "a myth."[78]

On the Continent, the sociologist Pierre Bourdieu carefully avoided using the term "the state" in most of his work throughout the 1970s, only beginning to theorize it in the lectures he delivered at the Collège de France in the 1980s and early 1990s.[79] In those lectures, Bourdieu took as his point of departure the classic Weberian definition of the state, but he spent most of his time deconstructing it, at least in the early lectures. The state, for Bourdieu, appears first as mere fiction, a myth—the most powerful of all myths. The "state" is the product of our collective imaginary and gains its force precisely from our imagining it together—from a consensus that we craft. "The State is this well-founded illusion, this place that exists, essentially, only because we believe it exists," Bourdieu lectured. It is "this illusory reality, collectively validated by the consensus," "this mysterious reality that exists by our collective belief in its existence."[80] We simply come to agree that there is an entity called "the state," which then exercises normative force over our everyday lives. It becomes the Archimedean point, what Bourdieu calls "the point of view of all points of view."[81]

Other critics focused on the overlap and intrusions of state and society. The sociologist Theda Skocpol, for instance, who originally "brought the state back in"—that is, refocused American academic interest on the state, drawing on Weber as a counterweight to Marx—soon found herself studying the interplay and interrelationship between the state and civic organizations, specifically women's groups.[82] Skocpol's original "'state-centered' theoretical frame of reference" would evolve, in later work such as *Protecting Soldiers and Mothers,* "into a fully 'polity-centered approach'": a more intersectional perspective that focused on women's organizations and their role in promoting a maternalist welfare system in the early twentieth century and, more generally, on issues of gender and identities.[83] The state, it turned out, had to be studied in relation with nonstate actors to understand how the interaction shaped our society.[84]

Others explored what work the boundary itself—the boundary between state and civil society—actually performs. In other words, rather than define the border, they analyzed how it functions, what it achieves. The political theorist Timothy Mitchell, for instance, took this approach, and, rather than deconstructing definitions of the state, showed how the discourse surrounding the division of state and society serves to distribute power.[85] "The boundary of the state (or political system) never marks a real exterior," Mitchell emphasized. "The line between state and society is not the perimeter of an intrinsic entity, which can be thought of as a free-standing object or actor. It is a line drawn internally, *within* the network of institutional mechanisms through which a certain social and political order is maintained."[86] It is precisely in the struggle over the boundaries of the state, in the effort to delineate society, in the exercise of pitting one against the other, that power circulates between institutions. From this perspective, "producing and maintaining the distinction between state and society is itself a mechanism that generates resources of power."[87] And it generates those resources because of the meanings, duties, and rights we associate with the public or private spheres.

Timothy Mitchell's intervention in his 1991 article "The Limits of the State" is a classic illustration of the critical move to deconstruct the state-society divide. As Mitchell argued, "the state" does not really have, and never has had, a fixed boundary. It is not a fixed object. Rather, its boundaries are endlessly contested and the very struggle over its boundaries is a key site for the circulation of power.[88] By drawing distinctions between the private sector and the state, for instance, multinational corporations can seek tax advantages that subsidize private investment, without necessarily politicizing the activities or subjecting them to public debate; at the same time, the state can pursue national interests under the guise of private investment. As Mitchell carefully showed, "the fact that [the multinational] can be said to lie outside the formal political system, thereby disguising its role in international politics, is essential to its strength as part of a larger political order."[89] (An excellent contemporary illustration, discussed earlier, would be the NSA compensating the private telecoms for storing bulk telephony metadata—the drawing of a fictitious line that keeps the metadata collection program alive, achieves the same goals, serves the interests of both parties, and yet supposedly does not raise any privacy concerns because the telecoms are nonstate actors). Mitchell's critique demonstrated brilliantly how institutional actors—governmental, corporate, civilian, et cetera—play with the boundaries of the state in order to advance their interests. Using a number of examples, Mitchell showed that "the state-society divide is not a simple border between two free-standing objects or domains, but a complex distinction internal to these realms of practice."[90] The critical enterprise here was to explore the discursive struggles over the boundaries of "the state" and how power circulates as a result. Or as Mitchell would write, "The task of a critique of the state is not just to reject such metaphysics, but to explain how it has been possible to produce this practical yet ghost-like effect."[91]

Mitchell's critique drew on a discursive analysis that had been championed by Foucault, though Foucault himself also directly addressed—and resisted—the concept of "the state." Foucault's most direct engagement occurred in the early 1970s, in the lectures published as *Penal Theories and Institutions* and *The Punitive Society*.[92] At the time, Foucault was writing against the backdrop of Louis Althusser's focus on state apparatuses—what Althusser referred to as *les appareils d'État*—and to this notion, Foucault would propose a more fluid concept. Foucault drew attention to the intermingling of governing and private enterprise.[93] He took as an example the sequestration of women laborers at the penitentiary-style private factory-convents that manufactured silk cloth in the Ain department of France in the nineteenth century. These were private enterprises that depended on the state and administrative regulations but themselves regulated practically every aspect of their workers' lives—their sexuality, their recreation, their spirituality, their religious faith, their habits, their vices, their moral upbringing. These were private enterprises that benefited from, meshed with, resisted, and manipulated state initiatives, such as laws requiring workers to carry with them at all times their *livrets* (annotated notebooks that functioned as identification and registration cards), so that the shop foreman as well as the policeman could monitor the workers, as could the bartender, the night attendant—in short, anyone who came into contact with them. The laws requiring mandatory *livrets* would extend to other domains, such as the workers' savings accounts. The *livrets* allowed the employer to annotate and comment for future employers and to evaluate each worker, facilitating his hold on his workers through a system of micropunishments that accompanied the women through the revolving doors of employment, debt, and the penal system.

Foucault elaborated a fluid conception of quasi-public, quasi-private institutions, multifaceted practices, and normalizing relations of power, all complexly tied up within "statelike" nodes. They are state-like in the sense that they harked back to the styles of governing that we were accustomed to associate with "governing." They are not the state, but they have statelike attributes. Foucault would jot down in his notes for his 1973 lectures, as if he were speaking directly to Althusser: "It is not a state apparatus, but rather an apparatus caught in a statelike knot. An infra-statelike system."[94] In other words, these multifaceted quasi-private, quasi-public institutions at times borrowed statelike governance methods and took on a statelike appearance. It is precisely this approach that would serve, in Foucault's research, "as a historical background to various studies of the power of normalization and the formation of knowledge in modern society."[95] A few years later, in this vein of studying fluid, multidimensional practices and institutions complexly tied up within statelike knots, Foucault would turn to the study of neoliberal governmentality.[96]

It is important to emphasize that the term "statelike"—*étatique* in French—serves as a referent. It describes the type of act we intuitively associate with the state. Different relations of power emerge over time— sovereign and majestic forms of power that create truth by marking the body, disciplinary and capillary forms of power that correct and individualize, or equilibrating forms of power that seek to balance and maximize certain objectives. These different forms of power overlap. They fill the space of social relations, are often instantiated in part by state actors, and become associated with governance. They are what we consider "statelike." As the writings of Timothy Mitchell suggest, "statelike" is something that we contest, that is deployed, that does work.

KNOTS OF STATELIKE POWER: that is what we face. A tenticular amalgam of public and private institutions that includes signals intelligence agencies, Netflix, Amazon, Microsoft, Google, eBay, Facebook, Samsung, Target, and others—a behemoth that spans government, commerce, and our intimate lives—all tied up in knots of statelike power. Economy, society, and private life melt into a giant data market for everyone to trade, mine, analyze, and target.

We are told, by this voyeuristic amalgam, that everything is being done for our benefit only—to protect us from terrorist attacks, to keep us safe from digital threats, to make our digital experience more pleasant, to show us products we desire, to better avoid spam and unwanted communications, to satisfy us better.... But there are other interests at stake, naturally—and little pretense about that. President Obama prominently mentioned in his speech on surveillance "our trade and investment relationships, including the concerns of America's companies," and he made sure to reference "malware that targets a stock exchange."[97] Clearly, there are significant profit motives—the cash reserves of the high-tech companies are staggering. There are deep competitive interests as well, in a seething world of corporate espionage. There are political and economic questions of dominance in the most valuable currency of all today, what has become the single most important primary resource: electronic communication capabilities in a digital age.

The prevalence of economic espionage in the archive of NSA documents leaked by Snowden is telling. As Glenn Greenwald showed us, "much of the Snowden archive revealed what can only be called economic espionage: eavesdropping and email interception aimed at the Brazilian oil giant Petrobras, economic conferences in Latin America, energy companies in Venezuela and Mexico, and spying by the NSA's allies—including Canada, Norway, and Sweden—on the Brazilian Ministry of Mines and Energy and energy companies in several other countries."[98] The NSA documents reveal targets that include, in addition, Russian companies such as Gazprom and Aeroflot.[99]

So is it really our interests that are at stake, or is it other ambitions? Will we, in the long run, be protected in our intimacy, in our sociality, in our political struggles—which at times may be more radical for some than for others? Hard to tell, is it not, when the drives of this tenticular oligarchy may include such pressing strategic global interests, such intense financial and commercial competition, not to mention world dominance, given our position, in President Obama's words, as "the world's only superpower."[100]

EIGHT

THE MORTIFICATION OF THE SELF

EXPOSED, WATCHED, RECORDED, predicted—for many of us, the new digital technologies have begun to shape our subjectivity. The inability to control our intimate information, the sentiment of being followed or tracked, these reinforce our sense of vulnerability. Our constant attention to rankings and ratings, to the number of "likes," retweets, comments, and shares, start to define our conception of self. For some of us, we depend increasingly on the metrics. We start judging and evaluating ourselves by the numbers. A sense of insecurity may begin to erode our self-confidence. The new platforms start to shape what we like about ourselves. The recommendations mold our preferences.

We have known for a long time that our selves are shaped by our interactions with others. From George Mead through Erving Goffman, we have learned that "the self is not something we are born with or something that is innate in us; instead, it is something we acquire through interaction with others."[1] It should come as no surprise, then, that the digital is reconfiguring our subjectivity. For many of us, the technologies we depend on for our own flourishing, for our professional advancement, for our personal development, have effects on us. The *doppelgänger* logic itself has a looping effect: it shapes our subjectivity on the basis of others' tastes, while shaping others' subjectivities on the basis of our own. These algorithms challenge our self-reliance, make us aware of our fragility, and produce a feeling of things being out of our personal control.

Research suggests that online visibility and the exposure and transparency of social media may have a damping effect on our willingness to voice our opinions and express ourselves, particularly when we think that we are in the minority—feeding into a "spiral of silence."[2] Studies show that being electronically watched at work by a supervisor has adverse effects on productivity, creativity, and stress levels: in more technical terms, "electronic performance monitoring can have adverse effects on employees' perceptions of their jobs' stressors and on their self-reported levels of physical and psychological strain."[3] There is even biometric data to support the findings: "Their performance of tasks suffers and they have elevated pulse rates and levels of stress hormones."[4] There is a large body of psychological research, associated with "objective self-awareness theory," suggesting that stimuli that prompt self-awareness (for example, mirrors, photos, autobiographical information, or now digital posts) may activate discrepancies between one's perceived self and social standards, and as a consequence lower self-esteem.[5] Other research suggests that the feeling that one is exposed to others or being watched is associated with lower self-esteem and heightened stress and anxiety: "privacy research in both online and offline environments has shown that just the perception, let alone the reality, of being watched results in feelings of low self-esteem, depression and anxiety."[6]

In a fascinating study of children at boarding school in the United Kingdom who are subject to these forms of digital surveillance, researchers found that the children express an "out-of-control" feeling. The students, particularly those in the English private school setting, were acutely aware of being watched, with CCTV and computer "watchware," and felt they were simply losing control:

> Pupils said that they were subject to "real-time" computer monitoring which included the teachers or IT people being able to

"see what we're doing on the computer" ... ; watching "what you type" ... ; watching "what's on everyone's screen" ... ; seeing "who's logged on to where" ... ; and taking "control of your computer" ... "The teacher can see it but also the IT people can do it as well. So they can write a message like 'Get off this site.' Or they'll just say 'go see X.'"[7]

Some of the pupils expressed concern that this would interfere with their schoolwork. One boy feared that "just by the touch of a button, they can shut down your computer. So your work could disappear and everything ... it's just a cause of worry that you're going to get a message and he's just going to shut your computer down and you're going to lose your work. It's just a constant threat of a teacher just killing your computer."

The research on these schoolchildren suggests that these feelings of vulnerability have significant effects on them as subjects. Such feelings act as a depressant, they wear on the soul, they undermine the children's self-confidence. The surveillance inhibits them, shakes them, makes them lose their sense of self.

A number of studies dating back to the 1970s—during and following a previous FBI surveillance scandal—documented the effects of omnipresent surveillance, including some of the famous Stanford University experiments of Philip Zimbardo, specifically the 1975 study Zimbardo conducted with Gregory White.[8] More recently, the journalist association PEN America confirmed a chilling effect on journalists resulting already from the NSA surveillance disclosures.[9] There is a loss of the self in the face of so much digital knowledge about the self. A feeling that one has no control over one's self. That everything can be known, will be known, is known. That there is nothing that can be kept from the other. Similar findings have surfaced in other industries that are subject to increasing digital monitoring of employees using real-time location systems, implantable radio-frequency identification devices, electronic onboard recorder devices, and electronic driver logs.[10]

Daniel Soar, in an essay in the *London Review of Books,* gives expression to this overwhelming sense of loss and lack of control:

> I know that Google knows, because I've looked it up, that on 30 April 2011 at 4.33 P.M. I was at Willesden Junction station, travelling west. It knows where I was, as it knows where I am now, because like many millions of others I have an Android-powered smartphone with Google's location service turned on. If you use the full range of its products, Google knows the identity of everyone you communicate with by email, instant messaging and phone, with a master list—accessible only by you, and by Google—of the people you contact most. If you use its products, Google knows the content of your emails and voicemail messages (a feature of Google Voice is that it transcribes messages and emails them to you, storing the text on Google servers indefinitely). If you find Google products compelling—and their promise of access-anywhere, conflagration and laptop-theft-proof document creation makes them quite compelling—Google knows the content of every document you write or spreadsheet you fiddle or presentation you construct. If as many Google-enabled robotic devices get installed as Google hopes, Google may soon know the contents of your fridge, your heart rate when you're exercising, the weather outside your front door, the pattern of electricity use in your home.
>
> Google knows or has sought to know, and may increasingly seek to know, your credit card numbers, your purchasing history, your date of birth, your medical history, your reading habits, your taste in music, your interest or otherwise (thanks to your searching habits) in the First Intifada or the career of Audrey Hepburn or flights to Mexico or interest-free loans, or whatever you idly speculate about at 3.45 on a Wednesday afternoon. Here's something: if you have an Android phone, Google can guess your home address, since that's where your phone tends to be at night. I don't mean that in theory some rogue Google employee could hack into your phone to find out where you sleep; I

mean that Google, as a system, explicitly deduces where you live and openly logs it as "home address" in its location service, to put beside the "work address" where you spend the majority of your daytime hours.[11]

This omnipresent knowledge deprives us of a secure space of our own, a place to feel safe, protected. The schoolchildren at one of the U.K. boarding schools in the study spoke about this in terms of "their expressed desire to have a 'backstage' area of 'emotional release.' ... Or as one pupil put it, in relation to the use of CCTV cameras in toilets or changing rooms, 'Why won't they let you sort yourself out, your face and stuff? ... It's somewhere where I can sort myself out'" (282). There is something about feeling that someone else has control over your personal life, your intimate self, your laptop or smartphone—the very extension of your person, with all your files, your photos, your tax forms, your intimate correspondence. It is destabilizing. It is corrosive. Some, like Erving Goffman, even suggest it is mortifying.

In his book *Nothing to Hide,* Daniel Solove explores these issues through the literary metaphor of Franz Kafka's novel *The Trial.*[12] It is not, he suggests, the Orwellian lens but Kafka that sheds the most light on our digital condition. And the problem that Kafka identified is, in Solove's words, that of "a suffocating powerlessness and vulnerability created by the court system's use of personal data and its denial to the protagonist of any knowledge of or participation in the process."[13] The real harm is not just that we inhibit our own behavior; it involves "bureaucratic ones—indifference, error, abuse, frustration, and lack of transparency and accountability."[14] These are what create the effects of powerlessness and vulnerability on human subjectivity.

· · ·

I recently walked into the Bloomberg building in Manhattan. I was there to speak at a University of Chicago reception. It was about 6:00 p.m. and already dark outside when I arrived. When I asked for directions to the function, I was sent to the reception area on the main floor. I soon realized the receptionist was a security guard, and that I needed to present some ID and get a badge so that I could get on the elevators. It then dawned on me that they wanted a photo of my face. They were going to place it on the temporary ID card. There was nothing I wanted less. But it was all done so seamlessly and quickly. The security guard just asked me to look behind him, where a camera lens was hidden somewhere on the wall in a bank of black glass. I could barely see the camera. I hadn't even noticed it coming into the lobby. Hadn't noticed it when I was showing my driver's license. But in a swift, rapid move, the security guard asked me, in an offhand way, to look behind him into the camera.

And *click*—it was done.

Next thing I knew, I had a badge with my photo on it and a personal Wi-Fi identifier and password. I could not resist. I did not resist. I could not challenge the security protocol. I was embarrassed to challenge it, so I gave in without any resistance. But it still bothers me today. Why? Because I had no control over the dissemination of my own identity, of my face. Because I felt like I had no power to challenge, to assert myself. And, really, who wants to be the person who is constantly challenging the hapless security guard? It's just his job. It sounds so paranoid.

. . .

I had been reading a fascinating article about how Facebook widgets track you even if you are not a Facebook member. I found the topic gripping. I should have been writing my book or preparing a lecture, but I couldn't stop reading about these new digital technologies. My insightful friends and my partner usually send any news articles about this to me. I am not a Facebook member, so I was getting slightly concerned.

I then started reading about how Google was vacuuming up data with its Google Street View technology. The article was on a website called AlterNet.org. All of a sudden, every time I would change pages on the article, a pop-up from Facebook would intrude on the bottom right of my screen, asking me to "like" AlterNet on Facebook. Then the same thing started happening at ProPublica.org as I was reading an article titled—get this—"Why Online Tracking Is Getting Creepier."[15] I kept getting hit with this pop-up window on the side of my screen: "Like ProPublica on Facebook. For investigations, news, data, discussions & more. 89,016 people like this. Sign up to see what your friends like."

I was surprised. But then I realized the AlterNet website naturally could track me, since I was visiting it, even if I didn't "like" Facebook. So could ProPublica. As well as Google, where I'd been searching before. And so can Facebook—as well as the NSA. I'm exposed, regardless.

IN HIS MASTERFUL ETHNOGRAPHY *ASYLUMS*, published in 1961, Erving Goffman minutely dissected how our surroundings shape us as subjects. Goffman gave us a phenomenology of the structure of the self—what he himself described as a "sociological version of the structure of the self."[16]

Goffman was studying, of course, the "analog": the asylum, the prison, the sanatorium, those institutions born in the early nineteenth century that contributed to a pervasive disciplinary power circulating through society. Goffman's research was based on ethnographic field work conducted in 1955–1956 at St. Elizabeth's Hospital in Washington, D.C., one of the largest asylums in the United States at the time, with over 6,000 patients. Goffman was writing during a period that produced much of the classic, critical literature on these institutions. It was several decades before the onset of our current crisis of massive overincarceration; but it was a time when people were beginning to question the concentration of patients that had built up in asylums and mental hospitals and would later take hold of the prison.[17] Goffman wrote on the asylum in 1961, as did Foucault, the same year, in his *History of Madness*, and as would David Rothman in 1971 in *The Discovery of the Asylum*, as well as R. D. Laing, David Cooper, Thomas Szasz, and others.[18] There was a growing awareness at the time of how psychiatric institutions and practices—and analog security measures more generally—shaped us as subjects. The experience in these institutions, Goffman showed, transformed the subject through a staged process that began with the mortification of the self.

Goffman's research would come to define the notion of the "total institution": a space where a large number of patients are cut off and separated from society for a long period of time and live together in a fully administered space, under the supervision of a separate group of guards and doctors who have open access to the free world. Goffman studied how that kind of institution affects patients, and what interested him deeply was this notion of the "mortification of the self"—the first of four stages of the shaping of subjectivity in the asylum. Mortification, according to Goffman, was followed by reorganization of the self, which led him to analyze all the influences that reorganize the subject (for example, the systems of punishments and privileges that reconstitute a subject). This was followed by a set of responses that are produced by the inmates themselves in reaction to this reorganization; and then finally by the production of cultural milieus that develop around these sets of responses, systems of power, and resistance.

These four stages are relevant to us today in the digital era because they demonstrate how power relations within an institution or larger system produce a moral experience. They offer a phenomenology of the analog carceral experience that may shed light on our digital condition. Goffman in fact called his study the analysis of the "moral career" of the patients. He interpreted these stages of their life course through the metaphor of a "career" by means of which, or through which, the patients become different moral agents. Goffman explored a sequence of changes in the life of an individual, of transformations of their subjectivity. And what is important is that it is a *moral* experience: the subject adjusts to his environment, is influenced by it, creates a new identity, transforms himself, all in relation to the forms of power that circulate around him.

The first step, then, the mortification of the self, happens in the asylum through a continual process of humiliation and degradation. These humiliating practices are carried out through a whole set of techniques. There is, for example, a rule in Benedictine monasteries that all private property must be seized from the young monk in order to prevent any form of attachment.[19] The monk must not even be allowed to attach himself to his bedding, his blanket, his mattress. Not even his room. The young monk shall be moved from room to room, displaced so that he cannot form any bonds and cannot develop an identity. There must be no association whatsoever with possessions or private property, because any attachment to property can create the possibility that the monk might develop an identity or a subjectivity that could comfort him. In the male prison, mortification happens through the loss of masculinity. This detachment serves to loosen the prisoner's identification with his conception of self. The detainees learn that their time and effort are worth nothing—that they are incapable of doing anything, of responding, of acting. The prison administration cuts the detainees' links to their prior selves, making it so that the subjects cannot and do not have the internal strength to respond or resist.[20]

Mortification operates through the contamination of the body and forced encounters with dirt and filth. The prisoners have to be dispossessed, stripped of their roles, so that they can maintain no control over their actions. Their actions must become so routinized and scheduled that they have no possibility of independence or choice. This is achieved through strict time schedules that dispossess inmates of any possible sense that they are acting for themselves.[21] It is a process, simultaneously, of social control: at 5:30 a.m., the inmates are awakened; on the count of one they wake up, on two they stand at attention, on three they make their bed. They dress by numbers, socks at two, at three, shoes, any noise is enough to send the inmate to the line ... A strict regulation, a regimentation, continual, pervasive, running across every aspect of time and space, is presented as being necessary for the security of the establishment. It is always justified on the basis of security. These are the practices that mortify the self, that detach it from its former identity. This is the deconstruction, piece by piece, of the subject.

The mortification of the self is followed by practices that reorganize and reconstruct the subject. These are the systems of punishments and privileges: a whole new set of norms and rules that are used to govern the individual, who is in the process of losing his own conception of self and taking on a new subjectivity. These are the "house rules" that determine who is privileged, how they may gain favor, what are the benefits, and what is the punishment.[22] The privilege system and the house rules distribute pleasure and mete out pain. The explicit and formal set of prescriptions lays out the main contours of permitted inmate conduct.

Goffman then turns to the different types of responses that inmates and patients formulate in reaction to these forms of reorganization. Some respond by withdrawing entirely, regressing to earlier stages of moral development, focusing obsessively on a momentary activity, or ceasing to participate in common life. Others refuse to cooperate and create lines of intransigence. Some invest themselves in their acts of resistance. Yet others become docile, or worse, take on the role of the disciplining superior. Some become fully "colonized": for them, the asylum turns into their ideal world.[23] They find their comfort zone, they believe they have "found a home, never having had it so good."[24] This is a completely different reaction, but one that also represents, in its extremity, a type of resistance. To adopt the total institution as the greatest place in the world is resistance to power; it is almost as if the inmate is appropriating the space to himself.

Conversion is another way to react: to try to be the perfect inmate, to adopt all the rules—not just to be happy or fully colonized, but to become an even more perfect detainee than the staff wants the inmate to be. Goffman gives the example of Chinese prisoner camps, where the detainees embraced communism and the communist way of life even more fervently than the guards. To convert is to become even more invested than what is expected or imposed by the guard staff— more than what the very ideology of the institution demands. It is to embrace the surveillance completely and become a part of it.

And, finally, at the fourth stage, the inmates form cliques within the institution to deal with their situation. They sort themselves by means of their various reactions and responses. Relations develop between particular inmates who express the same types of resistance. Cultural milieus emerge and crystallize around certain relations to time—to "doing time." Together, some inmates exploit, others abandon, and still others destroy time in the total institution. The control of time, it turns out, becomes one of the most important forms of power and resistance.

These processes of subjectivation reach deep into the structure of the self—much further down than the ideational realm. They penetrate below those layers of the self that can more easily be peeled away, like a mistaken conclusion or an ill-formed belief. The notion of false consciousness does not serve us well here; we need to explore, instead, the deeper regions of desire, with a full recognition that those very desires may be shaped and molded by the digital age itself.[25] As Deleuze and Guattari remind us:

> It was not by means of a metaphor, even a paternal metaphor, that Hitler was able to sexually arouse the fascists. It is not by means of a metaphor that a banking or stock-market transaction, a claim, a coupon, a credit, is able to arouse people who are not necessarily bankers.... There are socioeconomic "complexes" that are also veritable complexes of the unconscious, and that communicate a voluptuous wave from the top to the bottom of their hierarchy (the military-industrial complex). And ideology, Oedipus, and the phallus have nothing to do with this, because they depend on it rather than being its impetus. For it is a matter of flows, of stocks, of breaks in and fluctuations of flows; desire is present wherever something flows and runs, carrying along with it interested subjects—but also drunken or slumbering subjects—toward lethal destinations.[26]

The constant monitoring, the Netflix recommendations, Twitter's "popular accounts" and "find friends" do their work. They shape our digital selves. They constitute our distinctive "looking glass."[27] There is no authentic self down there, nor layers of false consciousness that can be peeled away. There is instead a deeply embedded self, shaped by these new digital technologies, that cannot easily be pried open. We are deeply invested—with "investments of desire," as Deleuze suggested—and these investments need to be explored. "We never desire against our interests," Deleuze explained, "because interest always follows and finds itself where desire has placed it. We have to be willing to hear the call of Reich: no, the masses were not deceived; at a particular time, they actually wanted a fascist regime!"[28]

Drawing on Deleuze and Guattari, who tried to move us beyond psychoanalysis to what they called "schizoanalysis," we might say that digital technology "liberates the flows of desire"—much like capitalism does in their work.[29] That it shapes and produces desires locked onto other desiring machines. Those other machines, we know them well today. We are glued to them. Inseparable. And we give ourselves up to them—in the process, giving ourselves away. This may also help to explain the self-destructive nature of some of the digital cravings. The objective of the material psychology that Deleuze and Guattari developed reached deep into the recesses of desire to understand how we might end up in obscene places. The "goal of schizoanalysis," they argued, is "to analyze the specific nature of the libidinal investments in the economic and political spheres, and thereby to show how, in the subject who desires, desire can be made to desire its own repression— whence the role of the death instinct in the circuit connecting desire to the social sphere."[30] The technologies that end up facilitating surveillance are the very technologies we crave. We desire those digital spaces, those virtual experiences, all those electronic gadgets—and we have become, slowly but surely, slaves to them. Slaves to them and to our desires, our desires for shares, clicks, friends, and "likes." Kevin Haggerty and Richard Ericson deploy the term "surveillant assemblage" to describe the broader trend of surveillance in virtually all sectors of life that "operates by abstracting human bodies from their territorial settings and separating them into a series of discrete flows. These flows are then reassembled into distinct 'data doubles' which can be scrutinized and targeted for intervention."[31] The surveillant assemblage becomes normalized and ubiquitous—and it has deeply normalizing effects on the self. It shapes us.

GOFFMAN'S ANALYSIS OF MORAL careers sheds light on our new condition of digital exposure. It may be possible, in fact, to identify the phenomenological steps of the structuration of the self in the age of Google and the NSA. To describe the moral careers that we digital subjects experience—just like the inmate in the asylum. There are striking parallels. The monk in his Benedictine monastery, who is deprived of his possessions and any form of attachment, has much in common with us digital subjects today, who are deprived of an intimate space of our own, of any anonymity or genuine privacy, who cannot control the dissemination of our secrets and of our most cherished moments. Many of us become increasingly detached even from our selves, as we know and get used to the fact that others can watch us. It is as if we begin to dissociate from our materiality and look at it from a distance. For many of us, especially for the youths in the British school study, we begin to feel like we are being watched in a demeaning way. In fact, some of the schoolchildren in the U.K. study explicitly refer to digital security as "big brother" (281). It begins to take on the same symbolic meaning as the custodial staff in Goffman's asylum—with features common to the first stage of mortification.

There is, for instance, humiliation and degradation when the CCTV cameras look into the girls' toilets or changing rooms, and watch them when they are undressing (281–282). Some of the pupils said that they "had heard stories making an explicit connection between 'surveillance' and 'voyeurism,' including firemen in helicopters using cameras to go around gardens looking at ladies sunbathing . . . and police officers radioing each other to say 'oh there's a MILF over here, come over here'" (282–283). Some of the pupils experienced this firsthand: being seen on camera and identified as not having had a top on in a park, even though they had a bikini top on—with the police saying repeatedly,

"We just saw you on camera" (283). We know, from Snowden's revelations, that this takes place all the time among young analysts working for the NSA, who share pictures of naked women or people in compromising positions.[32] Recall the trove of X-rated images gleaned by operation Optic Nerve.[33]

Then there is the opprobrium and increased surveillance for those who wear certain items of clothing: in the study of British schoolchildren, it was hoodies, tracksuits, or trainers (279). There is special attention to those who are not conforming, who display resistance, even the slightest resistance, to the mainstream norm. The schoolchildren said that what draws the most attention is wearing hoodies. It is, according to the upstanding children (the "angels"), "those who dress in 'subcultural' attire," or more simply the "chavs," who draw attention (279). The term "chavs" refers to "members of the working class who dress in tracksuits and baseball caps"—or, as the authors elaborate, "the term is also applied to members of the working class who are perceived by 'superordinate classes' to be 'aesthetically impoverished' due to their 'vulgar' and 'excessive' consumption habits" (280 n. 6). We focus more attention, more of our digital surveillance, on those whom we deem more vulgar—and that is often us, or at least our perception of ourselves.

There is, as well, the dimension of turning us into marketized objects of consumption. The *doppelgänger* logic constantly reminds us that we are hardly more than consuming things. "The working logics of these algorithms not only shape user practices," Tarleton Gillespie tells us, "but lead users to internalize their norms and priorities."[34] Many of us get "habituated" by "the code of commercial platform" to understand ourselves as targets of advertising, as objects of consumption; in this, "the principles of capitalism are embedded in the workings of search engines."[35] Targets of advertising, or simply targets. Some of the children say that "when they were in shops and supermarkets they sometimes wondered whether their actions could be misinterpreted by CCTV operators" (283).

The surveillance also affected the schoolchildren's relations with their teachers, parents, and others:

In terms of the "subjective impact" of these systems, some children explained how they impacted on patterns of "sociation" or "face-to-face" interaction with parents, teachers, and their peers. There was "no negotiation," for example, with the teacher for "late arrivals"; pupils could no longer buy their friends dinner due to the introduction of the "cashless" dinner programme; and the "automated text messages" could "really land you in it" with parents. (287)

These technologies have a different objectivity that undermines the possibility of negotiation, fudging, playing with facts and circumstances—something that is so important to human interaction.

To be sure, some research suggests that certain forms of social media—particularly those that are "nonymous" rather than "anonymous"—may promote self-esteem by allowing us to present ourselves in more flattering or self-pleasing ways, given that the narratives we construct are often those that make us more confident about ourselves. In one particular study, the researchers find that the opportunity to present more positive information about oneself and to filter out negative information—in other words, the selective self-presentation that is made possible by digitally mediated environments—has a positive influence on self-esteem.[36] They find that while an unedited view of the self (for instance, in a mirror) is likely to decrease self-esteem, the extra care involved in digital self-representation on online platforms may improve self-esteem.

But that does not seem to be the experience for many of the schoolchildren under digital supervision—especially those who feel less entitled.[37] Many of the schoolchildren sense that the digital platforms are trying to shape them. One of the pupils explains, "Everyone's watching each other. And they're all trying to make us all perfect" (283). As the U.K. study reports:

For some pupils these processes could lead to "self-policing" of the body." As one girl explained, "CCTV just encourages—you know—beauty, and everyone wanting to be perfect ... Like everybody wanting to be size 8 or really skinny and really, really, beautiful" . . . The emergence of a "surveillance society" reminded some of the pupils of a "futuristic" school play they had recently performed called "The Perfect Generation." "The cameras," as one girl explained, "are trying to control us and make us perfect." (287)

"The perfect generation": these technologies have effects on our subjectivity. They have "chilling effects," according to the pupils in the British study. They made the students "acutely aware that their actions were being monitored and led them to change 'legitimate' forms of behaviour or activities due to a concern that their actions could be misinterpreted by the 'surveyors'" (283).

Digital exposure is restructuring the self in the ways that Goffman identified. We are experiencing a moral transformation, undergoing a moral career, becoming different moral agents. For many of us, the virtual transparence has begun to mortify our analog selves—they are fading away like the image on an old Polaroid instant photo. Google, Facebook, Amazon, Scorelogix, the NSA—through their rankings, ratings, recommendations, scores, and no-fly-lists—are putting in place a new system of privileges and punishments that are restructuring our selves. We each react and respond differently to these incentives, and we sort ourselves out into different cliques—some who do not seem to care, others who feel targeted, some who resist, others who feel they've "never had it so good." This new digital age has not only given birth to our data doubles—our second bodies—but is having profound effects on our analog selves. In combination, it is radically transforming our subjectivity—even for those, perhaps even more, who believe they have nothing to fear.

SOMETIMES IT WOULD ALMOST APPEAR as if the whole of our digital society is divided into two different types of people: those who say they have nothing to hide and feel protected by the surveillance, and those who feel vulnerable and fear they will be the targets of digital surveillance. In *Asylums,* there is a haunting passage where Erving Goffman asks provocatively: "May this not be the situation in free society, too?"[38] Perhaps we should ask ourselves the inverse question today: What if our expository society is a total institution, too? Could it be that there are some people who see themselves as virtual prisoners of digital exposure, as potential targets, as the usual suspects, and others who, because of their privilege or for whatever other reason, feel that they are protected by the digital surveillance?

Perhaps it is society as a whole that has become the totalizing institution, with two categories of subjects: those who perceive themselves as under surveillance (the captive subjectivities) and those who believe they are being protected (the guardians of society). After all, there are many of us who believe, almost reflexively or unthinkingly, that the state is protecting our interests, that our self-interests line up with those of the NSA, Google, Microsoft, Instagram, and Facebook—that we are the beneficiaries. These may be our guardians of the expository society; the digital protects them. On the other side, there are those of us who always suspect we will be surveilled, vulnerable, exposed—and who question whether the system is there to protect us or detect us. These would be the inmates, the patients, those of us who are stuck in the expository society.

Mortification of the self, in our digital world, happens when subjects voluntarily cede their private attachments and their personal privacy, when they give up their protected personal space, cease monitoring their exposure on the Internet, let go of their personal data, and expose their intimate lives. The moral transformation happens when subjects no longer resist the virtual transparency of digital life. "I have nothing to hide." "It is no big deal." "Nothing to worry about."[39] That, paradoxically, may be the final stage of the mortification of the self.

NINE

THE STEEL MESH

A STRIKING FEATURE of our digital age is the utter contrast between the ethic of play and desire at the heart of our expository society and the crushing nature of our punitive practices. The surprising resilience of analog forms of punishment—of the iron bars and cinderblock cells—is entirely puzzling when juxtaposed with the virtual transparency that characterizes so much of our digital lives. In many respects, the contrast could not be greater. At one end, there is this new, capacious form of digital exposure that thrives on our willing embrace and deepest passions, on love and desire, and on anxiety too—on the time we spend on Xbox, on what we share on Facebook, on how we surf so freely with Explorer, Firefox, and Chrome. This is the space of open frontiers, of experimentation, of liberty, of expression, and of curiosity. At the other end, there is confinement and sequestration, shackles and chains, isolation cells and cages, with growing solitary confinement at one extreme and overcrowding and warehousing at the other. This is the space of bolted iron doors and heavy locks, of watchtowers and armed guards, of chain gangs and armored vans. Open, free surfing at one end; strapped down on the gurney used for lethal injection at the other end.

We hardly need to be reminded, but our liberal democracy in the United States still leads the world in imprisonment. Of all countries, our expository society, with all its virtual transparence, incarcerates the greatest number and percentage of its citizens.[1] About one in a hundred adults in America is behind bars, a total of more than 2.2 million people, and the number of people under correctional supervision is growing.[2] The prison and jail population in the United States is about the size of half the population of Denmark. The statistics are even worse when we disaggregate by demographics. In 2008, one out of nine young adult black men between the ages of twenty and thirty-four—or approximately 11 percent of that population—was incarcerated in prison or jail in the United States.[3] As of 2011, more than 2 million African American men were either behind bars or under correctional supervision (that is, had been arrested and processed by the criminal justice system and were on probation, on parole, or behind bars). That too represents about 11 percent of the total population of black men—one out of nine.[4] Not only that, but our prisons have reverted to the prerevolutionary function of extraction.[5] No longer to reeducate or to correct the inmate, no longer to improve or provide skills, no longer to honestly deter others, the prison has become dedicated to, or rather reverted to, the task of bare extraction. The prison has folded back on itself, returned to a procrustean age—to the dark age before the eighteenth-century reformers created the penitentiary. Our new digital existence is accompanied today by a vertiginous form of analog control: massive, archaic, and racialized overincarceration.[6]

It is surprising that our new digital transparence could coexist so seamlessly with such a massive, physical, brute, archaic regime of punishment. The forms of power are so staggeringly different. The freedom to surf the web, to stroll through any city in the world via Google Street View, to videochat with loved ones on the other side of the planet—it is hard to imagine a more radical contrast to the unplugged, unwired, locked-down, and isolated life in jails and prisons today. It is like the stark contrast between a brilliantly colorful, stimulating, and lustful digital life and the drab, gray, blunt, oppressive world of the fictional Oceania— nothing could be further apart. And yet the two extremes thrive.

As a first cut, we might imagine that we feel so free today in our liberal democracies—so liberated in our digital transparence—precisely because we have built our mirrored glass house without ever dismantling the iron cage of security at its inner core, the prison with its cinder blocks and iron bars. We allow ourselves the pleasure of digital surfing because we know, as we watch and are watched ourselves, that *some of us will be extracted*. Extracted from our digital life in common and returned to the "analog": to the prison, the iron bars, the cell block, the unplugged, the unwired. Returned to the closed chamber that is still there, at the heart of Philip Johnson's glass house.

Max Weber's notion of the iron cage—or rather, Talcott Parsons's somewhat idiosyncratic translation of Weber's expression *stahlhartes Gehäuse* as "iron cage"—captures well the analog condition of the iron bars and cinder blocks, of the archaic prison, still so dominant today despite the liquidity of our digital age.[7] The brute physicality of the penitentiary, the brick-and-mortar rationality of the eighteenth and nineteenth centuries—these are still present today throughout every facet of our massive overincarceration. The "analog" still survives deep inside the glass pavilion—and in such stark contrast to the fluid and ethereal digital age. For many of us, sadly, it may be the iron cage that comforts us, and serves as the condition of possibility for our digital wanderings.

BUT THERE IS MORE to it than that. An odd convergence is taking place today. At one end, our lived experience is gravitating dramatically from the analog to the digital. The digital self, the second body of today's liberal democratic citizen, is overtaking his analog physical existence and becoming far more permanent, durable, tangible, and demonstrable. But at the other end, the analog prison is gently sliding toward forms of digital correctional supervision—with more and more ankle bracelets and GPS tracking. As a result of the Great Recession of 2008, states and counties are doing everything they can to replace physical incarceration with digitized probation and supervision: with electronic bracelets and home monitoring, CCTV, biometric supervision, and all sorts of other digital technologies. Cost-cutting and efficiency measures are driving punitive practices into the virtual world.

The result is a paradoxical and surprising convergence of the radical freedom of digital life and the supervised existence of the parolee: the virtual transparence of digital exposure is beginning to mimic and replicate the new punitive form. Ordinary daily transparence has begun to map onto penal monitoring. Everyday life resembles the electronic bracelet and CCTV surveillance. The Apple Watch begins to function like the ankle bracelet. All is seen, all can be seen, all can be monitored—inside or out, wherever we are, free or supervised, we are permanently surveilled.

It is even possible to imagine a time in the not too distant future when there will be no need to incarcerate because we will all be watched so closely. We won't need to confine anymore, because we will be able to see everywhere—perhaps even control behavior from a distance. There exist already technologies to sedate and control. Psychotropic drugs serve this function, as do chemical castration and other forms of inhibitors. Would it be possible to imagine remote administration of such technologies once we can digitally see, track, watch, and follow everyone's most minor movements? Could it be an added feature of a smart watch? Not only will it take our pulse, it will, on remote orders, administer sedatives and chemical blockers . . . But that is merely science fiction. Perhaps.

AT ONE EXTREME, then, punishment is increasingly moving toward electronic monitoring and GPS-tracking. The growth of digital monitoring in the criminal justice system, especially of GPS devices, is striking. The *Journal of Offender Monitoring*—a journal dedicated to this burgeoning field—conducted a study charting the growth of electronic monitoring in the United States from 1999 to 2009.[8] According to that study, the number of people under GPS surveillance grew exponentially, up from 230 in 1999 to 91,329 in 2009. The annual rate of increase reached 95.6 percent in 2006 and 86.1 percent in 2007, growing practically eightfold from 2005 to 2009.[9] The other form of electronic monitoring, radio-frequency (RF) devices used to monitor home confinement orders, also continued to increase over the period, meaning that overall, the number of people on some form of electronic monitoring grew by a factor of more than 2.5 between 1999 and 2009, from approximately 75,000 to approximately 200,000.[10]

The same trends can be seen across the country. A study conducted in Florida tracks the use of electronic monitoring in that state from 1993–1994 through 2008–2009 and shows its use roughly doubling during that period, from 1,555 people in 1993–1994 to 3,177 in 2008–2009. Interestingly, the trends reflect a sharp decline in the number of people subject to RF monitoring, coinciding with the rise of the much more intrusive active GPS monitors. In 1993–1994, all of the 1,555 people subject to monitoring were monitored by RF (the only technology available then). GPS first came into use in 1997–1998, and within two years it had already eclipsed the number of people on RF devices (677 vs. 343). Finally, beginning in 2004–2005, the number of people on GPS devices increased dramatically, from 1,104 that year to 2,933 in 2008–2009, while the number of people on RF devices continued to fall over that same period, from 634 to 244.[11]

In North Carolina, the number of people under electronic monitoring (either satellite-based location monitoring or electronic home confinement) has increased dramatically from 875 in 2011 to 2,789 as of June 2014.[12] The Immigrations and Customs Enforcement (ICE) field office in New York dramatically increased the number of women monitored by GPS-enabled ankle bracelets, up roughly 4,000 percent in 2014 alone, from 18 to 719.[13] Even Vermont is turning to the use of electronic monitoring as a cost-effective alternative to pretrial detention in the state.[14]

One of the attractions of GPS monitoring for most counties and states—in addition to the reduced financial and human resources expended on incarceration and supervision of low-risk inmates—is the cost-saving factor of inmates being forced to pay part or all of the costs of the electronic monitoring devices themselves.[15] A recent study by National Public Radio and the Brennan Center shows that, with the exception of Hawaii and the District of Columbia, every state requires offenders to pay at least some portion of the cost associated with GPS monitoring.[16] This is part of a larger trend of requiring those on probation to pay for an increasing share of the cost of their supervision.

Looking ahead, there is compelling evidence of an even greater rise in electronic monitoring. The company with the single largest market share, BI, is set to have a banner year, at least in part due to its recent contract with ICE, which is predicted to be worth somewhere between $200 million and $300 million. The press releases on the company's website highlight the array of new contracts that the firm has entered into with county and municipal governments.[17] Meanwhile, SuperCom, an Israeli manufacturer of electronic monitoring devices, announced its expansion into the U.S. market in July 2014.

We are witnessing a sharp increase in the number of people subject to electronic monitoring in one form or another—from one end of the country to the other. Miami-Dade County in August 2014 introduced a program to use electronic monitoring as an alternative to pretrial detention, hoping to reduce the jail population by 10–20 percent; in November 2014, the Maine State Board of Corrections solicited proposals from counties to implement a pilot program of electronic monitoring for domestic abusers.[18] The second-largest county in Idaho, Canyon County, in July 2014 sent out a request for proposals for an electronic monitoring system as part of an alternative sentencing program.[19] More and more, local governments are seeking to implement such programs.

AT THE OTHER EXTREME, our free-world digital technologies more and more resemble correctional supervision. They in fact replicate the intrusiveness of digital carceral monitoring and share the same form of power: omnipresent, all-seeing. It is not an exaggeration to say that the reach of the new digital technologies into our ordinary lives is at least as powerful and intrusive as the capabilities of digital correctional monitoring, if not more so.

PRISM and UPSTREAM—as well as all the "people-based marketing" and "onboarding"—offer the signals and corporate intelligence services practically complete access to us. But even more than that, the NSA has found multiple other means of accessing devices and getting hold of the very functioning of those devices without a user's notice. In terms of our personal lives, we can be tracked in unparalleled ways—at least as much as correctional monitoring.

Two new ways of doing this seem to have emerged: physically inserting foreign devices into target computers that transmit data, and infiltrating computers through malware. It is almost as if we are implanting radio-frequency identification (RFID) devices into parolees. First, the NSA can gain access to computers using "radio pathways" even when the devices are not connected to the Internet. Radio pathway devices can be "physically inserted by a spy, a manufacturer or an unwitting user" into computers, emitting a radio frequency detectable by mobile NSA relay stations up to eight miles away. In addition to transmitting data from target computers to NSA stations, these devices also allow the NSA to plant malware in the target computers, as was the case in the 2010 U.S. cyberattack on Iran's nuclear enrichment facilities. The extent of operations abroad is staggering. A recent *New York Times* report notes that according to a "senior official, who spoke on the condition of anonymity," close to 100,000 such implants have been placed by the United States worldwide.[20] Much of this activity has ostensibly been used to track malware, a defensive move that one senior official interviewed compared to the activity of submarines that silently track one another. Such tracking has been used frequently on Chinese military targets, who have also been accused of using similar measures in U.S. government and industry devices.[21]

Second, malware now appears to be a very serious part of the U.S. intelligence toolkit. The NSA regularly uses malware to transmit data from users' devices to NSA databases without the user being aware of it. When the NSA is able to introduce the malware into another computer, the agency can, in its own words, "own" the machine; as Greenwald explains, it can "view every keystroke entered and every screen viewed."[22]

. . .

In November 2013 *Der Spiegel* broke a story, based on documents leaked by Edward Snowden, that concerned the NSA's Office of Tailored Access Operations (TAO), as well as its QUANTUMINSERT program.[23] TAO is "the NSA's top operative unit—something like a squad of plumbers that can be called in when normal access to a target is blocked." Their activity "ranges from counterterrorism to cyber attacks to traditional espionage." They have the ability to infiltrate, according to the information leaked to *Der Spiegel*, "servers, workstations, firewalls, routers, handsets, phone switches, SCADA systems, etc." The last of these, SCADAs, "are industrial control systems used in factories, as well as in power plants." In addition, TAO has developed the ability to infiltrate users of "virtually every popular Internet service provider," including "Facebook, Yahoo, Twitter and YouTube." While TAO has had difficulty infiltrating Google users, they have been able to see this information through their sources in GCHQ.[24]

Der Spiegel describes a "popular tool" of TAO called QUANTUMINSERT, which appears to be a way of gaining access to devices by covertly inserting malware into them. For instance, "GCHQ workers used this method to attack the computers of employees at partly government-held Belgian telecommunications company Belgacom, in order to use their computers to penetrate even further into the company's networks. The NSA, meanwhile, used the same technology to target high-ranking members of the Organization of the Petroleum Exporting Countries (OPEC) at the organization's Vienna headquarters." Perhaps most significantly, the NSA used the QUANTUMINSERT tool in order to infiltrate the SEA-ME-WE-4 underwater cable bundle, which connects "Europe with North Africa and the Gulf states and then continues on through Pakistan and India, all the way to Malaysia and Thailand."[25]

When these tactics fail, the TAO uses another NSA division, called ANT. *Der Spiegel* reports that the acronym is unclear but "presumably stands for Advanced or Access Network Technology." This division employs a whole catalogue of technologies that appear to follow roughly the same strategy as QUANTUMINSERT. The German newspaper adds, "In cases where [the TAO's] usual hacking and data-skimming methods don't suffice, ANT workers step in with their special tools, penetrating networking equipment, monitoring mobile phones and computers and diverting or even modifying data." ANT has "burrowed its way into nearly all the security architecture made by the major players in the industry—including American global market leader Cisco and its Chinese competitor Huawei, but also producers of mass-market goods, such as US computer-maker Dell."[26]

A major goal of these surveillance techniques is what ANT developers refer to as "persistence": the ability to insert into devices malware that continues to collect information even when changes have been made to those devices. "Persistent" malware continues to perform its function without any indication to the user of improper function. Targeted devices include not only computers but also routers and "the firmware in hard drives manufactured by Western Digital, Seagate, Maxtor and Samsung."[27] According to NSA documents leaked by Snowden, the practice is widespread: the NSA has, for instance, been able to infect well over 50,000 computers with one type of malware called Quantum Insertion, with the *New York Times* reporting that the number reached about 100,000.[28]

In some important respects, what TAO and ANT do holds different consequences for privacy than other activities revealed by earlier Snowden revelations. The NSA is not simply accessing bulk information available by means of a user's Internet activity in an anonymous or anonymizable way. The NSA's radio pathway surveillance is meant specifically to target activities that do not occur in the sphere of Internet-derived, bulk data. Devices must be covertly inserted into computers and phones on the factory floor, or later through a USB device. The creation of fake LinkedIn profiles, for instance, has been used by the GCHQ to infiltrate Belgacom "in order to tap their customers' telephone and data traffic."[29] These practices are different from the kinds of data skimming that might be performed by companies interested in consumer information. Rather, they target individual devices to listen to specific people.

DIGITAL SURVEILLANCE CAPABILITIES have reached new heights in ordinary digital existence and have begun to converge on carceral monitoring. It is difficult, in fact, to tell the two apart—and it is not even clear whether we should characterize these new digital surveillance techniques as "free-world" or "carceral." The Apple Watch and the ankle bracelet, the malware and the GPS tracking, begin to blend into one indistinguishable mass of digital monitoring as our lives begin to resemble that of a supervised parolee.

And as the technologies converge, they begin to feed increasingly into each other. They work together. Virtual transparence creates the possibility of seeing through populations, the way an X-ray or CAT scan does, to visualize pockets of resistance—and then it allows us to pluck them out of the system. The two different forms of power work together, as positive and negative—virtual transparence *and* massive incarceration, the wired *and* the unplugged—to fuel even more incarceration. The digital becomes a form of radiation.

There is growing evidence that this may be the direction we are headed. The *New York Times* revealed in September 2013 that the New York Police Department (NYPD) has begun to use social media to identify, arrest, and prosecute suspected members of crews, or youth gangs.[30] The program is called Operation Crew Cut, and it has a very simple strategy. According to the *Times,* "The strategy seeks to exploit the online postings of suspected members and their digital connections to build criminal conspiracy cases against whole groups that might otherwise take years of painstaking undercover work to penetrate. Facebook, officers like to say now, is the most reliable informer."[31]

At the center of the strategy is the NYPD's new "social media unit," which seeks to identify suspect youths and track them based on their online presence—their Facebook posts, their Twitter tweets, and the photos and videos that they upload. The strategy is very simple: "Officers follow crew members on Twitter and Instagram, or friend them on Facebook, pretending to be young women to get around privacy settings that limit what can be seen. They listen to the lyrical taunts of local rap artists, some affiliated with crews, and watch YouTube for clues to past trouble and future conflicts."[32]

By following what they say and whom they talk to, by tracing the network connections and watching the social interactions, NYPD officers can identify whom they want to go after and build cases against the youths. The social media unit complements the street intelligence both in the investigation and in the prosecution of criminal charges. "Though social media postings have emerged only recently as an element of prosecutions, those in the legal arena are fast learning that Facebook, MySpace and Twitter can help to pin down the whereabouts of suspects and shed light on motives."[33] Apparently this strategy is particularly effective despite the fact that it is widely publicized, because, as the NYPD officials explain, "an online persona is a necessary component of social life for the young crew members."[34] Earlier the *Times* reported that "as Twitter, Facebook and other forms of public electronic communication embed themselves in people's lives, the postings, rants and messages that appear online are emerging as a new trove for the police and prosecutors to sift through after crimes. Such sites are often the first place they go."[35]

While Ray Kelly, a former NYPD commissioner, was still heading the department, he made a joke at a news conference to the effect that he wanted to be "Facebook friends with all the city's criminal crews."[36] That may well be the direction we are headed. In at least one case, a federal agent with the U.S. Drug Enforcement Administration stole a suspect's identity, set up a fraudulent Facebook page in her name (unbeknownst to her), and posted suggestive photos that she had taken of herself (as well as photos of her infant son and niece) with her cell phone.[37] The Facebook page was a complete fabrication, intended simply to further investigation and identify other suspects. The woman in question, Sondra Arquiett, had agreed to plead guilty on a minor drug violation and was awaiting trial. The federal agent obtained her personal photos from a seizure and search of her mobile phone and then used those images to populate and give credibility to the Facebook page he created. The goal was to communicate with, identify, and locate other suspects. The United States government has filed legal briefs defending the actions of the federal agent, claiming that it was all legitimate investigation and legal. According to a news report, "The Justice Department is claiming, in a little-noticed court filing, that a federal agent had the right to impersonate a young woman online by creating a Facebook page in her name without her knowledge. Government lawyers also are defending the agent's right to scour the woman's seized cell phone and to post photographs—including racy pictures of her and even one of her young son and niece—to the phony social media account, which the agent was using to communicate with suspected criminals."[38]

Other local law enforcement agencies are also collecting digital data, creating data banks, and engaging in data mining for purposes of surveillance and investigation. Five municipalities in Virginia, for instance, have collaborated on building just such a data archive from subpoenaed records—note that no search warrant nor even probable cause is required for a subpoena—and fruits of investigations.[39] The five cities—Norfolk, Newport News, Hampton, Chesapeake, and Suffolk—have been building and sharing since 2012 what they call the Hampton Roads Telephone Analysis Sharing Network. It is made up of phone records subpoenaed by the local police agencies from telecommunication companies as well as information gleaned from mobile phones taken from suspects. The data is maintained in a "telephone analysis room" in Hampton and used by all of the agencies.[40]

According to *Wired* magazine, "The unusual and secretive database contains telecom customer subscriber information; records about individual phone calls, such as the numbers dialed, the time the calls were made and their duration; as well as the contents of seized mobile devices."[41] This data trove would supplement, of course, the ordinary means of law enforcement access to digital information: the directed subpoena to the telecom companies. This alone is considerable. Most telecoms are turning over troves of data in response to ordinary subpoena requests. As *Wired* recounts, AT&T received between January and June 2014 almost "80,000 criminal subpoenas for customer records from federal, state and local law enforcement agencies," and Verizon received "over 72,000 subpoenas from law enforcement during the same period."[42]

Kenneth Neil Cukier and Viktor Mayer-Schoenberger describe in their book *Big Data* the different ways in which datafication feeds into predictive policing: computer programs that data-mine criminal justice statistics in order to predict where crime will occur or who will commit it, Homeland Security research programs that analyze physiological patterns in order to identify potential terrorists. "If social workers could tell with 95 percent accuracy which teenage girls would get pregnant or which high school boys would drop out of school," Cukier and Mayer ask, "wouldn't they be remiss if they did not step in to help?"[43] And how easy it is to do just that, to predict delinquent outcomes based on texts, IMs, web searches, Internet gaming activities, et cetera. We know that in some cases Target can identify a young woman's pregnancy more effectively than her parent can. It is so frightfully easy to do with big data.

The trouble is that the allusions to crime and terrorism most often are overdetermined. They fail to specify the uncertainties and ambiguities; and, as a result, they have a way of silencing our questions, of neutralizing our doubts, and too easily justifying, for many, the surveillance of our phones and Internet activities. The possibility, or rather the probability, of overreaching should not be dismissed so quickly.

. . .

In September 2014, forty-three former members of Israel's elite military signals intelligence unit, Unit 8200, issued a public letter condemning the unit's exploitation of signals intelligence for political purposes.[44] Some of that intelligence may have been turned over to the unit by the NSA under an NSA agreement that has been in place since 2009 and that covers, by its own terms, practically all types of communication, including "unevaluated and unminimized transcripts, gists, facsimiles, telex, voice and Digital Network Intelligence metadata and content."[45] As James Bamford notes, the official NSA memorandum indicates that "the agency 'routinely sends' unminimized data."[46]

The forty-three signatories included former officers, current reservists, and former instructors of what is basically Israel's "equivalent of America's NSA or Britain's GCHQ."[47] In their open letter, they contend that signals intelligence is being used against innocent people for purely political motives: "The Palestinian population under military rule is completely exposed to espionage and surveillance by Israeli intelligence. It is used for political persecution and to create divisions within Palestinian society by recruiting collaborators and driving parts of Palestinian society against itself."[48] According to testimonials of the signatories, they were directed to focus on human weaknesses, on sexual affairs and financial problems, and on sexual preferences. As James Bamford explained in the *New York Times,* "In testimonies and interviews given to the media, they specified that data were gathered on Palestinians' sexual orientations, infidelities, money problems, family medical conditions and other private matters that could be used to coerce Palestinians into becoming collaborators or create divisions in their society."[49]

It seems that we in the United States are not immune to this behavior either. The *Huffington Post* published an NSA document dated October 3, 2012, leaked by Edward Snowden, that reveals that the NSA "has been gathering records of online sexual activity and evidence of visits to pornographic websites as part of a proposed plan to harm the reputations of those whom the agency believes are radicalizing others through incendiary speeches."[50] The NSA keyed in on any Internet activity that might harm a target's reputation, such as "viewing sexually explicit material online" and "using sexually explicit persuasive language when communicating with inexperienced young girls."[51] The targets included a "respected academic," a "well-known media celebrity," and a "U.S. person," all living abroad, probably in the Middle East.[52] As Bamford explains, "The document, from Gen. Keith B. Alexander, then the director of the N.S.A., notes that the agency had been compiling records of visits to pornographic websites and proposes using that information to damage the reputations of people whom the agency considers 'radicalizers'—not necessarily terrorists, but those attempting, through the use of incendiary speech, to radicalize others."[53] And with an estimated 1.2 million individuals populating the United States government's watch list and targeted for surveillance as potential threats or suspects, the potential for further abuse is high.[54]

As ordinary digital life and new forms of correctional surveillance begin to converge—as the Apple Watch, the ankle bracelet, and GPS tracking merge into each other—we face a new, generalized carceral condition marked by astounding levels of monitoring. This new condition has an uncanny relationship to earlier regimes of punishment. It is not just that there is an iron cage at the very heart of the digital age—though there is one for sure, comprised of massive over-incarceration, drone strikes, "black sites," and the Guantánamo Bay detention camp, among other things. But there is, eerily, something more. It is almost as if our iron cage today has been turned inside out and blankets us all.

The metaphor of the "iron cage" traces back to the Puritans and references the form of confinement that they invented: the penitentiary, that carceral gesture to penance imagined by the Quakers. The term "iron cage" was used in the seventeenth-century writings of the English Puritan preacher John Bunyan in his book *The Pilgrim's Progress,* written in 1678. The book was the fruit of Bunyan's own imprisonment and confinement from 1660 to 1672.[55] In the book, there is a man in despair, in the depths of dejection for having turned away from God and God from him. He is confined in "an iron cage," Bunyan wrote, and the misery he suffers there is a metaphor for the eternity of hell.[56]

The Harvard sociologist Talcott Parsons would read the metaphor into Max Weber's book *The Protestant Ethic and the Spirit of Capitalism* in preparing his 1930 English translation from the original German. As Peter Baehr notes, "Few concepts in the social sciences are more instantly recognizable than the 'iron cage.' Seemingly integral to the powerful denouement of *The Protestant Ethic and the Spirit of Capitalism,* the metaphor sums up, graphically and dramatically, the predicament of modern human beings trapped in a socioeconomic structure of their own making."[57] The term captures well Weber's genealogy of the rational, bureaucratic, and materialist world of modern capitalist production of the early twentieth century. In other respects, Parsons would underplay the Nietzsche in Weber.[58] But with the "iron cage," Parsons really got, in a truly Nietzschean way, the genealogy of morals at the heart of both German writers.

Talcott Parsons's reference to the Puritan preacher makes sense, even though it was a loaded translation of the German term *stahlhartes Gehäuse,* an expression that would translate more directly as "shell as hard as steel."[59] Most of the Anglo-Saxon interpretations would be guided by Parsons's adaptation—and many French and other readings as well, since many translators drew on Parsons's usage. To be sure, not all translators and commentators used the term "iron cage," some preferring "housing hard as steel" or "casing as hard as steel."[60] But the term, as Baehr comments, has taken on a life of its own.

Surprisingly, then, the metaphor of the iron cage was somewhat foisted onto Max Weber. But for our purposes here, it is perhaps a godsend. It allows us to see more clearly the cage in the mirrored glass pavilion—the punitive practices, the massive overincarceration, the omnipresent prison, the harsh punishment within the otherwise virtually transparent structure of the new digital age. It reminds us of the brick structure at the core of Philip Johnson's modern design. It puts the iron bars and cinder blocks back into the sequence—a sequence that has now reached a new stage in the Internet age.

The metaphor—in its more literal translation—allows us to recuperate, appropriate, and actualize Weber's original notion of a "shell as hard as steel" to better understand how the mirrored glass and digits work together: more of a tangled mesh, a webbed cloak, than iron bars. Steel is an alloy, a more modern material made by combining age-old iron with carbon or some other physical element. And there is something about a casing or a shell or a cloak that might be more appropriate than the prison cell or cage to describe our digital age. To be sure, the iron cage lives on, at the very heart of our mirrored glass house. But digital exposure takes on a different shape, a different form in this expository society. Virtual transparence and correctional monitoring work together more like a straitjacket, a casing, a shell made of some modern fabric, something like Teflon or woven Kevlar. We have graduated from the suit of armor, from the breastplate and gorget, from the pauldron and plackart, from the chain mail and iron plate of the analog age to the Kevlar jacket of digital times. Our expository society, it turns out, is a shell as hard as steel.

PART FOUR

Digital Disobedience

TEN

VIRTUAL DEMOCRACY

TODAY, DIGITS HAVE BECOME our pulse and our bloodstream. The cell phone in our pocket, the smart watch on our wrist, the MetroCard we swipe, the Kindle we read, the SMS texts—in short, every action we take in our daily routines, from the minute we wake up, is etched into the cloud to constitute a virtual self. Over the past ten to fifteen years, our digital self—the subject's second body—has taken on a life of its own and become more tangible than our analog self. We have built ourselves a mirrored glass pavilion: we expose ourselves to the gaze of others and embrace the virtual transparency with exhibitionist pleasure. We look in the glass as though we are looking in a mirror. We can see and be seen—and are seen, especially by a voyeuristic amalgam of NSA and FBI agents and subcontractors, social media, Silicon Valley firms, telecom companies, private consulting groups, hackers, advertisers, politicians, and ourselves. We watch and are watched, we knowingly strap surveillance devices on our bodies—and then some of us are arrested, some of us are disconnected, some of us are extracted.

How could this happen in a liberal democratic society? How is it possible that a democratic regime—a government *of the people*—could possibly engage in such massive and pervasive data collection, data mining, surveillance, and control of its own citizens, *of the people themselves*? Eben Moglen justifiably asks us whether "*any* form of democratic self-government, *anywhere,* is consistent with the kind of massive, pervasive surveillance into which the United States government has led not only its people but the world."[1] Could any self-respecting democratic polity knowingly allow the pervasive surveillance that we live in today? The answer, one would think, has to be no. In a state like the former German Democratic Republic, perhaps. Under the Stasi regime, certainly. In the film *The Lives of Others,* yes.[2] But in an advanced capitalist liberal democracy, in what we tout as a genuine democracy, is it even possible? On this question, the historian Quentin Skinner is surely right: "The current situation seems to me untenable *in a democratic society.*"[3]

Democratic theorists, for the most part, remain silent. Few have rolled up their sleeves to help understand our new digital condition.[4] Paradoxically, though, it may be precisely *because* we live in a liberal democracy—those of us in the United States, at least—that there is so little resistance. It may be that democratic practices themselves facilitate our complacency.

DEMOCRACIES MAY PRODUCE their own forms of apathy. Alexis de Tocqueville warned us about this in *Democracy in America,* particularly in the later volume from 1840, where he developed a theory about what he called "democratic despotism." Tocqueville's theory may have some bearing on our digital condition today. Democratic despotism, on Tocqueville's view, must be understood in contrast to the despotism of the Roman imperial leaders—an entirely different form of despotism, a tyranny of great leadership. In ancient Rome, despotic power expressed itself through a concentration of control in the hands of a few supreme arbiters, who possessed "an immense and unchecked power," Tocqueville wrote, "which allowed them to gratify all their whimsical tastes and to employ for that purpose the whole strength of the state."[5] Their despotism was spectacular and frequent, but it targeted mostly the few: "They frequently abused that power arbitrarily to deprive their subjects of property or of life; their tyranny was extremely onerous to the few, but it did not reach the many," Tocqueville suggested. "It was confined to some few main objects and neglected the rest; it was violent, but its range was limited."[6]

Tocqueville contrasted the notion of ancient despotism to the potential of despotism in a democracy, speculating that despotism would take an entirely different form in democracy. He warned that it might be "more extensive and more mild"; that it might "degrade men without tormenting them."[7] This, he suggested, would be a new condition, one that in a certain sense would require a new and different language regarding despotism and tyranny, for democratic power acts upon us differently than despotic leadership does. There's a stunning passage where Tocqueville described the way in which democratic power might act on the people: "It seeks, on the contrary, to keep [us] in perpetual childhood: it is well content that the people should rejoice, provided they think of nothing but rejoicing. For their happiness such a government willingly labors, but it chooses to be the sole agent and the only arbiter of that happiness; it provides for their security, foresees and supplies their necessities, facilitates their pleasures. . . . [W]hat remains, but to spare them all the care of thinking, and all the trouble of living?"[8]

This is a different form of despotism, one that might take hold only in a democracy: one in which people lose the will to resist and surrender with broken spirit. In striking language, Tocqueville suggested that this can compress, enervate, extinguish, and stupefy a people, "till each nation is reduced to nothing better than a flock of timid and industrious animals, of which the government is the shepherd."[9]

One need look no further than voting practices in the United States. David Graeber and others are undoubtedly right that democracy "is not necessarily defined by majority voting" and that it is better understood as "the process of collective deliberation on the principle of full and equal participation"; nevertheless, voting patterns are telling.[10] And in the United States, they are damning. Even in the most important national elections—in the United States, those are presidential elections—voter turnout since the Great Depression has fluctuated between 51 percent and 64 percent.[11] Turnout for midterm elections has consistently been below 50 percent since the 1930s, fluctuating be-tween 34 percent and 49 percent, as illustrated in Figure 10.1. The No-vember 2014 midterm election—just a few months after the Snowden leaks and in the heat of the NSA scandal—was "the lowest turnout election since 1942," with 35.9 percent of eligible voters and only 33.2 percent of the voting-age population going to the polls.[12] The Snowden revelations, it seems, fueled neither partisan divisions nor electoral turnout. There was a time, earlier in the republic, when voter turnout in the United States was consistently higher, but that time has passed.[13] Today, it seems, only about half the electorate—if even that—finds the time to vote in national elections. Walter Dean Burnham refers to this as "the disintegration of the American political order," a disinte-gration marked by the disillusionment of many Americans with ordi-nary party politics; "huge numbers of Americans are now wary of both major political parties and increasingly upset about prospects in the long term," Burnham and Thomas Ferguson note.[14] This may be true for 2014; but the low turnout has been a relative constant over the past eight decades.

Our democracy in the United States, it turns out, is not so much a democracy of voters as one of *potential* voters. It is not an actual de-mocracy so much as a *virtual* democracy—not in the sense commonly used by Internet enthusiasts or by what Evgeny Morozov calls "cyber-utopians," who believe that blogs and the web have reinvigorated democracy, but in the sense that our democracy is a figment of our imagination.[15] Virtual democracy has a potentiality, a capacity toward democratic rule, but one that is not actualized. It is through demo-cratic potentiality that the benefits of democracy are achieved. But it is also precisely what undermines real or timely resistance. Resistance, today, operates by catastrophe only.

What virtual democracy does is to provide a check on elected rep-resentatives, a check that essentially functions by the *threat* of demo-cratic vote: if our elected representatives exceed certain limits (and clearly those limits have a very high threshold), they may be voted out of office. There is no need for democratic action during ordinary times, no need even for that much democratic participation: even if the ma-

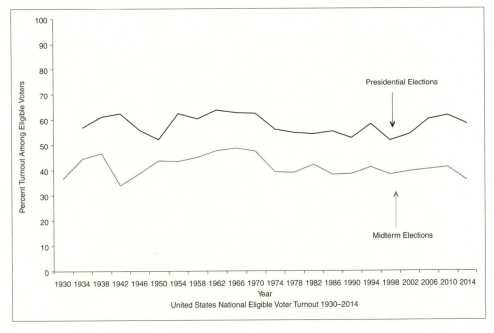

FIGURE 10.1 Voter turnout in federal elections in the United States (1930–2014). *Graph by* Bernard E. Harcourt. *Data source:* Michael P. McDonald, "National General Election VEP Turnout Rates, 1789–Present," United States Elections Project, http://www.electproject.org/national-1789-present.

jority does not vote in free elections, it *can,* and it is that potentiality that guarantees, in our imagination, the check on illegitimate exercises of authority.

Now, this may be entirely rational. Joseph Schumpeter, for instance, believed that it was entirely sensible to pay attention to political matters only when serious events arise. And many liberal thinkers have praised the way in which modern liberalism frees individuals to focus on their private lives and interests, and to avoid the public sphere. Whether it is justified or not, it has produced a form of politics that can best be described as "virtual democracy." There is a term in legal thought called "negotiating in the shadow of the law"—by which it is meant that legal actors in settlement negotiations always take into account how a judge might rule if the matter went to court. By analogy, we operate in the United States in the *shadow of democracy*—in the shadow

of a democratic vote. We rest on 50 percent turnout, but that is sufficient to keep alive the appearance and the threat of democratic rule.

DEMOCRATIC DESPOTISM IS DEEPLY LINKED to the apathy and
to the resulting complacency that facilitates and makes tolerable our
pervasive commercial and governmental surveillance today. In fact,
the problem of democratic despotism is, if anything, magnified in the
context of surveillance, policing, and punishment: in democracies,
these specific matters—surveillance, policing, and punishment—are
often viewed with distaste and often ignored as a result. They are considered necessities, but unpleasant ones. Democracies, for instance,
rarely if ever recognize that they might have political prisoners. They
tend to view all of their incarcerated as common-law detainees. Crime,
in a democracy, is not thought of as a political phenomenon.

This reflects and perpetuates a blindness that has plagued democratic theory for a long time. Throughout the history of democratic
theory there has been what Albert Dzur refers to as an *"invisibility* of
the problem of punishment," a "collective *non-seeing* or *dis-engagement*."[16]
This has been especially true during the recent period of massive overincarceration since the early 1970s, though it was equally true during the
earlier twentieth-century institutionalization in asylums and mental
hospitals—a phenomenon, in fact, that went virtually unnoticed in its
scale.[17] This blindness is not unique to the digital age; to the contrary, it
may have been born in tandem with democratic theory itself.

Returning to Tocqueville for a moment, we know well that *Democracy in America* set forth, for generations to come, the central problematics of democratic theory—the problem of majoritarianism, the
danger of a tyranny of the majority, the question of countermajoritarian
measures, and the matter of civic associations and town meetings, to
name only a few. Yet although the book itself was the direct result of
the birth of the penitentiary in the United States, there is one issue it
would ignore: punishment and the prison. Surprisingly—this is often
discounted—Tocqueville had come to the United States to visit the
new American penitentiary. Like his travel companion and coauthor,

Gustave de Beaumont, also a former public prosecutor, Tocqueville journeyed to the United States to study American prisons and report back on the democratic potential of those new institutions.[18] Their report, *The Penitentiary System in the United States and Its Application to France* (1833), as well as Tocqueville's continued writings and speeches on the topic during the 1840s, represented some of the very first theorizing of the prison and punishment by democratic political thinkers.[19]

But Tocqueville's orientation to the prison was marked by a staunch view of the criminal as a social enemy, and as a result, his political intervention regarding the penitentiary was strictly focused on administering prison populations, not democratizing prisoners as citizens. Tocqueville tellingly wrote in a parliamentary report of 1843: "It is necessary to recognize at this moment that there exists among us an organized society of criminals. All of the members of that society know one another; they support one another; they associate every day to disturb the public peace. They form a small nation in the midst of the large one. It is that society which should be scattered . . . and reduced, if possible, to a single being against all of the upright men united in defense of order."[20]

Tocqueville sought "to enlist prison reform in the antirevolutionary cause."[21] His reforms, which he gleaned from the experience of the United States, were, in his own words, "an ideal combination of 're-form' and 'repression.'"[22] They embraced a form of "despotism": Tocqueville admired the fact that the American prisons had produced "such complete obedience and useful work."[23] He wrote approvingly: "Although American society provides the example of the most extended liberty, the prisons of that same country offer the spectacle of the most complete despotism."[24] Tocqueville concluded: "Whence comes our near absolute repugnance toward using corporal punishment against prisoners? It is because one wants to treat men who are in prison as though they were still in society. Society and the prison are not, however, composed of the same elements. In general one could say that all of the inclinations of free men are toward the good, while all of the passions of the condemned criminals drive them violently toward evil. . . . 'The Rights of Man' . . . are not valid in prison."[25]

Civil rights had no place in prison, according to Tocqueville. Despotism was what was needed instead. Now, on Sheldon Wolin's reading, the question of the organization of prisons would become "one road" by which Tocqueville and other liberal thinkers would "find their way back to the state."[26] The prison was one way for these liberal thinkers, cautious of sovereign power, to turn to questions of administration, governmental planning, and public policy. It was, as Wolin suggests, "liberalism's way of reconciliation with the Old Regime by adopting and adapting its structure of power and its ideology of paternalism and benevolence."[27] But by the time that Tocqueville got back to the state and democratic theory, he had left behind any concern with punishment and the prison: Tocqueville was entirely silent about the prison in *Democracy in America*.

IT TURNS OUT, THEN, that democratic theory and the problems of surveillance and punishment may have been born together, but they were separated at birth. In fact, it is almost as if there was complete acoustic separation: by contrast to his writing in *Democracy in America*, Tocqueville did not tackle the questions of democracy and citizenship in *The Penitentiary System*. The criminal was viewed as an enemy, not as a potential citizen. And, by contrast to his writing in *The Penitentiary System*, Tocqueville would not tackle the question of punishment and prisons in *Democracy in America*. Democratic theory did not need to address punitive practices. The result is a distancing, a void: democratic theory did not give us useful tools to deal with these surveillance issues. If anything, it rendered those problems more tolerable because it placed surveillance, criminality, and punishment in the nonpolitical realm: the realm of the criminal as social enemy.[28]

Today there is beginning to be more awareness of excessive punishment and control in democratic discourse. Michelle Alexander's book *The New Jim Crow* has been on the *New York Times* best-seller list for over a year and a half and has contributed in part to this general awareness. Amy Lerman and Vesla Weaver have begun to document some of the more measurable effects of incarceration on citizenship and have shown how contact with the correctional system reduces participation in democratic politics, carrying with it a "substantial civic penalty": it produces a large, negative effect on "turning out to vote, involvement in civic groups, and trusting the government," even taking into account the possibility of selection bias.[29]

Despite all of this remarkable work, though, and some general awareness, it remains the case that the digital transparency and punishment issues are largely *invisible* to democratic theory and practice. They are not major topics of debate, they are not woven into theories of democracy, and they are not part of the political discourse. This is reflected, for instance, in the fact that President Obama's administration has never bothered to address the problem of overincarceration—and has been hesitant to challenge its security advisers on the question of NSA surveillance. Only on August 12, 2013, five years into the Obama administration, did Attorney General Eric Holder address the issue of overincarceration in a speech before the American Bar Association—but Holder focused predominantly on federal prison policy, which is tangential to the national problem.[30] President Obama himself only began to address the question of prisons in an interview in *Time* magazine in December 2012, but he has never made this a public issue.[31] His comments have been at best noncommittal, perhaps even unengaged.[32] This is equally true on the question of digital surveillance and the infamous NSA programs.

Paradoxically, then, the fact that we live in a democracy has facilitated, rather than hindered, our expository society. In our virtual democracy, the threat of democratic action has not slowed our digital exposure in any way. As a result, we will need to look elsewhere to locate digital resistance.

ELEVEN

DIGITAL RESISTANCE

IN HIS "POSTSCRIPT on the Societies of Control," published twenty years after *Anti-Oedipus,* Gilles Deleuze would return to the notion of the desire-producing machine, but this time as a metaphor for relations of power in society. Different forms of power at different historical periods are associated, Deleuze would argue, with different types of machines. There is a "match," Deleuze reports. A concordance. It is not that the type of machine determines the type of power, but rather that machines "express those social forms capable of generating them and using them."[1] So, Deleuze wrote, "the old societies of sovereignty made use of simple machines—levers, pulleys, clocks; but the recent disciplinary societies equipped themselves with machines involving energy, with the passive danger of entropy and the active danger of sabotage; the societies of control operate with machines of a third type, computers, whose passive danger is jamming and whose active one is piracy or the introduction of viruses."[2] These new machines, these computers, would match our times—our "societies of control."

Deleuze was insistent, though: do not fear these new machines, do not see only the dark side. "There is no need to ask which is the toughest or most tolerable regime, for it's within each of them that liberating and enslaving forces confront one another. . . . There is no need to fear or hope, but only to look for new weapons."[3]

"To look for new weapons": where shall we find them?

. . .

Jennifer Lyn Morone has incorporated herself. She is now a registered corporation with the state of Delaware: Jennifer Lyn Morone, Inc.[4] Establishing herself as a corporation allows Morone to "reclaim," in her words, "the whole process of resources, production, and ownership."[5]

Her logic goes something like this: while the corporate form was created in order "to provide a public benefit, such as building bridges and highways," today it primarily benefits the "owners," whose singular aim is profit. Corporations are now treated as individuals but have far more advantages than individuals in the flesh. For instance, they have certain tax advantages that individuals do not share, and can take advantage of certain loopholes that are unique to their corporate status. Plus, the individuals running the corporations—or should we say the corporations running the individuals—are barely liable even "if the company goes bankrupt or is accused of wrong-doing." At the same time, many companies sustain themselves on the basis of property they do not own: personal information, private data. Morone refers to this as "data slavery."[6]

Morone contends that her corporate self has "even more rights and benefits" than she does as an individual.[7] Incorporation transforms Morone's personal data into her property and allows her to profit from and control its use. As a corporation, Morone Inc. intends to derive revenue from three sources. First are Morone's "past experiences and present capabilities."[8] These are offered as biological, physical, and mental services such as genes, labor, creativity, blood, sweat, and tears. "Present capabilities" or services are, for now, mostly targeted toward people Morone knows personally, but they will also be marketed on her website, where customers will be able to fill out a calendar for her time and even check her status on the tasks concerned. The *Economist* notes that in addition to data, she will offer a range of biological services, from blood plasma at £30 ($50) and bone marrow donations ($5,100) to eggs at $170,000 apiece. Why so much? "Even though I'm older, which makes the eggs less valuable, they are more valuable to me as there is a limited supply and once this resource is gone, it's gone." And then there are mental services such as problem-solving (discounted if JLM gains something in return, such as knowledge); physical labour (she is a green-fingered gardener); and assets for which no pricing model yet exists: Ms Morone is still figuring out how to price "services" she currently gives away for free, such as compassion.[9]

Her website will include ads for particular goods and services Morone offers (these include everything from gardening to compassion; the latter can be traded to those friends who give her compassion in return, while more self-centered friends will receive an "invoice"), as well as advertising for goods and services of other people she knows.[10] Her website will also generate revenue, she notes, through "endorsements to promote events I might attend, clothes I might wear, restaurants I might eat at and products I might use."[11] Morone adds that another avenue is "the profitable but time consuming endeavor of pursuing intellectual property infringements," a course of action that in large part depends on whether her lawyer would be willing to work for "a percentage from cases won."[12]

A second source of revenue will be "selling future potential in the form of shares"; a third is "the accumulation, categorization, and evaluation of data that is generated as a result of Jennifer Lyn Morone's life."[13] Eventually those data will be generated through "a multi-sensor device that she will wear almost all the time" and a software program called Database of Me, or DOME, "which will store and manage all the data she generates."[14] As the *Economist* explains, "JLM's eventual goal is to create a software 'platform' for personal-data management; companies and other entities would be able to purchase data from DOME via the platform, but how they could use it would be limited by encryption or data-tagging. The software, then, would act as an automated data broker on behalf of the individual."[15] Morone's corporate status means that, in her words, "any data that I create that is linked to my name, IP address and appearance is copyrighted or trademarked and therefore subject to litigation if used without my permission. . . . So any photo I take, any email I write, any call, text, web search, cctv footage of me that is stored on someone else's, company's or government's server does not have the right to be there or to be used, sold, leased or traded."[16]

Morone describes the entire process as, at times, "schizophrenic"; the cluster of roles she must play do not all fit her self-image. She highlights this in her project's introductory video, in which she describes the project while wearing an ill-fitting suit and tie, adjusted in the back with visible clips.[17] On the other hand, as the clips suggest, she has embodied this role on her own terms: they "indicate that I am making this role fit me and not the other way around."[18]

When Morone is not working on the JLM Inc. project, she may be working on another one titled "Mouse Trap," a "hardly perceptible" mouse trap that operates all on its own and leaves the mouse physically unharmed but trapped in a clear cube—completely visible.[19] Now that's a frightening image of digital exposure.

. . .

Chelsea Manning and Edward Snowden could not live with their secrets—or, rather, with *our* secrets.[20] The loss of innocent civilian lives, hidden from public view by false claims of national security, was too much for Manning to keep to herself. The scope, depth, and intrusiveness of NSA surveillance were too much for Snowden.

On April 5, 2010, a decrypted video of a United States military Apache helicopter attack on unarmed civilians in Iraq, including two Reuters news staff, was made public on a WikiLeaks website and went viral.[21] The video, titled "Collateral Murder," was awful and heartwrenching. Shot from the gunsight of the Apache helicopter in 2007, it showed the copter circling and taking aim, shooting and killing unarmed civilians in a public square in eastern Baghdad. As the copter circles around, a father taking his children to school is seen trying to pull the injured men into his van, seemingly to get them medical assistance. The copter then fires mercilessly at the van, flipping it over with rounds of 30 mm ammunition, killing the father and severely injuring the children. "It's their fault for bringing their kids into a battle," the U.S. helicopter pilot is heard saying.[22] A battle of innocents it was, though. All told, twelve unarmed civilians were left dead. The official U.S. military position regarding the incident was that only insurgents were killed. The U.S. military claimed not to know how the Reuters news staff had died. The release of the "Collateral Murder" video was utterly shocking to many of us, especially to the uninitiated.

A few months later, in July 2010, the *Guardian,* the *New York Times,* and *Der Spiegel* began publishing a trove of U.S. military documents, including hundreds of thousands of war logs detailing U.S. incidents in Iraq and Afghanistan, and several months later, hundreds of thousands of confidential U.S. diplomatic cables. It was the largest cache of U.S. military and diplomatic documents leaked in history—and it had been made possible by the anonymous whistle-blower website set up by Julian Assange, WikiLeaks. Manning had been drawn to disclosing military secrets because of her feelings of revulsion at the conduct of the war in Iraq, but also, importantly, because of the condition of possibility that WikiLeaks created. According to reports, Manning read over the WikiLeaks list of the most sought-after pieces of secret government information, and identified and later leaked those documents that were in highest demand.[23]

Julian Assange had begun to develop WikiLeaks several years earlier, in 2006, to create a secure location where individual whistle-blowers could anonymously and securely upload secret documents that they felt the public should see. Assange referred to the site as "a web-based 'dead-letterbox' for would-be leakers."[24] WikiLeaks gained notoriety with its first major leak in July 2009, which concerned internal corporate documents proving bank fraud by the Icelandic bank Kaupthing, which was implicated in the Icelandic economic crisis.[25]

Assange, for many years, had been pursuing a radical approach of digital transparence, what some refer to as "a philosophy of transparence."[26] It represents a radical view that seeks to expose the truth no matter the consequence: a politically engaged practice of rendering public all governmental and commercial secrets that affect the public interest. As W. J. T. Mitchell suggests, Assange's project is to "turn the panopticon inside out."[27] The intention is to allow citizens to become the surveillers of the state and see directly into every crevasse and closet of the central watchtower, while rendering the public opaque and anonymous; to invert the line of sight so that it is we, the inmates, who cannot be seen; to place us behind the blinds in the central tower of the panoptic prison, so that our governments can anonymously be watched like the mice trapped in Jennifer Morone's clear cube.[28]

Assange's anonymous platform WikiLeaks represents an effort to institutionalize that inverted transparence. The idea is to allow any of us to anonymously and securely post documents to the WikiLeaks website in order for them to be accessible to everyone—a form of crowd-sourced transparence. "People determined to be in a democracy, to be their own government, must have the power that knowledge will bring—because knowledge will always rule ignorance," Assange asserts. "You can either be informed and your own rulers, or you can be ignorant and have someone else, who is not ignorant, rule over you."[29]

Daniel Ellsberg, who had leaked the Pentagon Papers in personal opposition to the Vietnam War in 1971, was a role model for Assange: someone who took personal responsibility and risked everything he had in life to expose the gruesome reality of war in order to foster public debate and democratic accountability. At a very early stage— it was on December 9, 2006, when WikiLeaks was just getting off the ground—Assange emailed Ellsberg, inviting him to get involved and lend his name to WikiLeaks. Assange wrote to Ellsberg that "fomenting a worldwide movement of mass leaking is the most effective political intervention."[30]

The idea of a "worldwide movement of mass leaking" captures well Assange's radical political vision. Assange views the world through a siege mentality: a political combat over secrecy and exposure that must be fought in the immediacy of the news cycle, under dire time constraints and with limited resources. Protecting the sources of leaks and ensuring that their information is immediately available—given that the whistle-blowers are often putting their liberty at stake—are of the highest concern to Assange.[31] This would lead to difficult choices about redacting leaked documents to protect named or identifiable individuals, given the overwhelming number of documents leaked, the limited resources of an essentially rogue outfit, and the U.S. government's refusal to do the work of redaction. In this context, Assange would stake out a clear political position. He would not remain neutral. He had views about the public interest that led him to privilege immediate exposure of secret information over the risk of harm to those exposed. He did so, in large part, because of an acute awareness of how secrecy shapes us as contemporary subjects. "Surveillance," Assange argues, "is another form of censorship. When people are frightened that what they are saying may be overheard by a power that has the ability to lock people up, then they adjust what they're saying. They start to self-censor."[32]

The U.S. government tried to shut down WikiLeaks by requesting that credit card companies stop processing donations from supporters. PayPal, MasterCard, Visa, Amazon, and the Swiss bank PostFinance all agreed to stop processing donations to WikiLeaks, leading the hacker collaborative Anonymous to initiate a series of DDoS (distributed denial of service) attacks against some of those companies in December 2010.[33] Meanwhile, Manning would be arrested in June 2010 and charged with violating the Espionage Act in a twenty-two-count indictment.[34] In February 2013, Manning pled guilty to ten of those twenty-two charges and faced a potential life sentence on the remaining counts at a highly publicized court-martial that was to begin on June 3, 2013.[35] Manning would ultimately be convicted on a total of seventeen counts, including espionage, and sentenced to thirty-five years in prison for leaking military secrets—in part, to deter others from doing the same.[36]

Two days later, on June 5, 2013, the *Guardian* published the first of several news articles on the top-secret NSA surveillance programs.[37] A few days after that, on June 9, 2013, the man who had leaked hundreds of thousands of classified top-secret NSA documents would reveal himself to the world: "My name is Ed Snowden. I'm twenty-nine years old. I worked for Booz Allen Hamilton as an infrastructure analyst for NSA in Hawaii."[38] Snowden, fully cognizant that he too could be sentenced to life imprisonment, would carefully turn over a colossal cache of secret NSA files to Glenn Greenwald, Laura Poitras, and Ewen MacAskill of the *Guardian*. In a calm and deliberate manner, Snowden explained why:

> When you're in positions of privileged access, like a systems administrator for the sort of intelligence community agencies, you're exposed to a lot more information on a broader scale then the average employee, and because of that, you see things that may be disturbing but, over the course of a normal person's career, you'd only see one or two of these instances. When you see everything, you see them on a more frequent basis, and you recognize that some of these things are actually abuses. And when you talk to people about them in a place like this, where this is the normal state of business, people tend not to take them very seriously and move on from them.
>
> But over time that awareness of wrongdoing sort of builds up and you feel compelled to talk about it. And the more you talk about it, the more you're ignored, the more you're told it's not a problem, until eventually you realize that these things need to be determined by the public and not by somebody who was simply hired by the government.[39]

The Snowden revelations received a lot of global media attention. In a survey in October and November 2014 of 23,376 Internet users in twenty-four countries—ranging from the "Five Eyes" to China, India, Indonesia, Egypt, Brazil, and Kenya—60 percent of those responding had heard of the Snowden affair.[40] This aggregated number, naturally, masks an internal differentiation, with 94 percent of respondents in Germany having heard about Snowden, 85 percent in China, and 76 percent in the United States.[41] Of those aware of the Snowden matter, 39 percent "have taken steps to protect their online privacy and security as a result of his revelations."[42] In the United States the figure was 36 percent, in the European countries surveyed it was 29 percent, and in the BRIC region it was 58 percent.[43] The largest effect has been in India (69 percent), Mexico (64 percent), China (62 percent), and Indonesia (61 percent).[44]

THERE IS TODAY a range of new weapons that we are using to challenge our virtual transparence and its pockets of obscurity. Some of us seek to diminish our own visibility. We fog up that plastic cube in which we are trapped. Some of us, like Manning and Snowden, create more transparence by leaking secrets. And some of us, like Assange, build platforms to throw sunlight on covert government and corporate operations. The interventions push in many different directions.

Along one dimension, there are ongoing efforts to securitize our personal information through self-help, education and awareness campaigns, and research leading to the development of more secure devices. The thrust here is to encrypt better, to create our own personal information management systems, to assemble our own secure servers, to retain our own information, to prevent exposing our data on the cloud. There are websites that share information about how to become more anonymous and avoid surveillance, and others that provide more secure technologies. The Electronic Frontier Foundation, for instance, has its own website, called I Fight Surveillance, that provides all kinds of information about encryption and security-audited free software, strengthening anonymity online, and other ways to resist virtual transparence.[45] Security in-a-Box, another organization, specifically targets the "digital security and privacy needs of advocates and human rights defenders."[46] (As you can imagine, the security needs among human rights activists around the world are acute.[47]) There are other freelance sites by activists and free software militants, such as Jason Self, a self-described "free software and free culture activist in Seattle, Washington," and Greycoder, who "test[s] privacy-friendly services, and explain[s] how to be private online."[48] These freelancers provide tips and services on how to protect personal data. The hacker collaborative Anonymous also posts advice and links to guides for protecting one's privacy and anonymity on the web.[49]

Apple has developed new iPhones and Google new Android devices that are more resistant to third-party data collection. These newer devices are equipped with more sophisticated encryption software, such that police and law enforcement officials are not able to access the data stored on the phone, even when they have a search warrant. These phones encrypt internal data, including emails, contacts, and photos, based on "a complex mathematical algorithm that uses a code created by, and unique to, the phone's user—and that Apple says it will not possess."[50] As a result, Apple itself cannot decrypt the data for the police, even if served with a warrant. According to Apple, it would take "more than 5 1/2 years to try all combinations of a six-character alphanumeric passcode with lowercase letters and numbers."[51] (Given the NSA's computing power, it's not entirely clear that this estimate is right.) In any event, Google is doing the same with its Android.[52]

Another company, a tech start-up by the name of Silent Circle, has developed a mobile device that is apparently even more secure. The device, called Blackphone, developed with a Madrid-based tech company called Geeksphone, apparently "allows users to make encrypted calls and send encrypted texts using Silent Circle software, and blocks third-party apps from accessing user data."[53] Others are working on developing integrated hardware and software that is more secure, through projects like the nonprofit FreedomBox foundation.[54] Start-ups are developing new Internet chat programs that are fully encrypted and anonymous, such as John Brooks's Ricochet.[55] Other anonymity and messaging systems include Tor, Pidgin, and Prosody.[56] Tor, for instance, is a gratis, open-source anonymity platform, originally funded by the United States Navy, that sends Internet traffic through a labyrinthine maze of encrypted networks in order to make it practically impossible for anyone to trace the communications back to their source. Tor has become extremely popular for those aggressively seeking anonymity, and it has been used extensively by human rights workers and whistle-blowers as well as those seeking to avoid law enforcement.[57]

Eben Moglen, who is involved in many of these projects and inspired an alternative to Facebook called Diaspora, is confident that there are ways to better secure personal communications.[58] He offers some quick tips:

> The first thing that you could do is to place your Facebook profile on a Web site where all the logs aren't in Mr. Zuckerberg's hands. You could do that by running a Diaspora pod or putting up an old-fashioned Web site.
>
> The second thing you could do is to have a mail server of your own, or a mail server with a few friends. The third thing you could do is to use the Firefox Web browser and to make sure that you're using the no-ad and no-script add-ons. If you do that, you're then 60 percent better.
>
> If you want to go 100 percent better, you can have a shell account somewhere—for example, on a server at The Washington Post—and you use a proxy to browse online so your browsing stream is mixed with the browsing stream of a thousand other people. You have ceased to be trackable in any meaningful way.[59]

These are relatively easy steps, and yet they can shelter personal data from the more immediate intrusive eyes of social media and data brokers. As Moglen suggests, "We did very little meaningful to your way of life, but we changed your privacy phenomenally."[60]

Naturally, none of this protects personal data from cable-splicing intelligence services. A lot of these efforts involve fending off the first or second layer of surveillance—the voyeuristic neighbor, the greedy advertisers, the amateur hacker—but not necessarily the NSA, the FBI, or large multinationals. As the Australian security journalist Patrick Gray suggests, "If the NSA is already targeting you, you're screwed. . . . But this is about stopping the wholesale violation of privacy and making it harder for people who shouldn't have access to this information from having access to it."[61] For this reason, some argue that the tech developments need to keep ramping up.[62] But it is not always possible to outfox the deep-pocketed signals intelligence agencies, especially when they have turned to undersea cable interception. Even the more sophisticated and wealthy Internet browser companies can find it hard at times to provide secure web encryption.[63]

It may be especially dangerous if we believe that we are secure when in fact we are not. *Wired* magazine recently revealed, for instance, that the FBI used an open-source code called Metasploit, available to anyone and used mostly by hackers, to infiltrate the Tor platform, which is supposed to ensure anonymity.[64] *Wired* describes the Metasploit app as "the most important tool in the hacking world," and one of the most well known, "an open-source Swiss Army knife of hacks that puts the latest exploits in the hands of anyone who's interested, from random criminals to the thousands of security professionals who rely on the app to scour client networks for holes."[65] The FBI, in what it called Operation Torpedo, apparently used Metasploit as the entryway to bypass Tor and reach genuine IP addresses. As *Wired* explains, "FBI agents relied on Flash code from an abandoned Metasploit side project called the 'Decloaking Engine' to stage its first known effort to successfully identify a multitude of suspects hiding behind the Tor anonymity network."[66] And, according to *Wired*, the FBI has only gotten better at it since then:

> Since Operation Torpedo, there's evidence the FBI's anti-Tor capabilities have been rapidly advancing. Torpedo was in November 2012. In late July 2013, computer security experts detected a similar attack through Dark Net websites hosted by a shady ISP called Freedom Hosting—court records have since confirmed it was another FBI operation. For this one, the bureau used custom attack code that exploited a relatively fresh Firefox vulnerability—the hacking equivalent of moving from a bow-and-arrow to a 9-mm pistol. In addition to the IP address, which identifies a household, this code collected the MAC address of the particular computer that [was] infected by the malware.[67]

These kinds of capabilities make it difficult for the less sophisticated and less wealthy digital users among us to imagine achieving real anonymity. Even many sophisticated experts are unsure that an effective privacy infrastructure will be realized during the next decade. In a Pew canvass of 2,511 Internet experts, builders, researchers, and innovators, a majority (55 percent) of the respondents said that they do not believe that there will be "a secure, popularly accepted, and trusted privacy-rights infrastructure by 2025 that allows for business innovation and monetization while also offering individuals choices for protecting their personal information in easy-to-use formats."[68] In other words, a majority of the surveyed experts are not optimistic about the ability to keep our personal information private, at least for a number of years to come.

This is precisely what fuels the view at the other extreme: assume total transparency of personal data, but invert the political economy. Privatize the data—truly privatize it for the benefit of individual users—and commercialize its exploitation. This would dramatically upend the financial incentives that are pushing our digital economy along. The idea here, in effect, is to truly marketize the personal data for the benefit of those who are producing them, to give the genuine holders of personal data the ability to control their dissemination and cash in on their value—in the vein of Jennifer Morone's incorporation project. Jaron Lanier takes this approach in his 2013 book *Who Owns the Future?*, where he sets forth a vision of a radically different information economy, one in which ordinary Internet users could be compensated anytime their data are used by others.[69]

This approach bears a family resemblance to the recent French digital taxation proposal, developed by Nicolas Colin and Pierre Collin, to tax corporations that use personal data—or more specifically, to tax certain data collection practices that generate revenue.[70] As Nicolas Colin writes, one of the biggest problems today is "our inability to add data as a primary economic category, just like goods and services . . . The reality may be that much of the value generated by the digital economy is not captured by official statistics."[71] If, however, we can get to the point where we tax the data practices, we should easily be able to reallocate the value extracted from those practices. Or alternatively, rather than turn to the fiscal system, as Colin and Collin do, it may be possible to redesign and reassign property rights to personal data; in effect, to use the property regime, rather than the tax regime, to privatize data.

Along another dimension, many of us are striving to render the internal operations of government, corporations, and the security apparatus itself more visible. This too comes in different variations. One approach is to turn the surveillance camera around: to aim the optic at those who are watching us. Frank Pasquale has long been an advocate of this kind of approach, which counters surveillance with even more surveillance.[72] William Simon also takes this approach in a provocative article titled "Rethinking Privacy" in the *Boston Review*. "The panopticon can be developed in ways Foucault never imagined," Simon argues, "to discipline the watchers as well as the watched."[73]

There are a number of creative projects along this dimension. Artist James Bridle started a project in 2013 called "Watching the Watchers," which collects "aerial photographs of military surveillance drones, found via online maps" such as Google Earth.[74] A collaborative called the Institute for Applied Autonomy has a web-based application called iSee that identifies and locates all of the CCTV cameras in urban environments and provides users with walking directions to avoid surveillance cameras.[75] The geographer, artist, and author Trevor Paglen documents the hidden world of governmental surveillance with photographs of secret sites, testing grounds, drone bases, and "black sites"—what he calls the "Blank Spots on the Map"—of the United States military and Department of Defense.[76] The filmmaker Laura Poitras has been documenting the construction of a massive surveillance data storage facility in the Utah desert, the NSA Utah Data Center, also called the Intelligence Community Comprehensive National Cybersecurity Initiative Data Center.[77] The art historian Ed Shanken teaches an entire course on the arts of countersurveillance, titled "Surveillance Culture: Privacy, Publicity, Art, Technology," which examines "how artists are responding by using surveillance technologies to look over 'big brother's' shoulder."[78]

Many of us, at political protests and ordinary police-civilian encounters, record undercover police officers who are engaging in surveillance or infiltrating protests, document the police filming of protests, or simply monitor the police—an increasingly common practice that has documented sometimes appalling, sometimes devastatingly upsetting uses of physical and lethal force by law enforcement.[79] The pupils discussed in the U.K. study also do a lot of this themselves:

> One pupil "set up a camera" to record her mum's "password on the laptop" . . . while some of the pupils at Girls Comprehensive said that they use "three-way telephone calls" which allows them to surreptitiously listen in to telephone conversations. This could be used as a convenient way for three people to arrange "to meet up" . . . or to listen in to telephone conversations when they had asked friends to "dump" a boyfriend: "X used to do it all the time and I would finish her boyfriend for her. She used to be on the other line." . . . Meanwhile some of the "techno-boys" . . . used their knowledge of [information and communication technologies] to monitor their peers and teachers."[80]

Another variant of this is the Open Data movement, which tries to render public information accessible on the web—a movement that got off the ground in spurts in the United States and in the United Kingdom in the mid-2000s. For instance, in Chicago, an early hacker site called Chicago Crimes published police crime data in 2005.[81] The Open Data movement has extended abroad to other liberal democracies. In 2011, in the wake of these developments, a French interministerial initiative started Étatlab, dedicated to the mission of open government data.[82]

Naturally, there are many other projects that favor a better regulation of personal and governmental data and that are located between some of the different extremes, or that mix and match strategies. Some take a more regulatory view overall, arguing for greater attention to the "architecture" of the Internet, in terms of code, law, and norms—in the vein of Larry Lessig's proposals in his updated *Code: And Other Laws of Cyberspace, Version 2.0*; others such as Jack Balkin argue for more robust congressional, executive, judicial, and technological oversight; and still others advocate structural compartmentalization of the data and privacy advocates along the lines enacted in the USA FREEDOM Act.[83] Daniel Solove urges us to lay a proper legal foundation for new technologies before they arrive.[84] Margo Schlanger recommends that we legally "infiltrate the NSA": that we insert and empower civil liberties ombudsmen within the institutional framework of intelligence services. "We need to add civil libertarians inside the surveillance state to nurture its civil liberties ecology," Schlanger writes.[85] Eben Moglen argues that intelligence should be placed under a rule-of-law regime and properly regulated. In his words, we must not only get our governments to protect us from the surveillance of foreign governments, but also force our own government to "subject its domestic listening to the rule of law."[86]

There is also a range of peaceful protest movements today challenging our loss of privacy and autonomy. Some link up with the Occupy movement or Occupy Central in Hong Kong—adding the problematics of surveillance to the call for renewed democratic mobilization. Others aim to awaken the sensibilities of the "average citizen," the person on the street—not the militant or the protester, but just the middle-class, suit-and-tie-wearing person who wants a private realm of his own. There are groups of "average data protesters" in Germany—such as the group Akkurater Widerstand (Accurate Resistance)—protesting the NSA's surveillance programs.[87] You can join them on Facebook here (how paradoxical): www.facebook.com/akkuraterwiderstand.

And then there are entirely different approaches that are more radical. There are online bazaars that operate on the Tor network (what people refer to as "a hidden part of the Internet," where it is far easier to remain anonymous), use a virtual currency called Bitcoin that is considered as anonymous as cash, and employ pseudonyms to protect people's identities.[88] There are orchestrated denial-of-service attacks intended to punish or deter—like those that the Anonymous collective conducted against MasterCard after it decided to stop servicing WikiLeaks.

The more radical forms of resistance can range from the very minor—lying about your personal information on websites—to more major forms of combat. Many of the teens danah boyd interviewed took the first route: they lied about their personal information on websites and "fabricated answers like name, location, age, and income to profile questions."[89] They playfully check "It's Complicated" as their relationship status on Facebook, and report their best friend as the person with whom they are "In a Relationship."[90] We can see a range of resistance at the British schools studied earlier, too. There are minor forms of resistance: some of the children use " 'blocking' tactics, like putting 'chewy and stuff over cameras' on the school buses" or "putting 'up your hood' to avoid being recorded by the school's CCTV cameras 'because you don't want them to see your face when you're twagging' (playing truant from school or a lesson)." Some of the parents resist as well, refusing, for instance, to give the school their cell phone numbers so that they don't receive "the 'automated text messages' sent out concerning pupil 'absence' and 'behaviour.' " There are some elaborate resistance strategies as well, with some pupils trying to stay one step ahead of the surveillance, to find a website before it is blocked, to use a computer that might not be controlled, to use proxy websites or a friend's account.[91] Some of the pupils engage in "avoidance tactics": "buying alcohol outside of the view of CCTV cameras," or "wearing hoodies up when 'twagging' school." And then there are more aggressive forms of resistance: "Council Estate pupils, on the other hand, returned the 'stare' at CCTV operators; 'threw stones' at the cameras; and 'swore at the cameras.' "[92] From avoidance techniques to throwing stones at the digital monitor or leaking troves of military and intelligence secrets: there are indeed a range of disruptive strategies, a lot of new weapons.

WITH DANIEL ELLSBERG and now Chelsea Manning and Edward Snowden, the term "whistle-blower" has become part of our lexicon today. The idea of the whistle-blower often goes hand in hand with notions of truth-telling and speaking truth to power. Recall that Daniel Ellsberg himself had titled a chapter of his memoirs "The Power of Truth."[93] Judith Ehrlich, the documentary filmmaker who codirected *The Most Dangerous Man in America: Daniel Ellsberg and the Pentagon Papers,* draws on these themes in her work. Ellsberg's resistance, in her words, was precisely about "speaking truth to power": "That one person, armed with a willingness to spend the rest of his life in prison and 7,000 pages of top secret documents in his safe, can make a huge difference is something that's shown by the film. . . . It's really a film about the power of truth-telling."[94] A recent (virtual) joint appearance of Snowden and Ellsberg has revived these themes and ideas. The headlines reflect it well. "Edward Snowden and Daniel Ellsberg speak truth to power at HOPE X," reads the caption at the *Daily Dot,* after the tenth "Hackers on Planet Earth" conference in New York City in July 2014, where they both appeared.[95]

Increasingly, many of us imagine these daunting acts—some of us might even say these treasonous acts—through a new lens: the courage of truth.[96] As forms of disobedience, these actions call for a level of self-awareness and self-sacrifice that is truly formidable. This is especially true today, in light of the more punitive reactions and increased criminalization. The Obama administration has aggressively pursued government officials who provide information to journalists on security issues without the approval of the administration, in what has been referred to as "an unprecedented crackdown."[97] Whereas all of the prior presidential administrations combined had prosecuted only three individuals for sharing secret information with journalists, the Obama administration has pursued eight criminal cases in six years—including in situations where there is no direct evidence of a leak.[98]

The cost of truth-telling has gone up, as has the amount of courage required.

On June 9, 2013, Edward Snowden presented himself to us all. He looked straight into the digital stream, his back fully exposed in the mirrored glass. Cognizant of the penalty he might have to pay—in decades, in a lifetime, or even perhaps with his bare life—he calmly told us: "If you realize that that's the world you helped create, and it's going to get worse with the next generation and the next generation, who extend the capabilities of this sort of architecture of oppression, you realize that you might be willing to accept any risk. And it doesn't matter what the outcome is, so long as the public gets to make their own decisions."[99]

TWELVE

POLITICAL DISOBEDIENCE

HERE IS A LIST of strong security questions, recommended to websites to prompt and verify us in order to render more secure our user IDs and passwords:

> What was your childhood nickname?
> In what city did you meet your spouse/significant other?
> What is the name of your favorite childhood friend?
> What street did you live on in third grade?
> What school did you attend for sixth grade?
> What was your childhood phone number including area code?
> What was the name of your first stuffed animal?
> In what city or town did your mother and father meet?
> Where were you when you had your first kiss?
> What was the last name of your third grade teacher?
> In what city or town was your first job?
> What is the name of the place your wedding reception was held?
> What is the name of a college you applied to but didn't attend?
> Where were you when you first heard about 9/11?
> What is the first name of the boy or girl that you first kissed?[1]

If you know my answers to these questions, I suspect, you know more of my secrets than my own companion of twenty-six years. You know my stories, my journey, my clues, my fondest memories. Perhaps you even know me better than what I can remember about myself—about my self, that self full of desire and disobedience ... desire and disobedience that is intercepted, recorded, mined, and then carved into the cold rock of the digital cloud of commercial and intelligence services.

WHAT, THEN, IS TO BE DONE—that is, beyond carefully analyzing, better understanding, and critically rethinking this new form of digital power that circulates today? The answer, I am afraid, is not simple or easy. It calls for courage and for ethical choice—for innovation and experimentation. In the end, it falls on each and every one of us—as desiring digital subjects, as parents and children, as teachers and students, as conscientious ethical selves—to do everything we can to resist the excesses of our expository society. Each one of us has a unique ability to intervene in our own way. And, as Eben Moglen reminds us, as he urges us: "Snowden's courage is exemplary. But he ended his effort because we needed to know now. We have to inherit his understanding of that fierce urgency."[2]

Resistance must come from within each and every one of us—and if it does, it will amount to a new form of leadership positioned against both Tocqueville's ancient forms of leaderful despotism and modern forms of democratic apathy. It would be, I believe, a form of leadership that could only properly be described as leaderless. Leaderless, but strenuously resistant. Something that might be imagined under the rubric of "political disobedience." It is a situated form of disobedience, one that is particularly appropriate as a type of resistance to democratic despotism, even though it may not necessarily be adequate to leaderful authoritarianism.[3]

What does such political disobedience offer? Most importantly, what it avoids is the recrystallization of oppressive structures or the return to repressive relations of power. For anyone sensitive to these issues, this alone should count in its favor. You may recall O'Brien in Orwell's *1984*, Winston's tormenter and eventually, I would argue, executioner. He had a few things to say about resistance, many of which I think were eerily prescient. "One does not establish a dictatorship in order to safeguard a revolution," he said. "One makes the revolution in order to establish the dictatorship."[4] What O'Brien is getting at, of course, is the intricate and tense relationship between leadership and despotism. But the coin that can be slotted between those two is precisely leaderlessness, which, equally importantly, produces knowledge differently and produces different knowledge.

The social movement initiated by Daniel Defert, Deleuze, Foucault, and others in 1970, the Prisons Information Group (GIP), was a deliberate attempt to create such a leaderless movement.[5] "There was no central organizing committee or symbolic figure, no charter, no statutes, no members, just activists."[6] The GIP was started specifically to allow voices to be heard that were not and could not be heard through traditional forms of leadership. The group embraced a leaderless model explicitly as an alternative to other forms of resistance—popular tribunals, accusatory methods, vanguard politics.[7] The idea was to produce a space, an opening, where people, voices, and discourses could be heard that were otherwise silenced. "The group demanded no ideological unity and provided no political guidance. Instead, it strove to 'break down the barriers required by power,' the so-called 'game of social hierarchies,' bringing together participants from various social strata: prisoners, lawyers, judges, family members, academics, etc."[8] The same was true with the Occupy Wall Street movement, which sought to create a space for innovative thought and genuine political resistance, rather than impose dogma. The idea, again, was to use gatherings and general assemblies to allow new ideas to emerge, new voices to be heard, new practices to originate. It was a movement, in David Graeber's words, "that resolutely refused to take the traditional path, that rejected the existing political order entirely as inherently corrupt, that called for the complete reinvention of American democracy."[9]

It is in this same sense that leaderless resistance, as a central element of political disobedience, can overcome modern-day democratic apathy, while avoiding leaderful despotism. But it requires each and every one of us to act bravely, with courage and conviction. It requires an ethic of the self.

Friendly critics of such leaderless movements, such as Jodi Dean and Doug Henwood, point to the family resemblance between leaderlessness and neoliberal individualism—suggesting that recent protest movements have more in common with the entrepreneurial ethic associated with neoliberalism than one might first think.[10] The ethical self, though, is by no means a recent by-product of neoliberalism. It far predates it. Seen from this angle, the emphasis on what we must do as ethical selves, each and every one of us—us digital subjects, with our desires and our disobedience—may be precisely what is necessary for us to begin to think of ourselves as *we*. Yes, as that *we* that has been haunting this book since page one.

"Revolutionaries often forget, or do not like to recognize, that one wants and makes revolution out of desire, not duty," Deleuze and Guattari remind us in *Anti-Oedipus*.[11] What a cruel reminder, given that it is precisely our desires and passions that have enslaved us, exposed us, and ensnared us in this digital shell as hard as steel. What a painful paradox. What a daunting prospect. That, I take it, is our greatest challenge today.

NOTES

THE EXPOSITORY SOCIETY

1. Amir Efrati, "'Like' Button Follows Web Users," *Wall Street Journal*, May 18, 2011; Asher Moses, "Facebook's Privacy Lie: Aussie Exposes 'Tracking' as New Patent Uncovered," *Sydney Morning Herald*, October 4, 2011.

2. Efrati, "'Like' Button Follows Web Users."

3. Ibid.

4. Ibid.

5. Byron Acohido, "Facebook Tracking Is under Scrutiny," *USA Today*, updated November 16, 2011.

6. Ibid.

7. Sara M. Watson, "What Happens When Online and Offline Boundaries Blur?," Al Jazeera America, October 7, 2014; see also Vindu Goel, "With New Ad Platform, Facebook Opens Gates to Its Vault of User Data," *New York Times*, September 28, 2014.

8. Erik Johnson, "Meet the New Atlas," September 29, 2014, http://atlassolutions .com/2014/09/29/meet-the-new-atlas.

9. Ibid.

10. Goel, "With New Ad Platform."

11. Johnson, "Meet the New Atlas."

12. Goel, "With New Ad Platform."

13. Watson, "What Happens When Online and Offline Boundaries Blur?"

14. Johnson, "Meet the New Atlas."

15. Yasha Levine, "The Psychological Dark Side of Gmail: Google Is Using Its Popular Gmail Service to Build Profiles on the Hundreds of Millions of People Who Use It," AlterNet, December 31, 2013.

16. Desmond Butler, Jack Gillum, and Alberto Arce, "US Secretly Created 'Cuban Twitter' to Stir Unrest," Associated Press, April 4, 2014.

17. Ibid.

18. Ibid.

19. Ibid.

20. Desmond Butler and Alberto Arce, "US Contractors Profiled 'Cuban Twitter' Responses," Associated Press, April 30, 2014.

21. Ibid.

22. Butler, Gillum, and Arce, "US Secretly Created 'Cuban Twitter' to Stir Unrest."

23. Ibid.

24. Ibid.

25. Ibid.

26. Ibid.

27. Ibid.

28. Ibid.

29. Ibid.

30. Glenn Greenwald and Ewen MacAskill, "NSA Prism Program Taps In to User Data of Apple, Google and Others," *Guardian,* June 6, 2013.

31. Ibid.; see also Glenn Greenwald, *No Place to Hide: Edward Snowden, the NSA, and the U.S. Surveillance State* (New York: Henry Holt, 2014), 153–157.

32. Glenn Greenwald, "XKeyscore: NSA Tool Collects 'Nearly Everything a User Does on the Internet,'" *Guardian,* July 31, 2013.

33. Glenn Greenwald, "Glenn Greenwald on Secret NSA Program to Crack Online Encryption," *Democracy Now!* interview with Amy Goodman and Juan Gonzalez, September 9, 2013, http://truth-out.org/news/item/18693-the-end-of-internet-privacy-glenn-greenwald-on-secret-nsa-program-to-crack-online-encryption.

34. Quoted in Greenwald, "XKeyscore."

35. Greenwald and MacAskill, "NSA Prism Program."

36. Greenwald, *No Place to Hide,* 109 (quoting Bart Gellman).

37. Greenwald and MacAskill, "NSA Prism Program."

38. Ibid.

39. Ibid.

40. Ibid.

41. Mark Zuckerberg post on Facebook, November 27, 2014, 5:28 p.m., www.facebook.com/zuck/posts/10101761674047881?pnref=story.

42. See, e.g., Anita Allen, *Unpopular Privacy: What Must We Hide?* (New York: Oxford University Press, 2011).

43. Julia Angwin, "Why Online Tracking Is Getting Creepier: The Merger of Online and Offline Data Is Bringing More Intrusive Tracking," *ProPublica*, June 12, 2014.

44. Greenwald, *No Place to Hide*, 92.

45. Andy Müller-Maguhn, Laura Poitras, Marcel Rosenbach, Michael Sontheimer, and Christian Grothoff, "Treasure Map: The NSA Breach of Telekom and Other German Firms," *Der Spiegel*, September 14, 2014.

46. Ibid.

47. Frances Ferguson, "Now It's Personal: D. A. Miller and Too-Close Reading," *Critical Inquiry* 41, no. 3 (2015): 521–540.

48. See Siva Vaidhyanathan, *The Googlization of Everything (and Why We Should Worry)* (Berkeley: University of California Press, 2011), 9, 26.

49. Jodi Dean, *The Communist Horizon* (New York: Verso, 2012).

50. None of this is to deny the ongoing ubiquitous forms of state repression that continue to saturate countries and regions around the world. I must emphasize that my analysis of our expository society does not displace analyses of state repression in liberal democracies, especially in the policing and prison context, nor of state repression in authoritarian regimes. None of this is to deny the continuing states of exception in both democratic and nondemocratic states. See generally Zygmunt Bauman, *Globalization: The Human Consequences* (Cambridge: Polity Press, 1998) (discussing bare life in California's Pelican Bay Prison); Didier Bigo, "Security, Exception, Ban and Surveillance," in David Lyon, ed., *Theorizing Surveillance: The Panopticon and Beyond*, 46–68 (Devon: Willan Publishing, 2006) (discussing surveillance as state of exception and the "ban-opticon"); Giorgio Agamben, *Homo Sacer: Sovereign Power and Bare Life* (Stanford, CA: Stanford University Press, 1998); and Hannah Arendt, *The Origins of Totalitarianism* (New York: Harcourt, Brace, 1951).

51. Kevin D. Haggerty and Richard V. Ericson, "The Surveillant Assemblage," *British Journal of Sociology* 51, no. 4 (2000): 611. There are other proper names for us digital subjects that I could use: "digital persons" (see Daniel J. Solove, *The Digital Person: Technology and Privacy in the Information Age* [New York: New York University Press, 2006]) and "digital selves" (see Shanyang Zhao, "The Digital Self: Through the Looking Glass of Telecopresent Others," *Symbolic Interaction* 28, no. 3 [2005], 387–405).

52. danah boyd, *It's Complicated: The Social Lives of Networked Teens* (New Haven, CT: Yale University Press, 2014), 5.

53. Ibid., 20.

54. Guy Debord, *La société du spectacle* (Paris: Buchet/Chastel, 1967).

55. Michel Foucault, *La société punitive: Cours au Collège de France (1972–1973)*, ed. Bernard E. Harcourt (Paris: Gallimard / Le Seuil, 2013).

56. Michel Foucault, *Discipline and Punish,* trans. Alan Sheridan (New York: Vintage, 1979), 217.

57. Gilles Deleuze, "Postscript on the Societies of Control," *October* 59 (Winter 1992): 6.

58. This is a rhetorical question posed by Google's CEO Eric Schmidt. Quoted in Vaidhyanathan, *The Googlization of Everything,* 45.

59. I am referring here, naturally, to the titles of the seminal books by Glenn Greenwald and Daniel Solove. See Greenwald, *No Place to Hide;* Daniel Solove, *Nothing to Hide* (New Haven, CT: Yale University Press, 2011).

60. Pierrette Poncela, "Chronique de l'exécution des peines: Les peines extensibles de la loi du 15 août 2014," *Revue de science criminelle* 3 (2014): 611–621.

61. See Alex Hern, "Samsung Rejects Concern over 'Orwellian' Privacy Policy," *Guardian,* February 9, 2015.

62. The quotation is from Foucault, *Discipline and Punish,* 82.

63. Kenneth Neil Cukier and Viktor Mayer-Schoenberger, *Big Data: A Revolution That Will Transform How We Live, Work, and Think* (New York: Houghton Mifflin Harcourt, 2013), 1–3.

64. A recent study has found that the use of body-worn video cameras by police officers may reduce the likelihood of force being used in police encounters. See Barak Ariel, William A. Farrar, and Alex Sutherland, "The Effect of Police Body-Worn Cameras on Use of Force and Citizens' Complaints against the Police: A Randomized Controlled Trial," *Journal of Quantitative Criminology,* November 2014.

65. Amy L. Gonzales and Jeffrey T. Hancock, "Mirror, Mirror on My Facebook Wall: Effects of Exposure to Facebook on Self-Esteem," *Cyberpsychology, Behavior, and Social Networking* 14, nos. 1–2 (2011): 79–83.

66. The *mouches* and *mouchards* of ancien régime France—the informants of the king's *lieutenant de police*—immediately became the eyes and ears of the revolutionary committees that would govern republican France, with their strict and unbending Comité de Sécurité. The very same was true in the early years of the American Republic with the Committee for Detecting and Defeating Conspiracies in the State of New York, of which John Jay was such a vital force. See generally George Pellew, *John Jay* (Boston: Houghton Mifflin, 1898), 60–63. Special thanks to Leo Ruyter-Harcourt and Henry Utset for excellent research on these questions.

67. Alexander Nazaryan, "The NSA's Chief Chronicler," *New Yorker,* June 10, 2013.

68. Today, for instance, practically all youths in the United States are glued to YouTube, posting on Facebook, or otherwise using digital media in sufficient amounts to be watched and surveilled around the clock. "By 2011, 95 percent of American teenagers had some form of access to the internet, whether at home or at

school," as danah boyd notes (*It's Complicated,* 193); 80 percent of high school students had cell phones (ibid., 3).

69. Megan Wollerton, "Website Spies on Thousands of People to Shed Light on Security Flaw," *CNET Magazine,* November 10, 2014.

70. "XKeyscore Presentation from 2008," *Guardian,* July 31, 2013.

71. Greenwald, "XKeyscore."

72. David Cole, "Can the NSA Be Controlled?," *New York Review of Books,* June 19, 2014, 17.

73. Greenwald, *No Place to Hide,* 98; see also Cole, "Can the NSA Be Controlled?," 17.

74. Quoted in Cole, "Can the NSA Be Controlled?," 16. As David Cole reminds us, through the words of the NSA general counsel, "Metadata absolutely tells you everything about somebody's life. If you have enough metadata, you don't really need content."

75. Eben Moglen, "Privacy under Attack: The NSA Files Revealed New Threats to Democracy," *Guardian,* May 27, 2014.

76. Vaidhyanathan, *The Googlization of Everything,* xiv.

77. Anthony Vidler, *The Architectural Uncanny: Essays in the Modern Unhomely* (Cambridge, MA: MIT Press, 1992), 220, 219.

78. Frank Pasquale refers to this as "the fading divide between 'state' and 'market.'" Pasquale, *The Black Box Society: The Secret Algorithms behind Money and Information* (Cambridge, MA: Harvard University Press, 2015), 48.

79. See, e.g., Ian Crouch, "So Are We Living in 1984?," *New Yorker,* June 11, 2013; see generally Chapter 1.

80. See, e.g., Solove, *Nothing to Hide,* 26; Pasquale, *The "Black Box" Society.*

81. On the surveillance state, see, e.g., Bill Keller, "Living with the Surveillance State," *New York Times,* June 16, 2013; see also generally Chapter 2. On the panopticon, see, e.g., Bruce Schneier, "Why the NSA's Defense of Mass Data Collection Makes No Sense," *Atlantic,* October 21, 2013; see also generally Chapter 3.

82. Glenn Greenwald, for instance, draws on Orwell, Bentham, and Foucault, and refers throughout and in the subtitle to "the U.S. surveillance state." See Greenwald, *No Place to Hide,* 174–176 and title page.

83. See "The President on Mass Surveillance" (editorial), *New York Times,* January 17, 2014; see also Barack Obama, "Speech on N.S.A. Phone Surveillance," *New York Times,* January 17, 2014.

84. See Chapter 2. While this book was already in page proofs, Congress passed the USA FREEDOM Act. *See* H.R.2048 - USA FREEDOM Act of 2015, Public Law No. 114-23, 114th Congress (2015–2016), passed on June 2, 2015, available at

https://www.congress.gov/bill/114th-congress/house-bill/2048/text. As a result, the Act can and will be treated only very briefly in this book.

85. President's Review Group on Intelligence and Communications Technologies, *Liberty and Security in a Changing World,* December 12, 2013, www.whitehouse .gov/sites/default/files/docs/2013-12-12_rg_final_report.pdf.

86. James Bamford, *The Shadow Factory: The Ultra-Secret NSA from 9/11 to the Eavesdropping on America* (New York: Doubleday, 2008), 199. In this regard, it is interesting to scrutinize what President Obama's advisers actually wrote in their report. Notice their language (my emphasis): "it would be *in the interests* of the providers and the government to agree on a voluntary system that *meets the needs of both.*" President's Review Group, *Liberty and Security in a Changing World.* They added that if such a mutually agreeable deal could not be worked out, "the government should *reimburse* the providers for the cost of retaining the data." Ibid. This would surely be a win-win solution for the surveillance-industrial complex.

Chapter One: GEORGE ORWELL'S BIG BROTHER

1. Jenny Hendrix, "NSA Surveillance Puts George Orwell's '1984' on Bestseller Lists," *Los Angeles Times,* June 11, 2013.

2. Charles Riley, "Sales of Orwell's '1984' Spike after NSA leak," CNN Money, June 12, 2013.

3. Dominic Rushe, "NSA Surveillance Goes beyond Orwell's Imagination—Alan Rusbridger," *Guardian,* September 23, 2013; "Orwell's Fears Refracted through the NSA's Prism," *Guardian,* June 9, 2013.

4. Nicholas Kristof, June 6, 2013, fhttps://twitter.com/NickKristof/status /342769970251448320.

5. Konrad Krawczyk, "NSA PRISM: 3 Ways Orwell's '1984' Saw This Coming," *International Business Times,* June 10, 2013.

6. Hendrix, "NSA Surveillance Puts George Orwell's '1984' on Bestseller Lists."

7. Griff Witte, "Snowden Says Spying Worse Than Orwellian," *Washington Post,* December 25, 2013. A video of Snowden's comments, aired Christmas Day on Britain's Channel 4, can be found at www.channel4.com/programmes/alternative -christmas-message/4od.

8. Glenn Greenwald, *No Place to Hide: Edward Snowden, the NSA, and the U.S. Surveillance State* (New York: Henry Holt, 2014), 174.

9. *Klayman v. Obama,* 957 F. Supp. 2d 1 (D.D.C. 2013), http://legaltimes.typepad .com/files/obamansa.pdf.

10. Richard Jinman, "NSA Spy Scandal Boosts Sales of Orwell's '1984'—Here's Why," *The Week,* June 12, 2013.

11. Ian Crouch, "So Are We Living in 1984?," *New Yorker,* June 11, 2013.

12. Ibid.

13. Regarding the third point, Crouch references Amy Davidson, "The N.S.A.-Verizon Scandal," *New Yorker,* June 6, 2013.

14. Crouch, "So Are We Living in 1984?"

15. Ibid.

16. Robert Colls, "Orwell and Ed: Snowden Saga Shows the One Thing '1984' Missed," *Salon,* December 17, 2013. Regarding the corporate interests, Colls links to Dan Roberts and Jemima Kiss, "Twitter, Facebook and More Demand Sweeping Changes to US Surveillance," *Guardian,* December 9, 2013.

17. Apple, "iPhone 5C," www.apple.com/iphone-5c/design.

18. George Orwell, *1984* (New York: Signet Classics, 1950), 140.

19. All references to George Orwell's *1984* are to the Signet Classics edition with an afterword by Eric Fromm. All references will be placed in parentheses after the quote or passage, without further identification.

20. Michel Foucault, *Discipline and Punish,* trans. Alan Sheridan (New York: Vintage, 1979), 19. On the actuarial in criminal justice, see generally Bernard E. Harcourt, "A Reader's Companion to 'Against Prediction': A Reply to Ariela Gross, Yoram Margalioth, and Yoav Sapir on Economic Modeling, Selective Incapacitation, Governmentality, and Race," *Law and Social Inquiry* 33 (2008): 273–277; Bernard E. Harcourt, *Against Prediction: Profiling, Policing, and Punishing in an Actuarial Age* (Chicago: University of Chicago Press, 2007).

21. See, e.g., William Blackstone, *Commentaries on the Laws of England,* vol. 4, chap. 1 (University of Chicago Press, 1979); Model Penal Code Sec. 1.13(2) (defining an "act" as a "bodily movement").

22. Crouch, "So Are We Living in 1984?"

23. See, e.g., "FBI Entrapped Suspects in Almost All High-Profile Terrorism Cases in US," RT, July 21, 2014, http://rt.com/usa/174484-hrw-fbi-sting-entrapment.

24. See, e.g., Thomas Szasz, *The Myth of Mental Illness: Foundations of a Theory of Personal Conduct* (New York: Hoeber-Harper, 1961); Michel Foucault, David Cooper, Jean-Pierre Faye, Marie-Odile Faye, and Marine Zecca, "Confinement, Psychiatry, Prison," in Michel Foucault, *Politics, Philosophy, Culture: Interviews and Other Writings, 1977–1984,* ed. Lawrence D. Kritzman, 178–210 (New York: Routledge, 1990).

25. Foucault, *Discipline and Punish,* 201.

26. On Admiral Poindexter's program, see James Bamford, *The Shadow Factory: The Ultra-Secret NSA from 9/11 to the Eavesdropping on America* (New York: Doubleday, 2008), 99–104. See generally Glenn Greenwald, "NSA Top Secret Program," *Guardian,* July 31, 2013; David E. Sanger and Thom Shankerjan, "N.S.A. Devises Radio Pathway into Computers," *New York Times,* January 14, 2014.

27. For a good discussion, see Marion Lewenstein, "Smokey Bear as Big Brother," in Peter Stansky, ed., *On Nineteen Eighty-Four* (New York: W. H. Freeman, 1983), 140ff.

28. Martha Nussbaum, "The Death of Pity: Orwell and American Political Life," in Abbott Gleason, Jack Goldsmith, and Martha Nussbaum, eds., *On Nineteen Eighty-Four: Orwell and Our Future* (Princeton, NJ: Princeton University Press, 2005), 281.

29. Cass Sunstein refers to this as the "political orthodoxy thesis": "Political orthodoxy is a consequence of sexual frustration, which governments can channel into marching and flag-waving and the Two Minutes Hate." Cass Sunstein, "Sexual and Political Freedom," in Gleason, Goldsmith, and Nussbaum, *On Nineteen Eighty-Four,* 237.

30. See, e.g., Joseph Goldstein and J. David Goodman, "Frisking Tactic Yields to Tighter Focus on Youth Gangs," *New York Times,* September 19, 2013.

31. Adam D. I. Kramer, Jamie E. Guillory, and Jeffrey T. Hancock, "Experimental Evidence of Massive-Scale Emotional Contagion through Social Networks," *Proceedings of the National Academy of Sciences* 111, no. 24 (2014): 8788–8790, epub ahead of print June 2, 2014, doi:10.1073/pnas.1320040111; see also Vindu Goel, "Facebook Tinkers with Users' Emotions in News Feed Experiment, Stirring Outcry," *New York Times,* June 29, 2014; Michael Roppolo, "Researcher Apologizes for Facebook Study in Emotional Manipulation," CBS News, June 30, 2014.

32. Adam D. I. Kramer's Facebook post explaining the study, dated June 29, 2014, https://www.facebook.com/akramer/posts/10152987150867796.

33. Kramer, Guillory, and Hancock, "Experimental Evidence of Massive-Scale Emotional Contagion."

34. Goel, "Facebook Tinkers with Users' Emotions."

35. Kramer, Guillory, and Hancock, "Experimental Evidence of Massive-Scale Emotional Contagion."

36. Ibid.

37. Ibid.

38. Ibid.

39. Ibid.

40. Ibid.

41. Ibid.

42. Ibid.

43. Ibid.

44. Ibid.

45. Varun Kacholia and Minwen Ji, "News Feed FYI: Helping You Find More News to Talk About," Facebook Newsroom, December 2, 2013, https://newsroom.fb.com/news/2013/12/news-feed-fyi-helping-you-find-more-news-to-talk-about; Roppolo, "Researcher Apologizes for Facebook Study in Emotional Manipulation"; Khalid El-Arini and Joyce Tang, "News Feed FYI: Click-Baiting," Facebook, August 25, 2014, http://newsroom.fb.com/news/2014/08/news-feed-fyi-click-baiting.

46. Goel, "Facebook Tinkers with Users' Emotions."

47. Kramer, Guillory, and Hancock, "Experimental Evidence of Massive-Scale Emotional Contagion."

48. Goel, "Facebook Tinkers with Users' Emotions."

49. Ibid. And as Vindu Goel of the *New York Times* notes: "Facebook is hardly the only Internet company that manipulates and analyzes consumer data. Google and Yahoo also watch how users interact with search results or news articles to adjust what is shown; they say this improves the user experience." Ibid.

50. Adam D. I. Kramer's Facebook post explaining the study, June 29, 2014, https://www.facebook.com/akramer/posts/10152987150867796.

51. See Albert Hirschman, *The Passions and the Interests: Political Arguments for Capitalism before Its Triumph* (Princeton, NJ: Princeton University Press, 1977); Bernard E. Harcourt, *The Illusion of Free Markets* (Cambridge, MA: Harvard University Press, 2011), 49–50.

52. Siva Vaidhyanathan, *The Googlization of Everything (and Why We Should Worry)* (Berkeley: University of California Press, 2011), 3.

53. Luc Boltanski and Eve Chiapello, *The New Spirit of Capitalism* (New York: Verso, 2007).

54. Gilles Deleuze and Félix Guattari, *Anti-Oedipus: Capitalism and Schizophrenia* (Minneapolis: University of Minnesota Press, 1983), 293.

55. Ibid., 296. Kevin Haggerty and Richard Ericson have done excellent work exploring the contributions of Gilles Deleuze and Félix Guattari in the context of surveillance, developing the notion of "surveillant assemblages" that abstract and separate human bodies into data flows. See Kevin D. Haggerty and Richard V. Ericson, "The Surveillant Assemblage," *British Journal of Sociology* 51, no. 4 (2000): 605–622.

56. Gilles Deleuze, "D for Desire," in Deleuze, *L'Abécédaire de Gilles Deleuze*. This is a TV series that was produced in France; see http://www.imdb.com/title/tt0408472/. It is available on YouTube and elsewhere. A transcript of the "D for Desire" episode is available at http://www.arpla.fr/canal2/archeo/cours_2005/desirDeleuze.pdf.

57. Ibid.

58. Deleuze and Guattari, *Anti-Oedipus*, 293.

59. Ibid., 29.

60. Ibid., 293.

61. Ibid.

62. Ibid.

63. Ibid., 104–105.

64. Ibid.

65. See also ibid., 104 ("It is not a question of ideology. . . . It is not an ideological problem, a problem of failing to recognize, or of being subject to, an illusion. It is a problem of desire, *and desire is part of the infrastructure*. Preconscious investments

are made, or should be made, according to the interests of the opposing classes. But unconscious investments are made according to positions of *desire* and uses of synthesis, very different from the interests of the subject, individual or collective, who desires"), 118–119, 344 ("the concept of ideology is an execrable concept that hides the real problems, which are always of an organizational nature").

66. Gilles Deleuze quoted in "Les intellectuels et le pouvoir" (discussion between Michel Foucault and Gilles Deleuze, March 4, 1972, *L'Arc* no. 49), in Michel Foucault, *Dits et écrits* (Paris: Gallimard, 1994), 2:306–315 ("Quarto" edition [Paris: Gallimard, 2001], 1:1182).

67. Haggerty and Ericson, "The Surveillant Assemblage."

68. Deleuze and Guattari, *Anti-Oedipus,* 26.

69. See, generally, Jaak Panksepp, *Affective Neuroscience: The Foundations of Human and Animal Emotions* (New York: Oxford University Press, 2004). For a more accessible introduction to the psychological literature, see Emily Yoffe, "Seeking: How the Brain Hard-Wires Us to Love Google, Twitter, and Texting. And Why That's Dangerous," *Slate,* August 12, 2009.

70. Deleuze and Guattari, *Anti-Oedipus,* 54.

71. Ibid., 26.

72. Michel Foucault, "Preface," xi–xiv, in Deleuze and Guattari, *Anti-Oedipus,* xiii.

73. Ibid.

74. Ibid.

75. Ibid.

76. Ibid., xiv.

77. Deleuze and Guattari, *Anti-Oedipus,* 34.

78. Ibid., 139.

79. Ibid., 109.

80. Ibid., 380.

Chapter Two: THE SURVEILLANCE STATE

1. Jack M. Balkin, "The Constitution in the National Surveillance State," *Minnesota Law Review,* Vol. 93, 1–25 (1980), 3–4; see, e.g., Bruce Schneier, "The Internet Is a Surveillance State," CNN, March 16, 2013.

2. Bill Keller, "Living with the Surveillance State," *New York Times.*

3. Julian Sanchez, "Dismantling the Surveillance State," *Daily Beast,* December 21, 2013.

4. "Time to Rein in the Surveillance State," ACLU, www.aclu.org/time-rein -surveillance-state-0.

5. Keller, "Living with the Surveillance State"; Thomas L. Friedman, "Blowing a Whistle," *New York Times,* June 11, 2013.

6. Conor Friedersdorf, "The Surveillance State Puts U.S. Elections at Risk of Manipulation," *Atlantic*, November 7, 2013.

7. See *American Civil Liberties Union v. Clapper*, 959 F. Supp. 2d 724 (S.D.N.Y. 2013), *reversed* Slip Opinion in Case 14-42 (2d Circuit, May 7, 2015); *Klayman v. Obama*, 957 F. Supp. 2d 1 (D.D.C. 2013); the President's Review Group on Intelligence and Communications Technologies, *Liberty and Security in a Changing World*, December 12, 2013.

8. *Clapper*, 959 F. Supp. 2d at 729. While this book was in page proofs, on May 7, 2015, the U.S. Court of Appeals for the Second Circuit entered an opinion reversing Judge Pauley's decision. In an opinion written by Judge Gerald E. Lynch, the three-judge panel declared the telephone metadata program illegal and unauthorized under Section 215 of the PATRIOT Act. Slip Opinion in Case 14-42, 74. The court, however, declined to grant a preliminary injunction to halt the metadata program, in light of Congress's immanent decision whether to reauthorize the program on June 1, 2015. Ibid., 95.

9. Ibid.

10. Ibid.

11. Ibid.

12. Ibid., 748.

13. Ibid., 757.

14. Ibid., 729, 733, 754.

15. Ibid., 754.

16. Ibid., 757.

17. *Klayman*, 957 F. Supp. 2d at 33.

18. Ibid., 29.

19. Ibid., 18.

20. Ibid., 19.

21. Ibid.

22. Ibid., 27.

23. Ibid.

24. Ibid., 33 n. 36.

25. *Clapper*, 959 F. Supp. 2d at 732–733.

26. Ibid., 757.

27. Ibid., 751 (emphasis added).

28. President's Review Group, *Liberty and Security in a Changing World*, 15.

29. Ibid., 16.

30. Ibid.

31. Ibid.

32. Ibid., 42.

33. Ibid., 56.

34. Ibid., 194.

35. Ibid., 194–195.

36. Ibid., 195.

37. Ibid., 204.

38. Ibid., 219.

39. Ibid., 18.

40. Ibid., 21.

41. Ibid., 28.

42. Barack Obama, "Remarks by the President on Review of Signals Intelligence," Department of Justice, January 17, 2014, www.whitehouse.gov/the-press-office /2014/01/17/remarks-president-review-signals-intelligence.

43. Ibid., para. 1.

44. Ibid., para. 17.

45. Ibid., para. 38.

46. Ibid.

47. Ibid., para. 9.

48. Ibid., para. 15.

49. Ibid., para. 22.

50. Ibid., para. 33.

51. Ibid., paras. 53–54.

52. Ibid.

53. Ibid., 40.

54. As noted earlier, while this book was in page proofs, on May 7, 2015, the U.S Court of Appeals for the Second Circuit reversed Judge Pauley's decision and held the NSA program illegal, but nevertheless declined to issue a preliminary injunction. See Slip Opinion in 14-42. Judge Gerald E. Lynch, writing for the three-judge panel, reasoned that Congress had to decide whether to reauthorize the NSA program on June 1, 2015, only a few weeks away, and that it was therefore more equitable to allow the program to remain in place in the meantime. "In light of the asserted national security interests at stake, we deem it prudent to pause to allow an opportunity for debate in Congress that may (or may not) profoundly alter the legal landscape," Judge Lynch wrote. Ibid., 95. In addition, on June 2, 2015, Congress passed the USA FREEDOM Act, which calls only for a restructuring of the Section 215 program—not for its termination.

55. President's Review Group, *Liberty and Security in a Changing World,* 14, 209, 212.

56. See *Mémoire sur la réformation de la police de France, soumis au roi en 1749 par M. Guillauté, illustré de 28 dessins de Gabriel de Saint-Aubin,* ed. Jean Seznec (Paris: Hermann, 1974), 22. I discussed this remarkable episode in Bernard E. Harcourt, *The Illusion of Free Markets* (Cambridge, MA: Harvard University Press,

2011), 152–153. I was first made aware of this memoir by Eric Heilmann at a conference in Louvain in 2009. See Eric Heilmann, "Comment surveiller la population à distance? La machine de Guillauté et la naissance de la police moderne," paper presented at the LISEC conference "Distance," http://hal.archives-ouvertes.fr/docs /00/12/55/74/HTML. Grégoire Chamayou posted about this marvelous invention as well in his webtext "Every Move Will Be Recorded," Max Planck Institute for the History of Science, www.mpiwg-berlin.mpg.de/en/news/features/feature14.

57. *Mémoire sur la réformation de la police de France,* 64.

58. "XKeyscore Presentation from 2008," *Guardian,* July 31, 2013; Glenn Greenwald, "XKeyscore: NSA Tool Collects 'Nearly Everything a User Does on the Internet,'" *Guardian,* July 31, 2013.

59. See, generally, Harcourt, *Illusion of Free Markets;* Gary Becker, François Ewald, and Bernard Harcourt, "American Neoliberalism: Michel Foucault's Birth of Biopolitics Lectures," seminar at the University of Chicago, May 9, 2012, http:// vimeo.com/43984248.

60. Pierre Bourdieu, *Sur l'État, cours au Collège de France (1989–1992)* (Paris: Le Seuil, 2012).

61. James Bamford, "They Know Much More than You Think," *New York Review of Books,* August 15, 2013; Glenn Greenwald and Ewen MacAskill, "NSA Prism Program Taps in to User Data of Apple, Google and Others," *Guardian,* June 6, 2013; Jacques Follorou, "Surveillance: la DGSE a transmis des données à la NSA américaine," *Le Monde,* October 30, 2013.

62. Glenn Greenwald, Ewen MacAskill, Laura Poitras, Spencer Ackerman, and Dominic Rushe, "Microsoft Handed the NSA Access to Encrypted Messages," *Guardian,* July 11, 2013; Glenn Greenwald, *No Place to Hide: Edward Snowden, the NSA, and the U.S. Surveillance State* (New York: Henry Holt, 2014), 112–115, 134–135; George Ellard, "Letter to Sen. Charles E. Grassley," September 11, 2013, www.nsa .gov/public_info/press_room/2013/grassley_letter.pdf; Marie-Louise Gumuchian and David Goldman, "Security Firm Traces Target Malware to Russia," CNN, January 21, 2014; Elizabeth A. Harris, Nicole Perlroth, Nathaniel Popper, and Hilary Stout, "A Sneaky Path into Target Customers' Wallets," *New York Times,* January 17, 2014; Kim Zetter, "Neiman Marcus: 1.1 Million Credit Cards Exposed in Three-Month Hack," *Wired,* January 24, 2014; Peter Maass and Laura Poitras, "Core Secrets: NSA Saboteurs in China and Germany," *Intercept,* October 10, 2014, https:// firstlook.org/theintercept/2014/10/10/core-secrets.

63. Dwight D. Eisenhower, "Eisenhower's Farewell Address," January 17, 1961, http://avalon.law.yale.edu/20th_century/eisenhower001.asp.

64. James Bamford, *The Shadow Factory: The Ultra-Secret NSA from 9/11 to the Eavesdropping on America* (New York: Doubleday, 2008), 199.

65. Greenwald and MacAskill, "NSA Prism Program Taps in to User Data."

66. Greenwald, MacAskill, Poitras, Ackerman and Rushe, "Microsoft Handed the NSA Access to Encrypted Messages."

67. Ibid.

68. Greenwald, *No Place to Hide,* 115.

69. Greenwald, MacAskill, Poitras, Ackerman and Rushe, "Microsoft Handed the NSA Access to Encrypted Messages."

70. Greenwald, *No Place to Hide,* 113; see also Greenwald, MacAskill, Poitras, Ackerman and Rushe, "Microsoft Handed the NSA Access to Encrypted Messages."

71. See Greenwald, MacAskill, Poitras, Ackerman and Rushe, "Microsoft Handed the NSA Access to Encrypted Messages."

72. Bamford, *The Shadow Factory,* 163–65.

73. Ibid., 163–64.

74. Ibid., 164–65.

75. With the end of military censorship in 1945, the intelligence agencies needed to enter into informal agreements with the telecommunications companies (ITT, RCA, Western Union) to get access to foreign government communications. Bamford, *The Shadow Factory,* 166–168. They did so under what was called Operation Shamrock, which the NSA would take over upon the agency's formation in 1952. Eventually, paper telegrams were replaced by computer discs, and the NSA was able to copy those as well under these informal agreements. This was disclosed in the early 1970s, resulting in the Church Commission investigation and report, and the passage of the Foreign Intelligence Surveillance Act in 1978. Bamford, *The Shadow Factory,* 168.

76. The scale in the United Kingdom was much smaller. "The general average of Warrants issued during the present century, does not much exceed 8 a-year," the investigation from 1844 revealed. "This number would comprehend, on an average, the Letters of about 16 persons annually." Jill Lepore, "The Prism: Privacy in an Age of Publicity," *New Yorker,* June 24, 2013.

77. Bamford, *The Shadow Factory,* 192.

78. Ibid., 197.

79. Ibid., 200.

80. Ibid.

81. Fulbright Association, "Admiral William O. Studeman, USN (Ret.)," http:// fulbrightevents.org/speaker/william-studeman.

82. Paracel, "William O. Studeman to Represent TRW on Paracel's Board of Directors," March 24, 1998, http://brandprosgroup.com/pages/first/websites/paracel /data/pr/032498.html.

83. Bamford, *The Shadow Factory,* 200.

84. Ibid., 200–201.

85. Ibid., 196.

86. Ibid., 199.

87. Greenwald, *No Place to Hide,* 101.

88. Ibid. (quoting Tim Shorrock).

89. Bamford, *The Shadow Factory,* 177–181.

90. Ibid., 181.

91. Bamford, "They Know Much More than You Think," 7.

92. Ibid., 10.

93. Greenwald, *No Place to Hide,* 105.

94. Ibid., 106.

95. Ibid., 105.

96. See Zygmunt Bauman, *Liquid Times: Living in an Age of Uncertainty* (Cambridge: Polity Press, 2007); Zygmunt Bauman, *Liquid Modernity* (Cambridge: Polity Press, 2000).

97. Hartmut Rosa raises these challenges in a provocative way in his work. See Hartmut Rosa, *Beschleunigung und Entfremdung: Entwurf einer kritischen Theorie spätmoderner Zeitlichkeit* (Frankfurt am Main: Suhrkamp, 2013); Sophia Grigull, *Hartmut Rosa und im Vergleich Zygmunt Bauman: Zur Konzeptualisierung von Autonomy spätmoderner Subjekte* (Berlin: Lit Verlag, 2014).

98. Eric Lichtblausept, "Spy Agencies Urge Caution on Phone Deal," *New York Times,* September 28, 2014.

99. Ibid.

100. Ibid.

101. Ibid.

102. Ibid.

103. Ibid.

104. Ibid.

105. Ibid.; see also Eric Lichtblausept, "Swedish-Owned Firm Defends Bid for Sensitive U.S. Phone Contract," *New York Times,* September 30, 2014.

106. Lichtblausept, "Swedish-Owned Firm Defends Bid."

107. David Pozen, "The Leaky Leviathan: Why the Government Condemns and Condones Unlawful Disclosures of Information," *Harvard Law Review* 127, no. 2 (2013): 512–635. On the larger question of leaks and state secrets, see Rahul Sagar, *Secrets and Leaks: The Dilemma of State Secrecy* (Princeton: Princeton University Press, 2013).

108. Greenwald, "Xkeyscore." Greenwald defines *metadata* as "the 'envelope' of a phone call or internet communication. For a phone call this could include the

duration of a phone call, the numbers it was between, and when it happened. For an email it would include the sender and recipient, time, but not the subject or content. In both cases it could include location information."

109. Jonathan Stray, "FAQ: What You Need to Know about the NSA's Surveillance Programs," ProPublica, August 5, 2013.

110. Greenwald, "XKeyscore."

111. Ibid.; see also Greenwald, *No Place to Hide,* 99.

112. Bamford, "They Know Much More than You Think," 9.

113. Ibid.

114. Ibid., 7.

115. Lindsay Goldwert, "How Do You Fix an Undersea Cable? Electrical Repairs on the Ocean Floor," *Slate,* February 5, 2008; Neil Jr., "Spy Agency Taps into Undersea Cable," ZDNet, May 23, 2001.

116. Neil Jr., "Spy Agency Taps into Undersea Cable."

117. Bamford, "They Know Much More than You Think," 6.

118. David Cole, "Can the NSA Be Controlled?," *New York Review of Books,* June 19, 2014, 17.

119. Bamford, *The Shadow Factory,* 13.

120. Ibid.

121. Ibid., 125.

122. Ibid., 127.

123. John Gilliom and Torin Monahan, *SuperVision: An Introduction to the Surveillance Society* (Chicago: University of Chicago Press, 2012).

124. Max Weber, "Politics as a Vocation," in *From Max Weber: Essays in Sociology,* ed. H. H. Gerth and C. Wright Mills (New York: Oxford University Press, 1946).

Chapter Three: JEREMY BENTHAM'S PANOPTICON

1. Glenn Greenwald, *No Place to Hide: Edward Snowden, the NSA, and the U.S. Surveillance State* (New York: Henry Holt, 2014), 175–177.

2. Ibid., 176.

3. William H. Simon, "Rethinking Privacy," *Boston Review,* September/October 2014, 60.

4. See generally Fabienne Brion and Bernard E. Harcourt, "The Louvain Lectures in Context," in Michel Foucault, *Wrong-Doing, Truth-Telling: The Function of Avowal in Justice* (Chicago: University of Chicago Press, 2014), 274–283; Bernard E. Harcourt, "Situation du cours," in Michel Foucault, *La société punitive: Cours au Collège de France, 1972–1973* (Paris: Gallimard / Le Seuil, 2013), 273–280.

5. See Michel Foucault, *Théories et institutions pénales: Cours au Collège de France, 1971–1972*, ed. Bernard E. Harcourt (Paris: Gallimard / Le Seuil, 2015), and Foucault, *La société punitive*.

6. Foucault, *Théories et institutions pénales* ("Pas d'introduction. La raison d'être de ce cours? Il suffit d'ouvrir les yeux. Ceux qui y répugneraient se trouveront dans ce que j'ai dit").

7. See Harcourt, "Situation du cours," 281–290.

8. The year before, in his 1972 lecture series *Penal Theories and Institutions*, Foucault had directly examined the question of state repression, focusing his lectures on the militarized response to the *nu-pieds* movement in Normandy in 1639. See Michel Foucault, "Penal Theories and Institutions," in *Ethics: Subjectivity and Truth* (New York: New Press, 1998), 7–12. The turn, in 1973, was to look at repression in society more generally—thus, the turn to panoptic power throughout society.

9. As a physician, Julius "worked on the staff of poorhouses and in the military." Norman Johnston, *Forms of Constraint: A History of Prison Architecture* (Urbana: University of Illinois Press, 2000), 180 n. 47. Julius was also a noted prison reformer. See Isidore Singer and Cyrus Adler, *The Jewish Encyclopedia: A Descriptive Record of the History, Religion, Literature, and Customs of the Jewish People from the Earliest Times to the Present Day* (New York: Funk and Wagnalls, 1916), 7:392.

10. Michel Foucault, "La vérité et les formes juridiques," in Foucault, *Dits et écrits* (Paris: Gallimard, 1994), 2:607 ("Quarto" edition [Paris: Gallimard, 2001], 1:1476).

11. N. H. Julius, *Vorlesungen über die Gefängniskunde-Kunde*, 2 vols. (Berlin: Stuhr, 1828); N. H. Julius, *Leçons sur les prisons, présentées en forme de cours au public de Berlin, en l'année 1827*, trans. H. Lagarmitte (Paris: F. G. Levrault, 1831), 384–385; discussed in Foucault, *La société punitive*, 24–25. Julius lectured in Berlin in 1827 "under the auspices of the newly formed *Verein für die Besserung der Stafgefangen*, a prisoners' aid society." These lectures were attended by Crown Prince Frederick William, who "became interested in prison reform" thereafter. See Johnston, *Forms of Constraint*, 180 n. 47.

12. Julius, *Leçons sur les prisons*, 385; discussed in Foucault, *La société punitive*, 25.

13. Foucault, *La société punitive*, 25.

14. Ibid.

15. Michel Foucault, *Surveiller et punir: La naissance de la prison* (Paris: Gallimard, 1975), 218.

16. Michel Foucault, *Discipline and Punish*, trans. Alan Sheridan (New York: Vintage, 1979), 216.

17. N. H. Julius, *Nord-amerikas sittliche Zustände, nach eigenen Anschauungen in den Jahren 1834, 1835 und 1836* (Leipzig: F. A. Brockhaus, 1839); N. H. Julius, *Du*

système pénitentiaire américain en 1836, trans. Victor Foucher (Paris: Joubert, 1837), 6 (underlined in the text).

18. Gustave de Beaumont et Alexis de Tocqueville, *Du système pénitentiaire aux États-Unis et de son application en France, suivi d'un appendice sur les colonies pénales et de notes statistiques* (Paris: H. Fournier Jeune, 1833); Alexis de Tocqueville, *Système pénitentiaire aux États-Unis et de son application en France,* 3rd ed. (Paris: Librairie de Charles Gosselin, 1845). See also Albert Krebs, "Julius, Nikolaus Heinrich," in *Neue Deutsche Biographie* (Berlin: Duncker and Humblot, 1974), 10:656–658.

19. His efforts in this regard were apparently "unmatched": "he founded and edited two of the central German-language journals of *Gefängniskunde,* and wrote textbooks and travel reports which became standard works on the subject." Thomas Nutz, "Global Networks and Local Prison Reforms: Monarchs, Bureaucrats and Penological Experts in Early Nineteenth-Century Prussia," *German History* 23, no. 4 (2005): 446; see also Bert Vanhulle, "Dreaming about the Prison: Édouard Ducpétiaux and Prison Reform in Belgium (1830–1848)," *Crime, History, and Societies* 14, no. 2 (2010): 111.

20. Nutz, "Global Networks and Local Prison Reforms," 454. After a self-financed 1825 trip to study British prison systems, Julius was hired by "the Prussian King Frederick William III . . . as an external inspector and adviser." Ibid., 448. Julius was subsequently hired by Frederick William IV after another largely self-financed 1834 trip to the United States to study innovations in American penitentiaries, a trip that produced the report *Nordamerikas sittliche Zustände.* Ibid., 448, 440.

21. Ibid., 454. Interestingly, however, those plans were never fully realized. This was due in part to the revolution of 1848 (after which Julius was stripped of his bureaucratic functions). Mostly, however, it was because of an "explosive increase" in prison population "after the new Prussian penal code, the *Strafgesetzbuch für die Preußischen Staaten,* came into force in 1851"—making implementation of solitary confinement for all prisoners impracticable. Nutz, "Global Networks and Local Prison Reforms," 455.

22. See Alessandro Stanziani, "The Traveling Panopticon: Labor Institutions and Labor Practices in Russia and Britain in the Eighteenth and Nineteenth Centuries," *Comparative Studies in Society and History* 51, no. 4 (October 2009): 715–741.

23. See Jeremy Bentham, *The Panopticon Writings,* ed. Miran Božovič (London: Verso, 1995); Jeremy Bentham, "Panopticon, or the Inspection-House," in *The Works of Jeremy Bentham,* ed. John Bowring (Edinburgh: William Tait, 1791), 4:37–173.

24. See Michel Foucault, "L'oeil du pouvoir," interview by J.-P. Barou and M. Perrot, in *Le panoptique,* ed. J.-P. Barou (Paris: Pierre Belfond, 1977), reproduced in Foucault, *Dits et écrits* (Paris: Gallimard, 1994), 3:190–207 ("Quarto" ed., 2:190–207); Michel Foucault, *Le pouvoir psychiatrique: Cours au Collège de France, 1973–1974,*

ed. Jacques Lagrange (Paris: Gallimard / Le Seuil, 2003), 352; Foucault, *La Société punitive*, 40 n. 3; *Les machines à guérir: Aux origines de l'hôpital modern*, ed. François Béguin, Blandine Barret-Kriegel, Bruno Fortier, Michel Foucault, and Anne Thalamy (Brussels: Pierre Mardaga, 1979).

25. See Béguin et al., *Les machines à guérir*, 48; on Bentham and the panopticon, see also Foucault, *La société punitive*, 78–79 n. 16.

26. Foucault, "L'oeil du pouvoir," 191.

27. See Foucault, *Discipline and Punish*, 6–7, 170–173, 183–192, and Foucault, *La société punitive*, 207–208 and 225 n. 1.

28. Foucault, *Discipline and Punish*, 187.

29. Foucault, *La société punitive*, 222–223.

30. Ibid.

31. Ibid., 198.

32. See Foucault, *Surveiller et punir*, 222–223.

33. Foucault, *Discipline and Punish*, 201.

34. See the "Notice" in the new *La Pléiade* complete works edition of Michel Foucault's *Surveiller et punir* (Paris: Gallimard, 2015).

35. Foucault, *Discipline and Punish*, 205.

36. Ibid., 200, 201, 204.

37. Ibid., 217.

38. Guy Debord, *La société du spectacle* (Paris: Buchet/Chastel, 1967).

39. Guy Debord, *The Society of the Spectacle* (New York: Zone Books, 1994), para. 3.

40. Ibid., para. 4.

41. Ibid., para. 215.

42. W. J. T. Mitchell, "The Spectacle Today: A Response to RETORT," *Public Culture* 20, no. 3 (Fall 2008): 573–582, revised version to appear in *Image Science* (forthcoming); see also W. J. T. Mitchell, *Picture Theory: Essays on Verbal and Visual Representation* (Chicago: University of Chicago Press, 1994), 326–328.

43. Mitchell, *Picture Theory*, 327; see also Jonathan Crary, *Techniques of the Observer: On Vision and Modernity in the Nineteenth Century* (Cambridge, MA: MIT Press, 1990), 18 (cited as well in Mitchell, *Picture Theory*, 327 n. 7).

44. Walking into the well of the last panopticon at Stateville Prison is an eerie experience today because the space has been turned on itself. With dirty, opaque Plexiglas covering the cell bars, it is difficult to see into the cells, but hard to ignore the calls and clamors from the prisoners. It feels almost as if surveillance has turned back into spectacle—in the most unfortunate and harrowing of spaces.

45. François Ewald, *"Omnes et singulatim:* After Risk," in *Carceral Notebooks* 7 (2011): 82.

46. Joseph Goldstein and J. David Goodman, "Seeking Clues to Gangs and Crime, Detectives Monitor Internet Rap Videos," *New York Times,* January 7, 2014.

47. See, e.g., G. W. Schulz, "Virginia Police Have Been Secretively Stockpiling Private Phone Records," *Wired,* October 20, 2014.

48. Nadia Prupis, "Big Brother 3.0: FBI Launches Facial Recognition Program," *Mint Press News,* September 17, 2014, www.mintpressnews.com/big-brother-3-0 -fbi-launches-facial-recognition-program/196657.

49. Georges Didi-Huberman, *Peuples exposés, peuples figurants.* L'oeil de l'histoire, vol. 4 (Paris: Les Éditions de Minuit, 2012), 11.

50. Ibid., 15 (*"sur-exposés* dans la lumière de leurs mise en spectacle").

51. Siva Vaidhyanathan, *The Googlization of Everything (and Why We Should Worry)* (Berkeley: University of California Press, 2011), 112.

52. Vaidhyanathan, *The Googlization of Everything,* 112.

53. Foucault, *Discipline and Punish,* 187.

54. Ibid., 201.

55. Vaidhyanathan, *The Googlization of Everything,* 112.

56. G. W. F. Hegel, "Preface," in *Philosophy of Right* (1820).

57. Foucault, *La société punitive,* 265.

58. Michel Foucault, *Histoire de la sexualité. 1, La volonté de savoir* (Paris: Gallimard, 1976).

59. Michel Foucault, *Sécurité, territoire, population: Cours au Collège de France (1977–1978),* ed. Michel Senellart (Paris: Gallimard / Le Seuil, 2004), 46.

60. Ibid., 7.

61. Ibid., 11–12.

62. Ibid., 15–19.

63. Jean Baudrillard, "The Precession of Simulacra," chap. 1 in *Simulacra and Simulations,* trans. Sheila Faria Glaser (Ann Arbor: University of Michigan Press, 1994). This discussion of the prison resonates entirely with Foucault's argument in *The Punitive Society* and *Discipline and Punish:* it is society as a whole that is panoptic, and the prison is just the metaphor.

64. Umberto Eco, *Travels in Hyperreality* (London: Pan, 1986); Louis Marin, *Utopics: Spatial Play,* trans. Robert A. Vollrath (Atlantic Highlands, NJ: Humanities, 1984), 239.

65. See, generally, Michael Sorkin, ed., *Variations on a Theme Park: The New American City and the End of Public Space* (New York: Hill and Wang, 1992); also Edward W. Soja, "Inside Exopolis: Scenes from Orange County," in Sorkin, ed., *Variations on a Theme Park,* 94–122.

66. George Ritzer, *The McDonaldization of Society: An Investigation into the Changing Character of Contemporary Social Life* (Newbury Park, CA: Pine Forge

Press, 1993); Alan Bryman, "Theme Parks and McDonaldization," in Barry Smart, ed., *Resisting McDonaldization* (London: Sage, 1999).

67. Steven Miles, *Spaces for Consumption: Pleasure and Placelessness in the Post-Industrial City* (Los Angeles: Sage, 2010), 142.

68. Ibid.

69. Ibid., 155. It is also a place of ritual and pilgrimage. See Alexander Moore, "Walt Disney World: Bounded Ritual Space and the Playful Pilgrimage Center," *Anthropological Quarterly* 53, no. 4 (1980): 207–218.

70. Keally McBride, *Collective Dreams: Political Imagination and Community* (University Park: Pennsylvania State University Press, 2005), 86ff.

71. Miles, *Spaces for Consumption*, 147.

72. Colin Campbell, *The Romantic Ethic and the Spirit of Modern Consumerism* (Oxford: Basil Blackwell, 1987), 78 (quoted in McBride, *Collective Dreams*, 114).

73. Gary S. Cross and John K. Walton, *The Playful Crowd: Pleasure Places in the Twentieth Century* (London: Sage, 2004), 167.

74. Ibid.

75. Miles, *Spaces for Consumption*, 154.

76. See McBride, *Collective Dreams*, 90; see also "Now Snowing—2013," available at http://celebrationtowncenter.com/events/now-snowing-2013.

77. McBride, *Collective Dreams*, 89.

78. Ibid., 116.

79. Ibid., 90.

80. Ibid.

81. Ibid., 93.

82. Ibid.

83. I am borrowing the title here, of course, of an excellent collection of essays on themed spaces in the United States edited by Michael Sorkin. See Sorkin, *Variations on a Theme Park*.

84. Special thanks to both W. J. T. Mitchell and Keally McBride, who encouraged me to explore further the shopping mall as a schema of neoliberalism. Tom Mitchell has also drawn my attention to the new show windows in Potsdamer Platz in Berlin that have mannequins with televisual eyes that move and follow you as you pass . . . an intriguing intersection of surveillance, spectacle, and exhibition in these themed spaces of consumption.

85. Angela Haggerty, "We Don't Do Clickbait, Insists BuzzFeed," *The Drum*, November 7, 2014, http://www.thedrum.com/news/2014/11/07/we-don-t-do-click bait-insists-buzzfeed.

86. Earlier, I thought this was the right way to understand our digital condition in this neoliberal security age. See Bernard E. Harcourt, "Digital Security in the

Expository Society: Spectacle, Surveillance, and Exhibition in the Neoliberal Age of Big Data," paper presented at the annual meeting of the American Political Science Association, August 8, 2014, http://ssrn.com/abstract=2455223. I am no longer convinced that this is right.

87. Quoted in Greenwald, *No Place to Hide,* 80.

88. Tim Shorrock quoted in ibid., 101.

89. I discuss this in *The Illusion of Free Markets* (Cambridge, MA: Harvard University Press, 2011); see also David Harvey, *A Brief History of Neoliberalism* (Oxford: Oxford University Press, 2005); Wendy Brown, *Undoing the Demos: Neoliberalism's Stealth Revolution* (New York: Zone Books, 2015).

90. Phillip Mirowski, *Never Let a Serious Crisis Go to Waste* (New York: Verso, 2013), 112. Alexander de la Paz, a PhD student in the political science department at Columbia University, offers a useful distinction, in his working paper "We 'Everyday Neoliberals'?" (2014), between genetic and analogical claims of neoliberal kinship. Copy in the author's possession.

91. See generally Michel Foucault, *Naissance de la biopolitique: Cours au Collège de France, 1978–1979,* ed. Michel Senellart (Paris: Gallimard/Le Seuil, 2004), specifically lectures of March 14, 21, and 28, 1979; Andrew Dilts, "From 'Entrepreneur of the Self' to 'Care of the Self': Neoliberal Governmentality and Foucault's Ethics," *Foucault Studies* 12 (October 2011): 130–146.

92. Mirowski, *Never Let a Serious Crisis Go to Waste,* 112–113.

93. One might productively study Facebook practices in relation to the practices of avowal of the self that Foucault explored in *Wrong-Doing, Truth-Telling.*

94. James Bamford, *The Shadow Factory: The Ultra-Secret NSA from 9/11 to the Eavesdropping on America* (New York: Doubleday, 2008), 102; Daniel Solove, *Nothing to Hide* (New Haven, CT: Yale University Press, 2011), 183–185.

95. Yasha Levine, "The Psychological Dark Side of Gmail," AlterNet, December 31, 2013, www.alternet.org/media/google-using-gmail-build-psychological-profiles-hundreds-millions-people.

96. Ibid.

97. Ibid.

98. Vaidhyanathan, *The Googlization of Everything,* 16.

99. Farhad Manjoo, "Google Unveils Inbox, a New Take on Email. And Possibly a Replacement for Gmail," *New York Times,* October 22, 2014.

100. Levine, "The Psychological Dark Side of Gmail."

101. The reference, naturally, is to Foucault, *History of Sexuality,* 1:138; see also Michel Foucault, *"Il faut défendre la société": Cours au Collège de France, 1975–1976,* ed. Mauro Bertani and Alessandro Fontana (Paris: Gallimard / Le Seuil, 1997).

Chapter Four: OUR MIRRORED GLASS PAVILION

1. See "The Roof Garden Commission: Dan Graham with Günther Vogt," video, Metropolitan Museum of Art, http://youtu.be/faR1Expe-O4. All quotations are from this video.

2. Ibid.

3. Ibid.

4. Karen Rosenberg, "Artists Hold up a Glass to a City's Changing Face: 'Hedge Two-Way Mirror Walkabout,' at the Met Museum," *New York Times,* April 29, 2014. Special thanks to Mia Ruyter for encouraging me to explore Dan Graham's artwork.

5. Other examples of Graham's work includes "Double Exposure" (1995), a mirror-glass structure-pavilion in Porto, Portugal, located in Serralves Park; his "Rooftop Urban Park Project," built on the rooftop of the Dia Art Foundation in the Chelsea neighborhood of New York, where it stood from 1991 to 2004; and his "Two-Way Mirror Pergola Bridge" (1990) in Clisson, France.

6. Tony Bennett, "The Exhibitionary Complex," *New Formation* 4 (Spring 1988). Special thanks to Irus Braverman.

7. Ibid., 87.

8. Ibid., 76.

9. Kim Kardashian published a book of selfies, titled *Selfish.* See Olivia B. Waxman, "Here's the Cover of Kim Kardashian's New Book of Selfies, *Selfish,*" *Time,* January 20, 2015. For more on selfies, see Alicia Eler, "Keeping up with the Selfies / Hyperallergic," April 28, 2014, http://aliciaeler.com/2014/04/28/keeping -selfies-hyperallergic.

10. See generally Emily Yoffe, "Seeking: How the Brain Hard-wires Us to Love Google, Twitter, and Texting. And Why That's Dangerous," *Slate,* August 12, 2009.

11. Spencer Ackerman and James Ball, "Optic Nerve: Millions of Yahoo Webcam Images Intercepted by GCHQ," *Guardian,* February 27, 2014; Associated Press, "British Spies Intercept Webcam Pictures, Report Says," *New York Times,* February 27, 2014.

12. Ackerman and Ball, "Optic Nerve."

13. Associated Press, "British Spies Intercept Webcam Pictures."

14. Ibid.; see also Ackerman and Ball, "Optic Nerve."

15. Ackerman and Ball, "Optic Nerve."

16. Ibid.; see also Associated Press, "British Spies Intercept Webcam Pictures."

17. Associated Press, "British Spies Intercept Webcam Pictures."

18. Julia Fioretti, "British Spy Agency Collected Images of Yahoo Webcam Chats: Guardian," Reuters, February 27, 2014.

19. Ackerman and Ball, "Optic Nerve."

20. Ibid.

21. Ibid.

22. Nick Hopkins, "GCHQ's Cover for Optic Nerve Provided by Legislation Introduced in 2000," *Guardian*, February 27, 2014.

23. Ibid.

24. Ackerman and Ball, "Optic Nerve."

25. Amy Davidson, "Are the NSA and GCHQ Trading Webcam Pictures?," *New Yorker*, February 28, 2014.

26. Ibid.

27. Ackerman and Ball, "Optic Nerve."

28. Spencer Ackerman, "Senators to Investigate NSA role in GCHQ 'Optic Nerve' Webcam Spying," *Guardian*, February 28, 2014.

29. Fioretti, "British Spy Agency Collected Images of Yahoo Webcam Chats."

30. Hopkins, "GCHQ's Cover for Optic Nerve."

31. Ackerman and Ball, "Optic Nerve."

32. You can see their selfie here: http://instagram.com/p/m3z66SlwW4.

33. Quoted in David Cole, "Can the NSA Be Controlled?," *New York Review of Books*, June 19, 2014, 16.

34. Michel Foucault, *Surveiller et punir: La naissance de la prison* (Paris: Gallimard, 1975), 203 n. 2; see also Michel Foucault, "La vérité et les formes juridiques," in Foucault, *Dits et écrits* (Paris: Gallimard, 1994), 2:607–609 ("Quarto" ed., 1:1475–1477).

35. Foucault, *Surveiller et puni*, 203 n. 2.

36. Ibid.

37. For a development of these themes, see Guillaume Tiffon, *La mise au travail des clients* (Paris: Economica, 2013), which explores this new form of surplus value that has come to dominate modern life. Elizabeth Emens has a marvelous book project on "admin" that explores this as well.

38. Karen Levy, "Data-Driven Dating: How Data Are Shaping Our Most Intimate Personal Relationships," *Privacy Perspectives*, December 17, 2013, www .privacyassociation.org/news/a/data-driven-dating-how-data-are-shaping-our -most-intimate-personal-relation.

39. Eben Moglen, quoted in Nancy Scola, "'People Love Spying on One Another': A Q&A with Facebook Critic Eben Moglen," *The Switch* (blog), *Washington Post*, November 19, 2014.

40. In a promising editorial, Colin Koopman refers to the dawning of a new age of "infopolitics," and he points usefully in the direction of other theorists who explore these new realms, such as Donna Haraway, who developed the notion of "the

informatics of domination," Grégoire Chamayou, who is working on "datapower," and Davide Panagia. See Colin Koopman, "The Age of Infopolitics," *New York Times,* January 26, 2014; Donna Haraway, "A Cyborg Manifesto: Science, Technology, and Socialist-Feminism in the Late Twentieth Century," in *Simians, Cyborgs and Women: The Reinvention of Nature* (New York; Routledge, 1991), 149–181.

41. For the conventional definition of literal transparence and the contrast with phenomenal transparence, see Colin Rowe and Robert Slutzky, "Transparency: Literal and Phenomenal," *Perspecta* 8 (1963): 45–54; and Colin Rowe and Robert Slutzky, "Transparency: Literal and Phenomenal . . . Part II," *Perspecta* 13/14 (1971): 287–301. For a contextualization of these texts, see Detlef Mertins, "Transparency: Autonomy and Relationality," *AA Files* 32 (Autumn 1996): 3–11; as Mertins suggests, for critiques of these texts, see Rosalind Krauss, "Death of a Hermeneutic Phantom," *Architecture+Urbanism* 112 (January 1980): 189–219; Robert Somol, "Oublier Rowe," *ANY* 7/8 (1994): 8–15. "Literal transparence" is defined by Rowe and Slutzky as "an inherent quality of substance, as in a glass curtain wall," Rowe and Slutzky, "Transparency: Literal and Phenomenal," 46; also as "the trompe l'oeil effect of a translucent object in a deep, naturalistic space," ibid., 48. Special thanks to Jonah Rowen.

42. Rowe and Slutzky, "Transparency: Literal and Phenomenal," 49.

43. Terence Riley, *LightConstruction* (New York: Harry N. Abrams, 1995), 11.

44. Anthony Vidler, *The Architectural Uncanny: Essays in the Modern Unhomely* (Cambridge, MA: MIT Press, 1992), 217.

45. Ibid., 220 (omitting "(as Pei himself admitted)").

46. See, generally, Rowe and Slutzky, "Transparency: Literal and Phenomenal," 48 (defining phenomenal space, based on certain cubist paintings, as "when a painter seeks the articulated presentation of frontally displayed objects in a shallow, abstracted space"); Rowe and Slutzky, "Transparency: Literal and Phenomenal . . . Part II," 288 (referring to phenomenal transparence as "the capacity of figures to interpenetrate without optical destruction of each other").

47. That is, for those who do not simply debunk the idea of phenomenal transparence as merely a hermeneutic phantom or as a ghost; see Krauss, "Death of a Hermeneutic Phantom." See also Reinhold Martin, *Utopia's Ghost: Architecture and Postmodernism, Again* (Minneapolis: University of Minnesota Press, 2010), 55–57; Reinhold Martin, *The Organizational Complex: Architecture, Media, and Corporate Space* (Cambridge, MA: MIT Press, 2003) (discussing the shallowness and thinness of the transparency metaphor).

48. Vidler, *The Architectural Uncanny,* 219.

49. Laura Kurgan, *Close Up at a Distance: Mapping, Technology, and Politics* (Cambridge, MA: MIT Press, 2013), 24.

50. See Marlborough Chelsea press release, "Broadway Morey Boogie," September 23, 2014–February 2015, http://www.broadwaymall.org/core/wp-content/uploads/2010/03/Broadway-Morey-Boogie-Press-Release.pdf

51. Ibid.

52. Michel Foucault, *Discipline and Punish,* trans. Alan Sheridan (New York: Vintage, 1979), 172.

53. Ibid., 207.

54. Ibid., 208 (emphasis added); see ibid., 249.

55. Ibid., 173.

56. Andy Müller-Maguhn, Laura Poitras, Marcel Rosenbach, Michael Sontheimer, and Christian Grothoff, "Treasure Map: The NSA Breach of Telekom and Other German Firms," *Der Spiegel,* September 14, 2014.

57. Siva Vaidhyanathan, *The Googlization of Everything (and Why We Should Worry)* (Berkeley: University of California Press, 2011), 2.

58. Ibid., 7.

59. Glenn Greenwald, *No Place to Hide: Edward Snowden, the NSA, and the U.S. Surveillance State* (New York: Henry Holt, 2014), 97.

60. Ibid.

61. Ibid.

62. Ibid., 92.

63. James Bamford, *The Shadow Factory: The Ultra-Secret NSA from 9/11 to the Eavesdropping on America* (New York: Doubleday, 2008), 129.

64. Ibid., 149.

65. Ibid.

66. Greenwald, *No Place to Hide,* 94.

67. Apple, "Introducing Apple Watch," www.apple.com/watch/films/#film-design.

68. All of the quoted passages that follow are from this video advertisement by Apple.

69. Charles Duhigg, "How Companies Learn Your Secrets," *New York Times,* February 16, 2012.

70. Ibid.

71. Frank Pasquale, *The Black Box Society: The Secret Algorithms behind Money and Information* (Cambridge, MA: Harvard University Press, 2015), 3–4, 156–160.

72. Ibid., 156.

73. Tim Wu, *The Master Switch: The Rise and Fall of Information Empires* (New York: Vintage, 2010).

74. See Andrew Norman Wilson, "Workers Leaving the Googleplex 2009–2011," video, http://www.andrewnormanwilson.com/WorkersGoogleplex.html; see

also Kenneth Goldsmith, "The Artful Accidents of Google Books," *New Yorker,* December 4, 2013. Thanks to Lucas Pinheiro for alerting me to this.

75. See Eve Kosofsky Sedwick, *Epistemology of the Closet* (Berkeley: University of California Press, 1990); Linda M. G. Zerilli, " 'Philosophy's Gaudy Dress': Rhetoric and Fantasy in the Lockean Social Contract," *European Journal of Political Theory* 4 (2005): 146–163.

76. Vaidhyanathan, *The Googlization of Everything,* 112.

77. Greenwald, *No Place to Hide,* 148.

78. Ibid., 149.

79. See generally Joseph Masco, *The Theater of Operations: National Security Affect from the Cold War to the War on Terror* (Durham, NC: Duke University Press, 2014) (exploring what might be called "governing through terrorism" and tracing its genealogy back to the Cold War).

80. Apple, "The Health and Fitness Film," available at http://www.apple.com /watch/films/#film-fitness.

81. Ibid.

82. Ibid.

83. Shanyang Zhao, "The Digital Self: Through the Looking Glass of Telecopresent Others," *Symbolic Interaction* 28, no. 3 (2005): 397.

84. These notions are developed further in Shanyang Zhao, Sherri Grasmuck, and Jason Martin, "Identity Construction on Facebook: Digital Empowerment in Anchored Relationships," *Computers in Human Behavior* 24, no. 5 (September 2008): 1816–1836.

85. John B. Thompson, *The Media and Modernity: A Social Theory of the Media* (Stanford, CA: Stanford University Press, 1995), 210 (quoted in Zhao, "The Digital Self," 397).

86. As a legal matter, the concept of the "mosaic" is particularly important and used to capture the danger, with our digital technology, of piecing together a composite knowledge of the individual with tiny pieces that would otherwise be irrelevant and useless—the danger, that is, to the right to privacy. Judge Richard Leon relies on the mosaic theory in *Klayman v. Obama,* 957 F. Supp. 2d 1 (D.D.C. 2013). David Pozen has an excellent discussion of the mosaic theory in his article "The Mosaic Theory, National Security, and the Freedom of Information Act, *Yale Law Journal* 115, no. 3 (December 2005): 628–679.

87. See, e.g., Mike Isaacs, "Nude Photos of Jennifer Lawrence Are Latest Front in Online Privacy Debate," *New York Times,* September 2, 2014; TMZ, " 'Kramer's' Racist Tirade—Caught on Tape," www.tmz.com/2006/11/20/kramers-racist-tirade-caught -on-tape.

88. See Foucault's opening discussion of nineteenth-century psychiatric practices (in particular the infliction of cold showers by Dr. Leuret on Patient A to make his patient avow that he was but is no longer mad) in the inaugural lecture of his 1981 Louvain lecture series, *Wrong-Doing, Truth-Telling: The Function of Avowal in Justice* (Chicago: University of Chicago Press, 2014), 11–12.

89. Jay Caspian Kang, "Moving Pictures: What's Really Radical about the Eric Garner Protests," *New York Times,* December 12, 2014.

90. Ibid.

91. See, e.g., Zhao, Grasmuck, and Martin, "Identity Construction on Facebook."

92. Ibid., 1829.

93. Ibid., 1830.

94. Hannah Seligson, "Facebook's Last Taboo: The Unhappy Marriage," *New York Times,* December 26, 2014.

95. The term is defined in different ways. TechAmerica Foundation's Federal Big Data Commission defines it as "a term that describes large volumes of high velocity, complex and variable data that require advanced techniques and technologies to enable the capture, storage, distribution, management, and analysis of the information." TechAmerica Foundation's Federal Big Data Commission, "Demystifying Big Data" (2013), 10, http://www.techamerica.org/Docs/fileManager.cfm?f=techamerica -bigdatareport-final.pdf. McKinsey and Company defines big data as "datasets whose size is beyond the ability of typical database software tools to capture, store, manage, and analyze." See James Manyika, Michael Chui, Brad Brown, Jacques Bughin, Richard Dobbs, Charles Roxburgh, and Angela Hung Byers, "Big Data: The Next Frontier for Innovation, Competition, and Productivity," McKinsey and Company, June 2011, 10, http://www.mckinsey.com/insights/business_technology/big_data _the_next_frontier_for_innovation. Another commentator defines it, broadly, as "an umbrella term. It encompasses everything from digital data to health data (including your DNA and genome) to the data collected from years and years of paperwork issued and filed by the government"; Higinio Maycotte, "The Evolution of Big Data, and Where We're Headed," *Wired,* March 26, 2014.

96. TechAmerica Foundation's Federal Big Data Commission, "Demystifying Big Data," 9, 11; see also SAS, "Big Data: What It Is and Why It Matters," http://www .sas.com/en_us/insights/big-data/what-is-big-data.html.

97. TechAmerica Foundation's Federal Big Data Commission, "Demystifying Big Data," 9, 11.

98. "The Digital Universe of Opportunities," EMC Infobrief with research and analysis by IDC, April 2014, http://www.emc.com/collateral/analyst-reports/idc -digital-universe-2014.pdf; also see Vernon Turner, John F. Gantz, David Reinsel, and Stephen Minton, "The Digital Universe of Opportunities: Rich Data and the

Increasing Value of the Internet of Things," IDC white paper, April 2014, http://idcdocserv.com/1678.

99. "The Digital Universe of Opportunities," slide 2.

100. Andrew McAfee and Erik Brynjolfsson, "Big Data: The Management Revolution," *Harvard Business Review* 62 (October 2012).

101. Martin Hilbert and Priscila López, "The World's Technological Capacity to Store, Communicate, and Compute Information," *Science* 332, no. 6025 (April 1, 2011): 60–65.

102. Ibid., 60.

103. Ibid., 63.

104. Manyika et al., "Big Data," 16–17.

105. Greenwald, *No Place to Hide,* 159.

106. "The Digital Universe of Opportunities," slide 5.

107. "The Digital Universe of Opportunities," slide 8.

108. Manyika et al., "Big Data," preface.

109. McAfee and Brynjolfsson, "Big Data."

110. Daniel Soar, "It Knows," *London Review of Books,* October 6, 2011, 3–6.

111. McAfee and Brynjolfsson, "Big Data."

112. See Pew Research Center, "The Web at 25," http://www.pewinternet.org/packages/the-web-at-25.

113. Mark Zuckerberg post on Facebook, November 27, 2014, 5:28 p.m., https://www.facebook.com/zuck/posts/10101761674047881.

114. "The Digital Universe of Opportunities," slide 4.

115. Manyika et al., "Big Data," 19, exhibit 7.

116. TechAmerica Foundation's Federal Big Data Commission, "Demystifying Big Data," 9.

117. Ibid.

118. See Manyika et al., "Big Data," 19, exhibit 7.

119. Maycotte, "The Evolution of Big Data."

120. Ibid.

121. Ibid.

122. Ibid.

123. Ibid.

124. Ibid.

125. "Big Data FAQs—A Primer," *Arcplan Business Intelligence Blog,* March 23, 2012, www.arcplan.com/en/blog/2012/03/big-data-faqs-a-primer.

126. Ibid.

127. Ibid.

128. Ibid.

129. World Economic Forum, "Big Data, Big Impact: New Possibilities for International Development," 2012, 2, http://www3.weforum.org/docs/WEF_TC_MFS_BigDataBigImpact_Briefing_2012.pdf.

130. Ibid.

131. Ibid., infographic on 5.

132. "All Too Much," *Economist,* February 25, 2010.

133. TechAmerica Foundation's Federal Big Data Commission, "Demystifying Big Data," 11.

134. Thomas H. Davenport, Paul Barth, and Randy Bean, "How 'Big Data' Is Different," *MITSloan Management Review* 54, no. 1 (Fall 2012): 22.

135. I discovered this passage in Vidler, *The Architectural Uncanny,* 218, along with a marvelous passage from André Breton, who writes, "As for me, I continue to inhabit my glass house." The Benjamin passage comes from "Surrealism," in *Reflections: Essays, Aphorisms, Autobiographical Writings,* ed. Peter Demetz (New York: Harcourt Brace Jovanovich, 1978), 180.

Chapter Five: A GENEALOGY OF THE NEW *DOPPELGÄNGER* LOGIC

1. See Jean de Coras, *Arrest mémorable du Parlement de Toloze, contenant une histoire prodigieuse de nostre temps, avec cent belles et doctes annotations, de monsieur maistre Jean de Coras* (Lyon: A. Vincent, 1561); Jean de Coras, *Arrest Memorable,* trans. Jeannette K. Ringold and Janet Lewis, in *Triquarterly* 55 (Fall 1982): 86–103; Natalie Zemon Davis, *The Return of Martin Guerre* (Cambridge, MA: Harvard University Press, 1985).

2. Davis, *The Return of Martin Guerre,* 34–35.

3. Ibid., 24. According to Davis, Martin Guerre left for Spain and eventually became a lackey in the household of a cardinal of the Roman Catholic Church, Francisco de Mendoza y Bobadilla, in Burgos, Spain. Davis believes that Martin then served the cardinal's brother, Pedro, and as part of his entourage, went into the Spanish army, losing his leg in battle against the French at Saint-Quentin on Saint Lawrence Day, August 10, 1557. Ibid., 26.

4. Coras, *Arrest mémorable du Parlement de Toloze,* 2.

5. Ibid.

6. The historian Robert Finlay took Davis to task—harshly, one might add—for her interpretation of the historical record, charging Davis with reinventing Bertrande: "No longer a dupe and victim, she has become a heroine, a sort of proto-feminist of peasant culture. This Bertrande de Rols," Finlay argues, "seems to be far more a product of invention than of historical reconstruction." Robert Finlay, "The Refashioning of Martin Guerre," *American Historical Review* 93, no. 3 (June 1988):

553–571. Davis responded in a forceful and persuasive essay, "On the Lame," *American Historical Review* 93, no. 3 (June 1988): 572–603 (June 1988), drawing on Michel de Montaigne's essay about the Guerre case, "Du démentir." Michel de Montaigne, "Du démentir," in *Oeuvres completes,* ed. A. Thibaudet and M. Rat (Paris: Gallimard, 1962), 649; Michel de Montaigne, *The Complete Works of Montaigne,* trans. Donald Frame (Stanford, CA: Stanford University Press, 1948), 505.

7. Coras, *Arrest mémorable du Parlement de Toloze,* 7.

8. Ibid., 3.

9. Ibid., 4.

10. Davis, *The Return of Martin Guerre,* 43.

11. Coras, *Arrest mémorable du Parlement de Toloze,* 6.

12. Ibid., 5.

13. Davis, *The Return of Martin Guerre,* 67; see also 55, 56, and 70.

14. Ibid., 70.

15. Coras, *Arrest mémorable du Parlement de Toloze,* 8.

16. See FBI, "Integrated Automated Fingerprint Identification System," www .fbi.gov/about-us/cjis/fingerprints_biometrics/iafis/iafis.

17. It is, for instance, extremely costly and laborious for the FBI to give its informants new identities. See Pete Earley, *Witsec: Inside the Federal Witness Protection Program* (New York: Bantam, 2003).

18. Bernard E. Harcourt, *Against Prediction* (Chicago: University of Chicago Press, 2007).

19. Ian Hacking, *The Taming of Chance* (New York: Cambridge University Press, 1990), 1–8.

20. Nikolas Rose, "At Risk of Madness," in Tom Baker and Jonathan Simon, eds., *Embracing Risk: The Changing Culture of Insurance and Responsibility* (Chicago: University of Chicago Press, 2002), 214.

21. Ibid., 214.

22. Naturally, the probabilistic turn was not uniformly deployed toward that aim. The cultural context matters, and as historian of science Deborah Coen demonstrates in her research on nineteenth-century imperial Austria, the probabilistic turn there promoted liberalism and tolerance, as opposed to the control of the Catholic Church. See Deborah Rachel Coen, *A Scientific Dynasty: Probability, Liberalism, and the Exner Family in Imperial Austria* (Ann Arbor, MI: UMI Dissertation Services, 2004).

23. Hacking, *The Taming of Chance,* 1.

24. Ibid.

25. Ibid.; see also Michel Foucault, *Abnormal: Lectures at the Collège de France, 1974–1975,* ed. Valerio Marchetti and Antonella Salomoni, trans. Graham Burchell

(London: Verso, 2004); Foucault, "Sixth Lecture: May 20, 1981," in Michel Foucault, *Wrong-Doing, Truth-Telling: The Function of Avowal in Justice* (Chicago: University of Chicago Press, 2014); Michel Foucault, "About the Concept of the 'Dangerous Individual' in 19th-Century Legal Psychiatry," *International Journal of Law and Psychiatry* 1 (1978): 1–18.

26. Hacking, *The Taming of Chance,* 105.

27. Ibid.

28. George B. Vold, *Prediction Methods and Parole: A Study of the Factors Involved in the Violation or Non-violation of Parole in a Group of Minnesota Adult Males* (Hanover, NH: Sociological Press, 1931), 103.

29. Andrew A. Bruce, Ernest W. Burgess, and Albert M. Harno, "A Study of the Indeterminate Sentence and Parole in the State of Illinois," *Journal of the American Institute of Criminal Law and Criminology* 19, no. 1, pt. 2 (May 1928): 284.

30. Ibid., 271.

31. See Ferris F. Laune, *Predicting Criminality: Forecasting Behavior on Parole,* Northwestern University Studies in the Social Sciences, no. 1 (Evanston, IL: Northwestern University Press, 1936), cover page.

32. Hacking, *The Taming of Chance,* 2.

33. Ernest Burgess, "Is Prediction Feasible in Social Work? An Inquiry Based upon a Sociological Study of Parole Records," *Social Forces* 7 (1928): 533–545.

34. Laune, *Predicting Criminality,* 2.

35. See Harcourt, *Against Prediction,* 181–182.

36. Vold, *Prediction Methods and Parole,* 70.

37. Daniel Glaser, "A Reformulation and Testing of Parole Prediction Factors" (PhD diss., Department of Sociology, University of Chicago, 1954), 268.

38. François Ewald, *L'État providence* (Paris: Bernard Grasset, 1986). There were also inflections of this in Foucault's "Sixth Lecture: May 20, 1981," in *Wrong-Doing, Truth-Telling,* and in Foucault, "About the Concept of the 'Dangerous Individual.'"

39. See also Robert Castel, *La gestion des risques—de l'anti-psychiatrie à l'après-psychanalyse* (Paris: Éditions de Minuit, 1981); Michel Foucault, *The Birth of Biopolitics: Lectures at the Collège de France, 1978–79,* trans. Graham Burchell (New York: Palgrave Macmillan, 2008).

40. Glaser, "A Reformulation and Testing of Parole Prediction Factors," 285.

41. See *Handbook for New Parole Board Members,* "Chapter 4: Parole Decision-making," 35, www.apaintl.org/documents/CEPPParoleHandbook.pdf.

42. Walter Webster Argow, "A Criminal Liability-Index for Predicting Possibility of Rehabilitation," *Journal of Criminal Law and Criminology* 26 (1935): 562. See also how Courtlandt Churchill Van Vetchen would open his study in 1935 with a bow to the insurance industry: "There is nothing new about the idea of evaluating

the probability of a future happening on the basis of past experience. Insurance, which depends upon actuarial ratings worked out with more or less precision, dates back at least to Roman times." Courtlandt Churchill Van Vechten Jr., "A Study of Success and Failure of One Thousand Delinquents Committed to a Boy's Republic" (PhD diss., Department of Sociology, University of Chicago; private ed., distributed by the University of Chicago Libraries, 1935), 19.

43. This is, with apologies, a play on words of Philip Mirowski's *Machine Dreams: Economics Becomes a Cyborg Science* (Cambridge: Cambridge University Press, 2001).

44. See generally Bernard E. Harcourt, "The Systems Fallacy," work in progress presented at Columbia Law School, December 6, 2013.

45. For a history of operations research, see Maurice W. Kirby, *Operational Research in War and Peace: The British Experience from the 1930s to 1970* (London: Imperial College Press, 2003); S. M. Amadae, *Rationalizing Capitalist Democracy: The Cold War Origins of Rational Choice Liberalism* (Chicago: University of Chicago Press, 2003).

46. Jennifer S. Light, *From Warfare to Welfare: Defense Intellectuals and Urban Problems in Cold War America* (Baltimore: Johns Hopkins University Press, 2003), 67. For an excellent history of the emergence of these techniques and rationalities, see Amadae, *Rationalizing Capitalist Democracy*.

47. James R. Schlesinger, "Quantitative Analysis and National Security," *World Politics* 15, no. 2 (1963): 314.

48. By contrast to the endlessly divisible "dividual" that, in Deleuze's "Postscript on the Societies of Control," *October* 59 (Winter 1992): 3–7, becomes endlessly reducible to data representations by means of digital technologies of control. See also Robert W. Williams, "Politics and Self in the Age of Digital Re(pro)ducibility," *Fast Capitalism* 1, no. 1 (2005).

49. Kevin D. Haggerty and Richard V. Ericson, "The Surveillant Assemblage," *British Journal of Sociology* 51, no. 4 (2000): 611.

50. Tom Vanderbilt, "The Science behind the Netflix Algorithms That Decide What You'll Watch Next," *Wired*, August 7, 2013.

51. Ibid.

52. Ibid.

53. "Recommender Systems: A Computer Science Comprehensive Exercise at Carleton College, Northfield, MN," http://www.cs.carleton.edu/cs_comps/0607/recommend/recommender/memorybased.html, referring to J. S. Breese, D. Heckerman, and C. Kadie, "Empirical Analysis of Predictive Algorithms for Collaborative Filtering," in *Proceedings of the Fourteenth Conference on Uncertainty in Artificial Intelligence* (San Francisco: Morgan Kaufmann, 1998).

54. Xavier Amatriain and Justin Basilico, "Netflix Recommendations: Beyond the 5 Stars (Part 1)," *Netflix Tech Blog,* April 6, 2012, http://techblog.netflix.com /2012/04/netflix-recommendations-beyond-5-stars.html.

55. Derrick Harris, "Twitter Open Sourced a Recommendation Algorithm for Massive Datasets," Gigaom, September 24, 2014, https://gigaom.com/2014/09/24 /twitter-open-sourced-a-recommendation-algorithm-for-massive-datasets.

56. Vanderbilt, "The Science behind the Netflix Algorithms."

57. Ibid.

58. Ibid.

59. Ibid.

60. Tarleton Gillespie, "The Relevance of Algorithms," in Tarleton Gillespie, Pablo Boczkowski, and Kirsten Foot, eds., *Media Technologies* (Cambridge, MA: MIT Press, forthcoming). Special thanks to Irus Braverman.

61. Julia Angwin, "Why Online Tracking Is Getting Creepier: The Merger of Online and Offline Data Is Bringing More Intrusive Tracking," ProPublica, June 12, 2014, www.propublica.org/article/why-online-tracking-is-getting-creepier.

62. Ibid.

63. Ibid.

64. Kevin Weil, "Experimenting with New Ways to Tailor Ads," *The Official Twitter Blog,* July 3, 2013, https://blog.twitter.com/2013/experimenting-with-new -ways-to-tailor-ads.

65. Facebook, "What Is Hashing? How Is Hashing Different than Encryption?," https://www.facebook.com/help/112061095610075.

66. Ibid.

67. See Facebook, "What Are Lookalike Audiences?," www.facebook.com/ help/164749007013531.

68. Facebook, "How Do I Create a Custom Audience?," https://www.facebook .com/help/170456843145568.

69. Facebook, "What Are Lookalike Audiences?"

70. Facebook, "What Does It Mean to Optimize for 'Similarity' or 'Reach' When Creating a Lookalike Audience?," https://www.facebook.com/help/1377 07129734780.

71. Clark Tibbitts, "Success and Failure on Parole Can Be Predicted," *Journal of Criminal Law and Criminology* 22, no. 1 (1931): 40.

72. Kenneth Neil Cukier and Viktor Mayer-Schoenberger, *Big Data: A Revolution That Will Transform How We Live, Work, and Think* (New York: Houghton Mifflin Harcourt, 2013), 68.

73. William Petty, *Political Arithmetick* (London: Clavel, 1691), xxiv.

74. Ibid., 31–32.

75. Richard Cantillon, *Essai sur la nature du commerce en général* (Paris: Institut national d'études démographiques, 1997), 186.

76. Ibid., 40.

77. Karl Marx and Friedrich Engels, *Capital, Volume 1* (Moscow: Progress, 1965), 54 n. 18, 57 n. 33, 179.

78. Alessandro Roncaglia, *Petty: The Origins of Political Economy* (Armonk, NY: M. E. Sharpe, 1985), 22.

Chapter Six: THE ECLIPSE OF HUMANISM

1. Kevin D. Haggerty and Richard V. Ericson, "The Surveillant Assemblage," *British Journal of Sociology* 51, no. 4 (2000): 616.

2. Bobbie Johnson, "Privacy No Longer a Social Norm, Says Facebook Founder," *Guardian,* January 10, 2010.

3. Centre for International Governance Innovation and IPSOS, "CIGI-IPSOS Global Survey on Internet Security and Trust," November 24, 2014, 3, www .cigionline.org/internet-survey.

4. Pew Research Center for the People and the Press, "Majority Views NSA Phone Tracking as Acceptable Anti-terror Tactic. Public Says Investigate Terrorism, Even if It Intrudes on Privacy," June 10, 2013, www.people-press.org/2013 /06/10/majority-views-nsa-phone-tracking-as-acceptable-anti-terror-tactic.

5. See D. A. Miller, *The Novel and the Police* (Berkeley: University of California Press, 1988), 162. Special thanks to Christopher Stolj for his discussion of Miller's classic study in our seminar with W. J. T. Mitchell, "Spectacle and Surveillance," Columbia University and University of Chicago, Monday, January 26, 2015. I will use these terms—*privacy, autonomy,* and *anonymity*—interchangeably and as shorthand to refer to the bundle of valuables that includes not only privacy and autonomy but also at times a certain anonymity (the ability to not be identified), some secrecy, and the right to be left alone and to not have to explain why. I subscribe here to a copious notion of privacy that has several dimensions. As Eben Moglen suggests—and I agree—our usual notion of "privacy" contains a number of aspects: "First is secrecy, or our ability to keep the content of our messages known only to those we intend to receive them. Second is anonymity, or secrecy about who is sending and receiving messages, where the content of the messages may not be secret at all. It is very important that anonymity is an interest we can have both in our publishing and in our reading. Third is autonomy, or our ability to make our own life decisions free from any force that has violated our secrecy or our anonymity. These three—secrecy, anonymity and autonomy—are the principal components of a mixture we call privacy.'" Eben Moglen, "Privacy under Attack: The

NSA Files Revealed New Threats to Democracy," *Guardian,* Tuesday May 27, 2014. In this respect, I also appreciate and agree with Daniel Solove's multidimensional view of privacy, along the lines of Wittgenstein's "family resemblances," as well as Jed Rubenfeld's intervention to make us see the special importance of anonymity in the digital age. See generally Daniel Solove, *Nothing to Hide* (New Haven, CT: Yale University Press, 2011), 24–25; Jed Rubenfeld, "We Need a New Jurisprudence of Anonymity," *Washington Post,* January 12, 2014.

6. Living a full and rewarding life resembled, far more than today, reading a novel—which, in the words of D. A. Miller, "takes for granted the existence of a space in which the reading subject remains safe from the surveillance, suspicion, reading and rape of others." Miller, *The Novel and the Police,* 162.

7. See generally Bernard E. Harcourt, *The Illusion of Free Markets* (Cambridge, MA: Harvard University Press, 2011).

8. Siva Vaidhyanathan, *The Googlization of Everything (and Why We Should Worry)* (Berkeley: University of California Press, 2011), xii.

9. Eric Schmidt, quoted in Vaidhyanathan, *The Googlization of Everything,* 45.

10. Jean-Paul Sartre, *Existentialism Is a Humanism,* trans. Carol Macomber (New Haven, CT: Yale University Press, 2007).

11. Mark Poster, *Existential Marxism in Postwar France* (Princeton, NJ: Princeton University Press, 1975), 90.

12. Chris Thornhill, "Humanism and Wars: Karl Jaspers between Politics, Culture and Law," in Helmut Wautischer, Alan M. Olson, and Gregory J. Walters, eds., *Philosophical Faith and the Future of Humanity* (New York: Springer, 2012), 299–319.

13. Chris Butler, *Henri Lefebvre: Spatial Politics, Everyday Life and the Right to the City* (New York: Routledge, 2012), 14, 21.

14. François Mauriac's intimate journals are a testament to these humanist interventions. See François Mauriac, *Oeuvres autobiographiques complètes,* ed. François Durand (Paris: Gallimard / Bibliothèque de la Pléiade, 1990).

15. In fact, humanism was so hegemonic at the time that many found it suffocating. See Foucault, "Interview with André Berten," in Michel Foucault, *Wrong-Doing, Truth-Telling: The Function of Avowal in Justice* (Chicago: University of Chicago Press, 2014), 264–265. See also "Francois Mauriac (1885–1970)," in Lawrence D. Kritzman, ed., *The Columbia History of Twentieth-Century French Thought* (New York: Columbia University Press, 2006), 607–608.

16. Carl J. Richard, *The Battle for the American Mind: A Brief History of a Nation's Thought* (Lanham, MD: Rowman and Littlefield, 2006), 311–312.

17. Andrew R. Heinze, *Jews and the American Soul: Human Nature in the Twentieth Century* (Princeton, NJ: Princeton University Press, 2004), 273–274.

18. Hannah Arendt, *The Human Condition* (Chicago: University of Chicago Press, 1998 [1958]); see generally Seyla Benhabib, *The Reluctant Modernism of Hannah Arendt* (New York: Rowman and Littlefield, 2003); Lewis P. Hinchman and Sandra K. Hinchman, "In Heidegger's Shadow: Hannah Arendt's Phenomenological Humanism, *Review of Politics* 46, no. 2 (April 1984): 205–206.

19. Deborah Nelson, *Pursuing Privacy in Cold War America* (New York: Columbia University Press, 2002), 9.

20. Ibid.

21. *Papachristou v. City of Jacksonville*, 405 U.S. 156 (1972).

22. Ibid., 156.

23. Ibid.

24. Ibid., 163.

25. Ibid., 164.

26. Ibid., 164 n. 7 (quoting Henry David Thoreau, *Excursions* [1893], 251–252).

27. *Griswold v. Connecticut*, 381 U.S. 479, 484 (1965).

28. *Katz v. United States*, 389 U.S. 347 (1967).

29. *Chimel v. California*, 395 U.S. 752 (1969). Deborah Nelson documents and traces the expansion in the scope of privacy—from Samuel Warren and Louis Brandeis's famous 1890 law review article ("The Right to Privacy," *Harvard Law Review* 4, no. 5 [December 15, 1890]: 193) through the impact of feminism in the 1970s. She demonstrates well how the "universalist" conception of the privacy-bearing subject—originally formulated by Warren and Brandeis, developed in Justice Douglas's 1957 *The Right to Privacy*, and culminating in the "zone of privacy" invoked in the *Griswold* decision—was gradually particularized but extended to other groups. It eventually "expanded beyond the home" and the "domestic autonomy" of the heterosexual marital bond. See Nelson, *Pursuing Privacy in Cold War America*, 21.

30. *United States v. United States District Court*, 407 U.S. 297 (1972) ("The danger to political dissent is acute where the Government attempts to act under so vague a concept as the power to protect 'domestic security.' Given the difficulty of defining the domestic security interest, the danger of abuse in acting to protect that interest becomes apparent").

31. This is reflected, for instance, in the Second Circuit's opinion on the legality of the NSA's bulk telephony metadata program under section 215 of the USA PATRIOT Act in *ACLU v. Clapper*, issued when this book was in page proofs. In its opinion dated May 7, 2015, the Second Circuit ruled that the NSA program was illegal, but declined to issue a preliminary injunction, and concluded its opinion on precisely this theme of the balancing of interests: "This case serves as an example of the increasing complexity of balancing the paramount interest in protecting the

security of our nation—a job in which, as the President has stated, "actions are second-guessed, success is unreported, and failure can be catastrophic," Remarks by the President on Review of Signals Intelligence—with the privacy interests of its citizens in a world where surveillance capabilities are vast and where it is difficult if not impossible to avoid exposing a wealth of information about oneself to those surveillance mechanisms. Reconciling the clash of these values requires productive contribution from all three branches of government, each of which is uniquely suited to the task in its own way." *ACLU v. Clapper*, Slip Opinion in 14-42 (2d Circuit, May 7, 2015), 96–97.

32. *United States v. Adel Daoud*, 755 F.3d. 479, 483 (7th Cir., June 16, 2014).

33. Wendy Brown does a brilliant job of tracing this history and its consequences in her book *Undoing the Demos: Neoliberalism's Stealth Revolution* (Cambridge, MA: Zone Books, 2015).

34. Naomi Klein, *The Shock Doctrine: The Rise of Disaster Capitalism* (New York: Metropolitan Books, 2007); David Harvey, *A Brief History of Neoliberalism* (Oxford: Oxford University Press, 2005); Noam Chomsky, *Profit over People: Neoliberalism and Global Order* (New York: Seven Stories Press, 1999).

35. See Klein, *The Shock Doctrine*; see also "Cheney/Halliburton Chronology," *Halliburton Watch*, www.halliburtonwatch.org/about_hal/chronology.html.

36. Often, the neoconservative rhetoric is in fact not faithful to the original writings of Hayek and others. See Bernard E. Harcourt, "How Paul Ryan Enslaves Friedrich Hayek's *The Road to Serfdom*," *Guardian*, September 12, 2012, http://www.theguardian.com/commentisfree/2012/sep/12/paul-ryan-enslaves-friedrich-hayek-road-serfdom. And see, generally, Harcourt, "The Chicago School," chap. 6 in *The Illusion of Free Markets*, 121–150.

37. See also *United States v. Jacobsen*, 466 U.S. 109 (1984); *United States v. Miller*, 425 U.S. 435, 443 (1976).

38. *Kyllo v. United States*, 533 U.S. 27, 38 (2001) ("The Agema Thermovision 210 might disclose, for example, at what hour each night the lady of the house takes her daily sauna and bath—a detail that many would consider 'intimate' ").

39. *California v. Greenwood*, 486 U.S. 35 (1988); *Maryland v. King*, 569 U.S. ____ (2013).

40. Those traditional conceptions of privacy—privacy as self-definition, privacy as individual dignity, and privacy as freedom—are all deeply humanist. For a discussion of those conceptions, see generally Robert Post, "Three Concepts of Privacy," *Georgetown Law Journal* 89 (2001): 2087–2098; Daniel J. Solove, "A Taxonomy of Privacy," *University of Pennsylvania Law Review* 154, no. 3 (2006): 477–566.

41. See generally Nelson, *Pursuing Privacy in Cold War America*; Arthur Miller, *The Assault on Privacy* (Ann Arbor: University of Michigan Press, 1971); Jon

L. Mills, *Privacy: The Lost Right* (Oxford: Oxford University Press, 2008); John C. Lautsch, "Computers, Communications and the Wealth of Nations: Some Theoretical and Policy Considerations about an Informational Economy," *Computer/Law Journal* 4, no. 1 (1983): 101–132; Daniel Solove, "A Brief History of Information Privacy Law," George Washington University Law School Public Law Research Paper No. 215, 2006, http://scholarship.law.gwu.edu/faculty_publications/923; Secretary's Advisory Committee on Automated Personal Data Systems, *Records, Computers and the Rights of Citizens* (Washington, DC: U.S. Department of Health, Education and Welfare, 1973).

42. Tim Hwang and Karen Levy, "'The Cloud' and Other Dangerous Metaphors," *Atlantic*, January 20, 2015.

43. Alan Rusbridger and Ewen MacAskill, "I, Spy: Edward Snowden in Exile," *Guardian*, July 19, 2014; see also Dustin Volz, "Snowden: NSA Employees Are Passing around Nude Photos," *National Journal*, July 17, 2014.

44. Ibid.

45. Ibid.

46. Alina Selyukh, "NSA Staff Used Spy Tools on Spouses, Ex-Lovers: Watchdog," Reuters, September 27, 2013.

47. Solove, *Nothing to Hide*.

Chapter Seven: THE COLLAPSE OF STATE, ECONOMY, AND SOCIETY

1. For treatments of these earlier dilemmas, see, e.g., Edward A. Shils, *The Torment of Secrecy: The Background and Consequences of American Security Policies* (Glencoe, IL: Free Press, 1956); Morton Grodzins, *The Loyal and the Disloyal* (Chicago: University of Chicago Press, 1956).

2. Max Weber, *Economy and Society*, 2 vols. (Berkeley: University of California Press, 1978).

3. Evgeny Morozov, "Who's the True Enemy of Internet Freedom—China, Russia, or the US?," *Guardian*, January 3, 2015.

4. Matt Williams, "Apple Blocks 'Objectionable' App That Reports Deaths from US Drone Strikes," *Guardian*, August 30, 2012.

5. Josh Begley, "Drones + iPhone App," August 21, 1990, http://vimeo.com/47976409.

6. Williams, "Apple Blocks 'Objectionable' App." See also Lorenzo Franceschi-Bicchierai, "After 5 Rejections, Apple Accepts App That Tracks U.S. Drone Strikes," Mashable, February 7, 2014.

7. Franceschi-Bicchierai, "After 5 Rejections, Apple Accepts App."

8. Williams, "Apple Blocks 'Objectionable' App."

9. Josh Begley, Dronestre.am, http://dronestre.am (accessed December 23, 2014).

10. Franceschi-Bicchierai, "After 5 Rejections, Apple Accepts App."

11. Ibid.

12. Ibid.

13. Ibid.

14. "Obama Administration Lifts Blanket Ban on Media Coverage of the Return of Fallen Soldiers," National Security Archive, George Washington University, February 26, 2009, http://nsarchive.gwu.edu/news/20090226.

15. Ibid.

16. Elisabeth Bumiller, "U.S. Lifts Photo Ban on Military Coffins," *New York Times,* December 7, 2009.

17. See http://drones.pitchinteractive.com. For additional reporting on the drone wars, see Steve Coll, "A Reporter at Large: The Unblinking Stare: The Drone War in Pakistan," *New Yorker,* November 24, 2014.

18. Austin Ramzy, "Facebook Deletes Post on Tibetan Monk's Self-Immolation," *New York Times,* December 27, 2014.

19. Ibid.

20. Ibid.

21. *Benjamin Joffe et al. v. Google, Inc.,* 746 F.3d 920 (9th Cir. 2013). As the Ninth Circuit noted in note 1, "Google may have also used its software to capture encrypted data, but the plaintiffs have conceded that their wireless networks were unencrypted."

22. Ibid., 923.

23. Ibid.

24. Ibid. As Google explains in its brief to the United States Supreme Court: "In addition to collecting identifying information about Wi-Fi networks, Google's Street View cars also collected so-called 'payload data' that was sent over unencrypted Wi-Fi networks if the data was being broadcast at the moment the Street View cars passed within range of the networks. App. 4a. Google did not use any of this data in any product or service. Upon learning of the collection of payload data, Google took its Street View cars off the road and segregated the payload data the cars had collected." *Google, Inc. v. Joffe et al.,* Petition for Writ of Certiorari, United States Supreme Court, No. 13-____, 5, www.documentcloud.org/documents/1100394 -google-v-joffe.html.

25. *Google, Inc. v. Joffe et al.,* Petition for Writ of Certiorari, United States Supreme Court, No. 13-____.

26. Ibid., 22.

27. Ibid.

28. Charles Duhigg, "How Companies Learn Your Secrets," *New York Times*, February 16, 2012.

29. Ibid.

30. See Ira S. Rubinstein, R. D. Lee, and P. M. Schwartz, "Data Mining and Internet Profiling: Emerging Regulatory and Technological Approaches," *University of Chicago Law Review* 75 (2008): 273 ("Once a person discloses information to a third party, as she does when requesting a URL or when running search queries, she relinquishes any reasonable expectation of constitutional privacy she has in that information. . . . Individuals have a right to keep their information secluded, but once they share it with others [i.e., reveal it via the Internet], privacy rights end"). See also James Risen and Laura Poitras, "N.S.A. Gathers Data on Social Connections of U.S. Citizens," *New York Times*, September 28, 2013.

31. Duhigg, "How Companies Learn Your Secrets."

32. Marc Parry, "Big Data on Campus," *New York Times*, July 18, 2012.

33. Regarding the trucking industry, see Chris Baraniuk, "Haulin' Data: How Trucking Became the Frontier of Work Surveillance," *Atlantic*, November 18, 2013, and, more generally, Karen Levy, "The Automation of Compliance: Techno-legal Regulation in the U.S. Trucking Industry" (PhD diss., Sociology Department, Princeton University, 2014); regarding tracked nurses, see Jill A. Fisher and Torin Monahan, "Evaluation of Real-Time Location Systems in Their Hospital Contexts," *International Journal of Medical Informatics* 81 (2012): 705–712, and Torin Monahan and Jill A. Fisher, "Implanting Inequality: Empirical Evidence of Social and Ethical Risks of Implantable Radio-Frequency Identification (RFID) Devices," *International Journal of Technology Assessment in Health Care* 26, no. 4 (2010): 370–376.

34. Sarah O'Connor, "Amazon Unpacked: The Online Giant Is Creating Thousands of UK Jobs, So Why Are Some Employees Less than Happy?," *Financial Times Magazine*, February 8, 2013.

35. Simon Head, "Worse than Wal-Mart: Amazon's Sick Brutality and Secret History of Ruthlessly Intimidating Workers," *Salon*, February 23, 2014.

36. Ibid.

37. Detlev Zwick and Janice Denegri Knott, "Manufacturing Customers: The Database as a New Means of Production," *Journal of Consumer Culture* 9, no. 2 (2009): 226.

38. Ibid., 228.

39. Ibid., 223; see also Greg Elmer, *Profiling Machines: Mapping the Personal Information Economy* (Cambridge, MA: MIT Press, 2004), 55.

40. Zwick and Knott, "Manufacturing Customers," 273.

41. Statement of Sen. John D. (Jay) Rockefeller IV at the Senate Hearings on December 18, 2013, "What Information Do Data Brokers Have on Consumers, and

How Do They Use It?," U.S. Senate Committee on Commerce, Science, and Transportation; see also Yasha Levine, "You'll Be Shocked at What 'Surveillance Valley' Knows about You," AlterNet, January 8, 2014, www.alternet.org/media/what-surveillance-valley-knows-about-you.

42. Levine, "You'll Be Shocked."

43. Frank Pasquale, "The Dark Market for Personal Data," *New York Times,* October 17, 2014.

44. Elizabeth Dwoskin, "Data Broker Removes Rape-Victims List after Journal Inquiry," *Digits* blog, *Wall Street Journal,* December 19, 2013.

45. The website is preserved here: https://www.evernote.com/shard/s1/sh/235f0aab-785b-4a10-bf3b-f39b3fd0dec7/ed697225af44a19f18240183df03cd0f.

46. Dwoskin, "Data Broker Removes Rape-Victims List."

47. Statement of Sen. John D. (Jay) Rockefeller IV at the Senate Hearings on December 18, 2013.

48. Ibid.

49. Charles Duhigg, "Bilking the Elderly, with a Corporate Assist," *New York Times,* May 20, 2007; Frank Pasquale discusses and develops this in his book *The Black Box Society: The Secret Algorithms behind Money and Information* (Cambridge, MA: Harvard University Press, 2015).

50. Duhigg, "Bilking the Elderly, with a Corporate Assist."

51. Natasha Singer, "Mapping, and Sharing, the Consumer Genome," *New York Times,* June 16, 2012.

52. Ibid.

53. Ibid.

54. Lois Beckett, "Everything We Know about What Data Brokers Know about You," ProPublica, June 13, 2014.

55. Ibid.

56. Julia Angwin, "Why Online Tracking Is Getting Creepier: The Merger of Online and Offline Data Is Bringing More Intrusive Tracking," ProPublica, June 12, 2014.

57. Pam Dixon and Robert Gellman, "The Scoring of America: How Secret Consumer Scores Threaten Your Privacy and Your Future," World Privacy Forum, April 2, 2014, www.worldprivacyforum.org/wp-content/uploads/2014/04/WPF_Scoring_of_America_April2014_fs.pdf.

58. Ibid., 6.

59. Ibid., 9.

60. Ibid., 47.

61. Ibid., 50.

62. Ibid., 84.

63. Emily Steel, "Companies in Scramble for Consumer Data." *Financial Times,* June 12, 2013.

64. Emily Steel, "Financial Worth of Data Comes In at under a Penny a Piece," *Financial Times,* June 12, 2013.

65. Steel, "Companies in Scramble for Consumer Data." See also Emily Steel, Callum Locke, Emily Cadman, and Ben Freese, "How Much Is Your Personal Data Worth?," *Financial Times,* June 12, 2013.

66. Steel, "Financial Worth of Data."

67. Steel, "Companies in Scramble for Consumer Data."

68. Natasha Singer, "Shoppers Who Can't Have Secrets," *New York Times,* May 1, 2010. See also Stephanie Clifford and Quentin Hardy, "Attention, Shoppers: Store Is Tracking Your Cell," *New York Times,* July 14, 2013.

69. Levine, "You'll Be Shocked."

70. Jen Wieczner, "How the Insurer Knows You Just Stocked Up on Ice Cream and Beer," *Wall Street Journal,* February 25, 2013.

71. Alexander Furnas, "Everything You Wanted to Know about Data Mining but Were Afraid to Ask," *Atlantic,* April 3, 2012.

72. David Lyon, "Surveillance Technology and Surveillance Society," in T. Misa, P. Brey, and A. Feenberg, eds., *Modernity and Technology* (Cambridge, MA: MIT Press, 2003), 172.

73. Glenn Greenwald, *No Place to Hide: Edward Snowden, the NSA, and the U.S. Surveillance State* (New York: Henry Holt, 2014), 81; transcript of President Obama's interview with PBS host Charlie Rose on June 17, 2013, available at Buzzfeed, "President Obama Defends NSA Spying," www.buzzfeed.com/buzzfeedpolitics/president-obama-defends-nsa-spying.

74. Ibid.

75. See Greenwald, *No Place to Hide,* 129; transcript of President Obama's interview with PBS host Charlie Rose on June 17, 2013; see also Louis Jacobson, "Barack Obama Says the Foreign Intelligence Surveillance Court 'Is Transparent,'" Politifact.com, June 21, 2013, www.politifact.com/truth-o-meter/statements/2013/jun/21/barack-obama/barack-obama-says-foreign-intelligence-surveillanc.

76. See generally Bloomberg.com.

77. Eben Moglen, "Privacy under Attack," *Guardian,* May 27, 2014.

78. David Easton, *The Political System: An Inquiry into the State of Political Science* (New York: Knopf, 1953), 112.

79. Rémi Lenoir, "L'état selon Pierre Bourdieu," *Sociétés contemporaines* 87 (2012–2013): 126.

80. Pierre Bourdieu, *Sur l'état* (Paris: Le Seuil, 2012), 25.

81. Ibid., 53. By way of example, Bourdieu focuses on the calendar and daylight savings time, which produce effects of reality on social existence. Ultimately, in his later lectures in 1990–1991, Bourdieu crystallizes this thought into a model of symbolic force that he uses to supplement Weber's definition: "Max Weber says that the State is the monopoly over the legitimate use of violence. And I correct that, by saying: it is the monopoly over the legitimate use of physical and symbolic violence"; ibid., 545. His early deconstruction of the notion of the state, though, is perhaps more useful: the state as the unspoken glue that binds us together.

82. See Theda Skocpol, "Five Minutes with Theda Skocpol: 'Even Those on the American Centre-Left Are Now Viewing Europe in a Negative Sense Because of Austerity,'" London School of Economics and Political Science, November 7, 2013, http://blogs.lse.ac.uk/europpblog/2013/05/13/interview-theda-skocpol-climate-change-tea-party-state-theory-europe. "The work that I did with colleagues that was embodied in the book, *Bringing the State Back In,* was in many ways drawing on the traditions of Max Weber, and other—especially German—theorists who in dialogue with Marxists argued that we need to take the organisations of the state more seriously in their own right."

83. Theda Skocpol, *Protecting Soldiers and Mothers: The Political Origins of Social Policy in the United States* (Cambridge, MA: Harvard University Press, 1995), viii, x.

84. Skocpol, "Five Minutes with Theda Skocpol."

85. Timothy Mitchell, "The Limits of the State: Beyond Statist Approaches and Their Critics," *American Political Science Review* 85, no. 1 (March 1991): 77–96.

86. Ibid., 90.

87. Ibid.

88. Ibid., 77–96.

89. Ibid., 90.

90. Ibid.

91. Ibid., 91.

92. Stephen Sawyer has an excellent paper, "Foucault and the State," which he presented at the Neubauer Collegium at the University of Chicago on May 16, 2014; the audio is at https://www.youtube.com/watch?v=RMisoCoJUko.

93. Ibid.

94. Ibid., 215 n. a ("Ce n'est pas un appareil d'État, c'est un appareil pris dans le noeud étatique. Un système infra-étatique").

95. Michel Foucault, *Discipline and Punish,* trans. Alan Sheridan (New York: Vintage, 1979), 308. In the original French, it is a footnote on the last page. See *Surveiller et punir: La naissance de la prison* (Paris: Gallimard, 1975), 315 n. 1.

96. Michel Foucault, *The Birth of Biopolitics: Lectures at the Collège de France, 1978–79,* trans. Graham Burchell (New York: Palgrave Macmillan, 2008); see also Wendy Brown, *Undoing the Demos: Neoliberalism's Stealth Revolution* (New York: Zone Books, 2015).

97. Barack Obama, "Speech on N.S.A. Phone Surveillance," transcript, January 17, 2014, *New York Times.*

98. Greenwald, *No Place to Hide,* 134–135.

99. Ibid., 135.

100. Obama, "Speech on N.S.A. Phone Surveillance."

Chapter Eight: THE MORTIFICATION OF THE SELF

1. Shanyang Zhao, "The Digital Self: Through the Looking Glass of Telecopresent Others," *Symbolic Interaction* 28, no. 3 (2005): 392.

2. Keith Hampton, Lee Rainie, Weixu Lu, Maria Dwyer, Inyoung Shin, and Kristen Purcell, "Social Media and the 'Spiral of Silence,'" Pew Research Internet Project, August 26, 2014, www.pewinternet.org/2014/08/26/social-media-and-the-spiral-of-silence.

3. M. J. Smith, P. Carayon, K. J. Sanders, S-Y. Lim, and D. LeGrande, "Employee Stress and Health Complaints in Jobs with and without Electronic Performance Monitoring," *Applied Ergonomics* 23, no. 1 (February 1992): 26; see also G. L. Rafnsdóttir and M. L. Gudmundsdottir, "EPM Technology and the Psychosocial Work Environment," *New Technology, Work and Employment* 26, no. 3 (2011): 210–221; Kate Murphy, "We Want Privacy, but Can't Stop Sharing," *New York Times,* October 4, 2014.

4. Murphy, "We Want Privacy, but Can't Stop Sharing."

5. See, e.g., Shelley T. Duval and Robert A. Wicklund, *A Theory of Objective Self-Awareness* (New York: Academic Press, 1972); Paul J. Silvia and Shelley T. Duval, "Objective Self-Awareness Theory: Recent Progress and Enduring Problems," *Personality and Social Psychology Review* 5 (2001): 230–241.

6. Murphy, "We Want Privacy, but Can't Stop Sharing." Glenn Greenwald chronicles some of the social science data as well in his book *No Place to Hide: Edward Snowden, the NSA, and the U.S. Surveillance State* (New York: Henry Holt, 2014), 178–181.

7. Michael McCahill and Rachel Finn, "The Social Impact of Surveillance in Three UK Schools: 'Angels,' 'Devils' and 'Teen Mums,'" *Surveillance and Society* 7, nos. 3–4 (2010): 273–289, quotation p. 278. All further references to this article will be in parentheses.

8. Gregory L. White and Philip G. Zimbardo, "The Chilling Effects of Surveillance: Deindividuation and Reactance," Department of Psychology, Stanford University, 1975.

9. PEN American Center, "Chilling Effects: NSA Surveillance Drives U.S. Writers to Self-Censor," November 12, 2013, www.pen.org/sites/default/files/Chilling%20Effects_PEN%20American.pdf.

10. See generally Jill A. Fisher and Torin Monahan, "Evaluation of Real-Time Location Systems in Their Hospital Contexts," *International Journal of Medical Informatics* 81 (2012): 705–712; Torin Monahan and Jill A. Fisher, "Implanting Inequality: Empirical Evidence of Social and Ethical Risks of Implantable Radio-Frequency Identification (RFID) Devices," *International Journal of Technology Assessment in Health Care* 26, no. 4 (2010): 370–376; Karen Levy, "The Automation of Compliance: Techno-legal Regulation in the U.S. Trucking Industry" (Ph.D. diss., Department of Sociology, Princeton University, 2014).

11. Daniel Soar, "It Knows," *London Review of Books,* October 6, 2011, 3–6.

12. Daniel Solove, *Nothing to Hide* (New Haven, CT: Yale University Press, 2011), 26.

13. Ibid., 27.

14. Ibid.

15. Julia Angwin, "Why Online Tracking Is Getting Creepier: The Merger of Online and Offline Data Is Bringing More Intrusive Tracking," ProPublica, June 12, 2014, www.propublica.org/article/why-online-tracking-is-getting-creepier.

16. Erving Goffman, *Asylums* (Piscataway, NJ: Transaction, 2007), xxii.

17. For an analysis of mass institutionalization in asylums and later in prisons, see Bernard E. Harcourt, "From the Asylum to the Prison: Rethinking the Incarceration Revolution," *Texas Law Review* 84 (2006): 1751–1786; Bernard E. Harcourt, "An Institutionalization Effect: The Impact of Mental Hospitalization and Imprisonment on Homicide in the United States, 1934–2001," *Journal of Legal Studies* 40 (2011): 39–83.

18. See Michel Foucault, *History of Madness* (London: Routledge, 2006); David Rothman, *The Discovery of the Asylum: Social Order and Disorder in the New Republic* (New York: Little, Brown, 1971); R. D. Laing and Aaron Esterson, *Sanity, Madness, and the Family: Families of Schizophrenics* (London: Routledge, 1964); David Cooper, ed., *Psychiatry and Anti-psychiatry* (London: Paladin, 1967); Thomas S. Szasz, *The Myth of Mental Illness* (New York: Hoeber-Harper, 1961).

19. Goffman, *Asylums,* 19.

20. Ibid., 33.

21. Ibid., 38–39.

22. Ibid., 48.

23. Ibid., 62.

24. Ibid.

25. In this regard, see the exchange between Steven Lukes, "In Defense of 'False Consciousness,'" *University of Chicago Legal Forum*, 2011, 19–28, and Bernard E. Harcourt, "Radical Thought from Marx, Nietzsche, and Freud, through Foucault, to the Present: Comments on Steven Lukes's 'In Defense of False Consciousness,'" *University of Chicago Legal Forum*, 2011, 29–51.

26. Gilles Deleuze and Félix Guattari, *Anti-Oedipus: Capitalism and Schizophrenia* (Minneapolis: University of Minnesota Press, 1983), 104–105.

27. Zhao, "The Digital Self," 402–403.

28. Gilles Deleuze, quoted in "Les intellectuels et le pouvoir" (discussion between Michel Foucault and Gilles Deleuze, March 4, 1972, *L'Arc*, no. 49), in Foucault, *Dits et écrits* (Paris: Gallimard, 1994), 2:306–315 (Quarto ed. [Paris: Gallimard, 2001], 1:1182); I am giving a slightly modified translation based on *Language, Counter-Memory, Practice: Selected Essays and Interviews by Michel Foucault*, ed. Donald F. Bouchard (Ithaca, NY: Cornell University Press, 1980). And to which Foucault would respond: "It is possible that the struggles now taking place and the local, regional, and discontinuous theories that derive from these struggles and that are indissociable from them, stand at the threshold of our discovery of the manner in which power is exercised." Foucault and Deleuze, "Les intellectuels et le pouvoir."

29. Deleuze and Guattari, *Anti-Oedipus*, 139.

30. Ibid., 105

31. Kevin D. Haggerty and Richard V. Ericson, "The Surveillant Assemblage," *British Journal of Sociology* 51, no. 4 (2000): 606.

32. Alan Rusbridger and Ewen MacAskill, "I, Spy: Edward Snowden in Exile," *Guardian*, July 19, 2014; see also Dustin Volz, "Snowden: NSA Employees Are Passing around Nude Photos," *National Journal*, July 17, 2014.

33. Associated Press, "British Spies Reportedly Snoop on Video Chats, X-Rated Webcam Pics," CBS News, February 27, 2014.

34. Tarleton Gillespie, "The Relevance of Algorithms," in Tarleton Gillespie, Pablo Boczkowski, and Kirsten Foot, eds., *Media Technologies* (Cambridge MA: MIT Press, forthcoming).

35. Ibid.

36. Amy L. Gonzales and Jeffrey T. Hancock, "Mirror, Mirror on My Facebook Wall: Effects of Exposure to Facebook on Self-Esteem," *Cyberpsychology, Behavior, and Social Networking* 14, nos. 1–2 (2011): 79–83.

37. Notice that it is only one of the more entitled young boys in the U.K. study who says about CCTV outside of school, "I never notice them but I don't really mind because I'm not doing anything wrong. . . . So, I've got nothing to be worried about" (279).

38. Goffman, *Asylums,* 320.

39. According to a survey conducted in October and November 2014, more than a third of American Internet users (36 percent) "share personal information with private companies online all the time and say 'it's no big deal'"; about 35 percent of American Internet users believe that "the chance of their personal information being compromised [on the Internet] is so small that it's not worth worrying about." See Centre for International Governance Innovation and IPSOS, "CIGI-IPSOS Global Survey on Internet Security and Trust," survey carried out between October 7 and November 12, 2014, www.cigionline.org/internet-survey.

Chapter Nine: THE STEEL MESH

1. "Highest to Lowest," International Centre for Prison Studies, 2014, www .prisonstudies.org/highest-to-lowest.

2. See, generally, Bernard E. Harcourt, *The Illusion of Free Markets* (Cambridge, MA: Harvard University Press, 2011).

3. "One in 100: Behind Bars in America 2008," Pew Research Center, February 2008, www.pewtrusts.org/en/research-and-analysis/reports/2008/02/28/one-in -100-behind-bars-in-america-2008. On page 5 of this study, the authors state that one in nine black men between the ages of twenty and thirty-four are incarcerated. This statistic was also cited in Adam Liptak, "1 in 100 U.S. Adults behind Bars, Study Shows," *New York Times,* February 28, 2008, which notes that the Pew report's finding is higher than the official Justice Department statistics because the latter "calculates the incarceration rate by using the total population rather than the adult population as the denominator. Using the department's methodology, about one in 130 Americans is behind bars." According to the 2011 Justice Department report, the incarceration rate of black men twenty to thirty-four years old in 2011 was roughly 6.367 percent (note that the 2011 data from the Justice Department only include incarceration rates per 100,000 for sentences of one year or more).

4. Laura M. Maruschak and Erika Parks, "Probation and Parole in the United States, 2011," Bureau of Justice Statistics, November 2012, 3, 7, www.bjs.gov/content /pub/pdf/ppus11.pdf. The total population of black men in 2011 was 19,463,223, according to U.S. Census estimates, and 2,144,787 divided by 19,463,223 is 0.11, so that's 11 percent of the black male population in 2011, about one of every nine individuals.

5. The prison today recalls the premodern models of the seventeenth and eighteenth centuries, depicted and described by Foucault in his lecture of March 21, 1973; see Michel Foucault, *La société punitive: Cours au Collège de France (1972–1973),* ed. Bernard E. Harcourt (Paris: Gallimard/Le Seuil, 2013), 213–214.

6. I agree with Loïc Wacquant that we should differentiate between rates of incarceration among different demographic groups and would be well advised to refer to mass incarceration as "the hyperincarceration of (sub)proletarian African American men from the imploding ghetto"; Loïc Wacquant, "Class, Race and Hyperincarceration in Revanchist America," *Daedalus* 139, no. 3 (2010): 74. However, I also agree with Marie Gottschalk that the problem of overincarceration affects practically every demographic group, even if at different levels. See Gottschalk, comments at Harvard University, December 5, 2013, www.youtube.com/watch?v =sOl503npe30. To keep matters straight, I will refer to the problem of "massive overincarceration."

7. Peter Baehr, "The 'Iron Cage' and the 'Shell as Hard as Steel': Parsons, Weber, and the *Stahlhartes Gehäuse* Metaphor in *The Protestant Ethic and the Spirit of Capitalism*," *History and Theory* 40, no. 2 (2001): 153–169.

8. See Matthew DeMichele and Brian Payne, *Offender Supervision with Electronic Technology: Community Corrections Resource*, 2nd ed., Office of Justice Programs, Bureau of Justice Assistance, Department of Justice, 2009, www.appa-net .org/eweb/docs/APPA/pubs/OSET_2.pdf.

9. Ibid., 17.

10. Ibid.

11. William Bales, Karen Mann, Thomas Blomberg, Gerry Gaes, Kelle Barrick, Karla Dhungana, and Brian McManus, "A Quantitative and Qualitative Assessment of Electronic Monitoring," Office of Justice Program, National Institute of Justice, U.S. Department of Justice, January 2010, www.ncjrs.gov/pdffiles1/nij /grants/230530.pdf.

12. Michael D. Abernethy, "Someone's Watching: Electronic Monitoring on the Rise, Better Technology and Newer State Laws Driving," *Times-News* (Burlington, NC), June 21, 2014.

13. Tessa Stuart, "There Has Been a 4,000% Increase in Ankle-Monitoring of Women in New York This Year," *Village Voice*, September 5, 2014.

14. Peter Hirschfeld, "State Eyes Electronic Monitoring as Alternative to Incarceration," Vermont Public Radio, June 17, 2014.

15. Trent Cornish and Jay Whetzel, "Location Monitoring for Low-Risk Inmates—A Cost Effective and Evidence-Based Reentry Strategy," *Federal Probation* 78, no. 1 (June 2014): 19–21.

16. Joseph Shapiro, "As Court Fees Rise, the Poor Are Paying the Price," NPR, May 19, 2014; for a graph comparing the common fees that different state courts are charging to defendants and offenders, see "State-by-State Court Fees," NPR, May 19, 2014.

17. See http://bi.com/news.

18. Associated Press, "Electronic Monitoring in Domestic Violence Cases to Expand in Maine," *Portland Press Herald,* November 20, 2014.

19. See Addendum No. 1 to "Canyon County's Request for Proposals, Electronic Monitoring System—Alternative Sentencing Program," July 16, 2014, www.canyon county.org/getattachment/Elected-Officials/Commissioners/Legal-Notices-and -RFPs/Electronic-Monitoring-System-Alternative-Sentencin/Addendum-No-1 -to-Electronic-Monitoring-System-RFP.pdf.aspx.

20. David E. Sanger and Thom Shanker, "N.S.A. Devises Radio Pathway into Computers," *New York Times,* January 14, 2014. See also Floor Boon, Steven Derix, and Huib Modderkolk, "NSA Infected 50,000 Computer Networks with Malicious Software," Nrc.nl, November 23, 2012, www.nrc.nl/nieuws/2013/11/23/nsa-infected -50000-computer-networks-with-malicious-software.

21. Sanger and Shanker, "N.S.A. Devises Radio Pathway into Computers."

22. Glenn Greenwald, *No Place to Hide: Edward Snowden, the NSA, and the U.S. Surveillance State* (New York: Henry Holt, 2014), 117.

23. "Inside TAO: Documents Reveal Top NSA Hacking Unit," *Der Spiegel,* December 29, 2013. Other articles referenced in this one include "Quantum Spying: GCHQ Used Fake LinkedIn Pages to Target Engineers," *Der Spiegel,* November 11, 2013; "Oil Espionage: How the NSA and GCHQ Spied on OPEC," *Der Spiegel,* November 11, 2013.

24. "Inside TAO: Documents Reveal Top NSA Hacking Unit."

25. Ibid.

26. Jacob Applebaum, Judith Horchert, and Christian Stöcker, "Shopping for Spy Gear: Catalog Advertises NSA Toolbox," *Der Spiegel,* December 29, 2013.

27. Ibid.

28. Greenwald, *No Place to Hide,* 117–118; Sanger and Shanker, "N.S.A. Devises Radio Pathway into Computers."

29. Boon, Derix and Modderkolk, "NSA Infected 50,000 Computer Networks."

30. Joseph Goldstein and J. David Goodman, "Frisking Tactic Yields to Tighter Focus on Youth Gangs," *New York Times,* September 18, 2013.

31. Ibid.

32. Ibid.

33. Joseph Goldstein, "In Social Media Postings, a Trove for Investigators," *New York Times,* March 2, 2011.

34. Ibid.

35. Ibid.

36. Goldstein and Goodman, "Frisking Tactic Yields to Tighter Focus on Youth Gangs."

37. Chris Hamby, "Government Set up a Fake Facebook Page in This Woman's Name," BuzzFeed, October 6, 2014.

38. Ibid.

39. See, e.g., G. W. Schulz, "Virginia Police Have Been Secretively Stockpiling Private Phone Records," *Wired,* October 20, 2014.

40. Ibid.

41. Ibid.

42. Ibid.

43. Kenneth Neil Cukier and Viktor Mayer-Schoenberger, "The Rise of Big Data: How It's Changing the Way We Think about the World," *Foreign Affairs,* May–June 2013.

44. Peter Beaumont, "Israeli Intelligence Veterans Refuse to Serve in Palestinian Territories," *Guardian,* September 12, 2014; James Bamford, "Israel's N.S.A. Scandal," *New York Times,* September 16, 2014; Anshel Pfeffer, "Unit 8200 Refuseniks Shed Light on Ethics of Israel's Intel Gathering," *Haaretz,* September 15, 2014.

45. NSA memorandum agreement cited in Bamford, "Israel's N.S.A. Scandal"; the memorandum can be read at www.theguardian.com/world/interactive/2013/sep/11/nsa-israel-intelligence-memorandum-understanding-document. See generally Glenn Greenwald, Laura Poitras, and Ewen MacAskill, "NSA Shares Raw Intelligence Including Americans' Data with Israel," *Guardian,* September 11, 2013.

46. Bamford, "Israel's N.S.A. Scandal."

47. Beaumont, "Israeli Intelligence Veterans Refuse to Serve."

48. Ibid.

49. Bamford, "Israel's N.S.A. Scandal"; see also Beaumont, "Israeli Intelligence Veterans Refuse to Serve."

50. Glenn Greenwald, Ryan Grim, and Ryan Gallagher, "Top-Secret Document Reveals NSA Spied on Porn Habits as Part of Plan to Discredit 'Radicalizers,'" *Huffington Post,* January 24, 2014; Bamford, "Israel's N.S.A. Scandal"; see also Greenwald, *No Place to Hide,* 187.

51. Greenwald, Grim, and Gallagher, "Top-Secret Document Reveals NSA Spied."

52. Ibid.

53. Bamford, "Israel's N.S.A. Scandal."

54. Ewen MacAskill, "Greenwald Is Said to Have Second US Intelligence Source," Readersupportednews.org, October 12, 2014, http://readersupportednews.org/news-section2/318-66/26357-greenwald-is-said-to-have-second-us-intelligence-source; *Citizenfour,* dir. Laura Poitras, Weinstein Company (2014).

55. Baehr, "The 'Iron Cage' and the 'Shell as Hard as Steel,'" 158.

56. Ibid., 159.

57. Parsons, for instance, would mistranslate Weber's final reference to "the last men"—one could hardly gesture more directly to Nietzsche—as a bland historical reference to "the last stage of this cultural development." Ibid., 153.

58. Max Weber, *The Protestant Ethic and the Spirit of Capitalism,* trans. Talcott Parsons (London: George Allen and Unwin, 1930), 182.

59. Baehr, "The 'Iron Cage' and the 'Shell as Hard as Steel,'" 154.

60. Ibid., 157 n. 21 (quoting Arthur Mitzman's book on *The Iron Cage* and Eric Matthews).

Chapter Ten: VIRTUAL DEMOCRACY

1. Eben Moglen, "Privacy under Attack," *Guardian,* May 27, 2014 (my emphasis).

2. For the most piercing research and analysis of the Stasi regime in East Germany, see Andreas Glaeser, *Political Epistemics: The Secret Police, the Opposition, and the End of East German Socialism* (Chicago: University of Chicago Press, 2011). The film referred to here is Florian Henckel von Donnersmarck's *The Lives of Others* (2006), which presents a chilling fictionalized representation of life in the former GDR.

3. Quentin Skinner and Richard Marshall, "More Liberty, Liberalism and Surveillance: A Historic Overview," Opendemocracy.net, July 26, 2013 (emphasis added), www.opendemocracy.net/ourkingdom/quentin-skinner-richard-marshall/liberty-liberalism-and-surveillance-historic-overview.

4. As Albert W. Dzur notes in his insightful article "Repellent Institutions and the Absentee Public: Grounding Opinion in Responsibility for Punishment," in Jesper Ryberg and Julian V. Roberts, eds., *Popular Punishment: On the Normative Significance of Public Opinion* (Oxford: Oxford University Press, 2014), there is silence from the democratic theorists on these questions. The only close response that can be heard is from the moderate civil libertarians, like Thomas Friedman and Bill Keller: NSA surveillance is necessary to avoid another 9/11, because there could be nothing worse for civil liberties in this country than another terrorist attack on American soil. Or, in the words of Bill Keller, "the gravest threat to our civil liberties is not the N.S.A. but another 9/11-scale catastrophe that could leave a panicky public willing to ratchet up the security state, even beyond the war-on-terror excesses that followed the last big attack." Bill Keller, "Living with the Surveillance State," *New York Times,* June 16, 2013. In other words, we need to cede our civil liberties to protect them.

5. Alexis de Tocqueville, *Democracy in America,* 317.

6. Ibid.

7. Ibid.

8. Ibid., 318.

9. Ibid., 319.

10. Ibid.

11. See the United States Elections Project website at www.electproject.org/home (data available at "National General Election VEP Turnout Rates, 1789–Present," www.electproject.org/national-1789-present); see also Michael P. McDonald, "Turnout in the 2012 Presidential Election," *Huffington Post,* February 11, 2013.

12. See United States Elections Project, "2014 General Election Turnout Rates," www.electproject.org/Election-Project-Blog/2014generalelectionturnoutrates.

13. See United States Elections Project, "National General Election VEP Turnout Rates, 1789–Present."

14. Walter Dean Burnham and Thomas Ferguson, "Americans Are Sick to Death of Both Parties: Why Our Politics Is in Worse Shape than We Thought," AlterNet, December 18, 2014, www.alternet.org/americans-are-sick-death-both-parties-why -our-politics-worse-shape-we-thought.

15. Evgeny Morozov, *The Net Delusion: The Dark Side of Internet Freedom* (New York: Public Affairs, 2011), xiii (defining cyberutopianism as "a naive belief in the emancipatory nature of online communication that rests on a stubborn refusal to acknowledge its downside").

16. Dzur, "Repellent Institutions and the Absentee Public."

17. See generally Bernard E. Harcourt, "An Institutionalization Effect: The Impact of Mental Hospitalization and Imprisonment on Homicide in the United States, 1934–2001," *Journal of Legal Studies* 40, no. 1 (January 2011): 39–83; Bernard E. Harcourt, "From the Asylum to the Prison: Rethinking the Incarceration Revolution," *Texas Law Review* 84 (2006): 1751–1786.

18. See generally Sheldon S. Wolin, *Tocqueville between Two Worlds: The Making of a Political and Theoretical Life* (Princeton, NJ: Princeton University Press, 2001), 383.

19. Ibid., 385. Wolin contends that this could be generalized to political theory more generally. As he suggests, since Plato's *Laws,* most of the major political thinkers prior to Tocqueville had not theorized punishment much. Contemporary punishment theorists have excavated more of the punishment theory from the works of early modern political theory and would likely contest Wolin's generalization. Andrew Dilts has done important work on the central role of punishment theory in John Locke's work. See Andrew Dilts, "To Kill a Thief: Punishment, Proportionality, and Criminal Subjectivity in Locke's Second Treatise," *Political Theory* 40, no. 1 (February 2012): 58–83. Alice Ristroph has written important work on the central role of the right of self-defense to Hobbes's political theory. See Alice

Ristroph, "Respect and Resistance in Punishment Theory," *California Law Review* 97 (2009): 601. Keally McBride has written an important book on punishment and political theory, *Punishment and Political Order* (Ann Arbor: University of Michigan Press, 2007). See also Susan Meld Shell, "Kant on Punishment," *Kantian Review* 1 (March 1997): 115.

20. Cited in Wolin, *Tocqueville between Two Worlds,* 388.

21. Ibid.

22. Ibid., 389.

23. Ibid., 391.

24. Quoted in ibid., 394.

25. Quoted in ibid. Tocqueville's writings on the prison would have been a perfect illustration for Foucault of the disciplinary turn in the early nineteenth century: "The punishment is simultaneously the mildest and the most terrible that has ever been invented." Quoted in ibid., 395.

26. Ibid., 386.

27. Ibid., 387.

28. Foucault develops this theme in *La société punitive: Cours au Collège de France, 1972–1973* (Paris: Gallimard / Le Seuil, 2013), 34–39, 75 n. 2.

29. Vesla Weaver and Amy Lerman, "Political Consequences of the Carceral State," *American Political Science Review* 104, no. 4 (November 2010): 817, 827.

30. Holder called for reforms to the federal prison system that would have no real effect on mass incarceration: the elimination of some mandatory minimum federal sentencing policies for nonviolent offenders, as well as "compassionate release" or the early release of senior citizen and ill federal inmates (essentially, people who present no risk of harm to others, and cost a lot to incarcerate). See "Holder's Remarks at American Bar Association as Prepared for Delivery," *Talking Points Memo,* August 12, 2013, http://livewire.talkingpointsmemo.com/entry/read-ag-eric-holders-remarks-at-american-bar; the video is at www.c-spanvideo.org/program/DrugOff. See also Carri Johnson, "With Holder in the Lead, Sentencing Reform Gains Momentum," National Public Radio, August 7, 2013.

31. See Richard Stengel, Michael Scherer, and Radhika Jones, "Setting the Stage for a Second Term," *Time,* December 19, 2012.

32. President Obama balances his discussion of prisons with strong law-and-order rhetoric, and minimizes the problem of massive overincarceration, in contrast to his focus on the economy. See Stengel, Scherer and Jones, "Setting the Stage for a Second Term."

Chapter Eleven: DIGITAL RESISTANCE

1. Gilles Deleuze, "Postscript on the Societies of Control," *October* 59 (Winter 1992): 6.

2. Ibid.

3. Ibid., 4.

4. See Jennifer Lyn Morone, "Jennifer Lyn Morone, Inc.," Project for the Royal Academy of Art, di14.rca.ac.uk/project/jennifer-lyn-morone-inc.

5. Ibid.

6. Regine, "Jennifer Lyn Morone™ Inc., the Girl Who Became a Corporation," We Make Money Not Art, June 23, 2014, http://we-make-money-not-art.com /archives/2014/06/jennifer-lyn-morone-inc.php.

7. Ibid.

8. Morone, "Jennifer Lyn Morone, Inc."

9. P.H., "The Incorporated Woman," *Schumpeter* blog, *Economist,* June 27, 2014.

10. Similar, she notes, to Google's AdSense system. Regine, "Jennifer Lyn Morone™ Inc."

11. Ibid.

12. Ibid.

13. Morone, "Jennifer Lyn Morone, Inc."

14. P.H., "The Incorporated Woman."

15. Ibid.

16. Regine, "Jennifer Lyn Morone™ Inc."

17. Jennifer Lyn Morone, "Jennifer Lyn Morone, Inc." (video), June 15, 2014, http://vimeo.com/98300179.

18. Regine, "Jennifer Lyn Morone™ Inc."

19. Jennifer Lyn Morone, "Mouse Trap," April 14, 2013, available at http://vimeo .com/64022517.

20. Special thanks to Léonard Ruyter-Harcourt for his research assistance on WikiLeaks.

21. WikiLeaks released the video, along with transcripts and other supporting documents on April 5, 2010. The video and information are still available at the web-site http://collateralmurder.com. See also David Leigh and Luke Harding, *WikiLeaks: Inside Julian Assange's War on Secrecy* (London: Guardian Books, 2011), 65ff.

22. Leigh and Harding, *WikiLeaks,* 65.

23. See *We Steal Secrets: The Story of WikiLeaks,* dir. Alex Gibney (Jigsaw/Focus World, 2013).

24. "Profile: Julian Assange," BBC News, August 18, 2014, available at www.bbc .com/news/world-11047811.

25. Simon Bowers, "Icelandic Bank Kaupthing's Top Executives Indicted over Market Rigging," *Guardian,* March 19, 2013.

26. Leigh and Harding, *WikiLeaks,* 3.

27. W. J. T. Mitchell commentary on the film *We Steal Secrets* in seminar "Spectacle and Surveillance," Columbia University and University of Chicago, January 26, 2015.

28. See Jennifer Morone, JLM Inc. (website), http://jenniferlynmorone.com.

29. Julian Assange quoted in Michael Hastings, "Julian Assange: The *Rolling Stone* Interview," *Rolling Stone,* January 18, 2012.

30. Leigh and Harding, *WikiLeaks,* 47.

31. See Andy Greenberg, "WikiLeaks' Assange Warns Sources against 'Direct-to-Newspaper' Leak Projects," *Forbes,* February 8, 2011.

32. Hastings, "Julian Assange"; see also *We Steal Secrets,* dir. Gibney.

33. Esther Adley and Josh Halliday, "Operation Payback Cripples MasterCard Site in Revenge for WikiLeaks Ban," *Guardian,* December 8, 2010.

34. Cora Currier, "Charting Obama's Crackdown on National Security Leaks," ProPublica, July 30, 2013.

35. Ibid.; Erin Lahman, "Bradley Manning Trial: An American Hero Gets Court-Martialed on June 3," *PolicyMic,* May 29, 2013, http://mic.com/articles/45063/bradley-manning-trial-an-american-hero-gets-court-martialed-on-june-3.

36. Paul Lewis, "Bradley Manning Given 35-Year Prison Term for Passing Files to WikiLeaks," *Guardian,* August 21, 2013.

37. Mirren Gidda, "Edward Snowden and the NSA Files—Timeline," *Guardian,* August 21, 2013.

38. The video of the June 9, 2013, interview is available at the *Guardian* website: www.theguardian.com/world/video/2013/jun/09/nsa-whistleblower-edward-snowden-interview-video. For a rough transcript, see http://mic.com/articles/47355/edward-snowden-interview-transcript-full-text-read-the-guardian-s-entire-interview-with-the-man-who-leaked-prism.

39. Ibid.

40. Centre for International Governance Innovation and IPSOS, "CIGI-IPSOS Global Survey on Internet Security and Trust," November 24, 2014, www.cigionline.org/internet-survey.

41. Ibid., 10.

42. Ibid., 3.

43. Ibid., 11.

44. Ibid.

45. See the I Fight Surveillance website at https://ifightsurveillance.org.

46. See the Security in-a-Box website at https://securityinabox.org.

47. Danny O'Brien, "13 Principles Week of Action: Human Rights Require a Secure Internet," Electronic Frontier Foundation, September 18, 2014, www.eff.org /deeplinks/2014/09/human-rights-require-secure-internet.

48. Jason Self, "Avoiding Surveillance," http://jxself.org/avoiding-surveillance .shtml; the Greycoder website is www.greycoder.com.

49. See, e.g., SAMURAI_LUCY, "Protecting Your Online Privacy—Basic Info," July 7, 2014, http://pastebin.com/nFoM3aVf.

50. David E. Sanger and Brian X. Chen, "Signaling Post-Snowden Era, New iPhone Locks out N.S.A.," *New York Times,* September 26, 2014.

51. Ibid.

52. Craig Timberg and Greg Miller, "FBI Blasts Apple, Google for Locking Police out of Phones," *Washington Post,* September 25, 2014.

53. Amrita Jayakumar, "As Mobile Device Privacy Hits the Spotlight, Silent Circle Zips Ahead," *Washington Post,* October 12, 2014.

54. For information, visit http://freedomboxfoundation.org.

55. Kim Zetter, "Middle-School Dropout Codes Clever Chat Program That Foils NSA Spying," *Wired,* September 17, 2014.

56. Ibid.

57. Kevin Poulsen, "The FBI Used the Web's Favorite Hacking Tool to Unmask Tor Users," *Wired,* December 15, 2014.

58. See Jim Dwyer, *More Awesome Than Money: Four Boys and Their Quest to Save the World from Facebook* (New York: Viking, 2014), 18–31.

59. Eben Moglen, quoted in Nancy Scola, "'People Love Spying on One Another': A Q&A with Facebook Critic Eben Moglen," *Washington Post,* November 19, 2014.

60. Ibid.

61. Zetter, "Middle-School Dropout Codes Clever Chat Program That Foils NSA Spying."

62. Eben Moglen, "Privacy under Attack: The NSA Files Revealed New Threats to Democracy," *Guardian,* May 27, 2014 ("From now on," Moglen suggests, "the communities that make free software crypto for everyone else must assume that they are up against 'national means of intelligence'").

63. See Thomas Fox-Brewster, "Why Web Giants Are Struggling to Stop Snoops Spying on Thousands of Websites," *Forbes,* September 10, 2014.

64. Poulsen, "The FBI Used the Web's Favorite Hacking Tool to Unmask Tor Users."

65. Ibid.

66. Ibid.

67. Ibid.

68. Lee Rainie and Janna Anderson, "The Future of Privacy," Pew Research Center, December 18, 2014, www.pewinternet.org/2014/12/18/future-of-privacy.

69. Jaron Lanier, *Who Owns the Future?* (New York: Simon and Schuster, 2013).

70. See Pierre Collin and Nicolas Colin, "Mission d'expertise sur la fiscalité de l'économie numérique," January 18, 2013, www.economie.gouv.fr/rapport-sur-la -fiscalite-du-secteur-numerique.

71. Nicolas Colin, "Corporate Tax 2.0: Why France and the World Need a New Tax System for the Digital Age," *Forbes,* January 28, 2013.

72. See, e.g., Frank Pasquale, "Restoring Transparency to Automated Authority," *Journal on Telecommunications and High Technology Law* 9 (2011): 235.

73. William H. Simon, "Rethinking Privacy," *Boston Review,* October 20, 2014.

74. See James Bridle, "Watching the Watchers," http://shorttermmemoryloss .com/portfolio/project/watching-the-watchers. The full series can be seen on Flickr: www.flickr.com/photos/stml/sets/72157632721630789. Special thanks to Lucas Pinheiro for introducing me to this work.

75. Institute for Applied Autonomy, iSee (web-based app), www.appliedautonomy .com/isee.html. Thanks to Rebecca Zorach for introducing me to this work, and to Trevor Paglen's.

76. Trevor Paglen, *Blank Spots on the Map: The Dark Geography of the Pentagon's Secret World* (New York: Penguin, 2009); see also generally "Negative Dialectics in the Google Era: A Conversation with Trevor Paglen," *October* 138 (Fall 2011): 3–14; Rebecca Zorach, ed., *Art against the Law* (Chicago: University of Chicago Press, 2015). A number of artists and activists are also involved in projects to expose black sites, military spaces of secrecy, and the carceral geography of penal repression. See, e.g., Allan Sekula, "Polonia and Other Fables," show documenting "black sites" in Poland, http://x-traonline.org/article/allan-sekula-polonia-and-other-fables; Josh Begley, "Prison Map," project documenting the geography of incarceration in the United States, http://prisonmap.com/about.

77. See Whitney Museum exhibition for spring 2016, http://whitney.org/Exhi bitions/LauraPoitras.

78. Special thanks to W. J. T. Mitchell for sharing Ed Shanken's syllabus.

79. The citizen video capturing the homicide of Eric Garner by NYPD officer Daniel Pantaleo in Staten Island on July 17, 2014, is the latest in a series of these. See generally Anna North, "What the Eric Garner Video Couldn't Change," *New York Times,* December 4, 2015.

80. Michael McCahill and Rachel Finn, "The Social Impact of Surveillance in Three UK Schools: 'Angels,' 'Devils' and 'Teen Mums,'" *Surveillance and Society* 7, nos. 3–4 (2010): 285.

81. Pierre Manière and Sylvain Rolland, "L'open data: mirage ou Eldorado?" *La Tribune,* June 27, 2014.

82. Ibid.

83. Lawrence Lessig, *Code: And Other Laws of Cyberspace, Version 2.0* (New York: Basic Books, 2006); Jack M. Balkin, "The Constitution in the National Surveillance State," *Minnesota Law Review,* Vol. 93, 1–25 (1980), 21–25; H.R.2048 - USA FREEDOM Act of 2015, Public Law No. 114-23, 114th Congress (2015–2016), passed on June 2, 2015, available at https://www.congress.gov/bill/114th-congress/house-bill/2048/text.

84. Daniel Solove, *Nothing to Hide* (New Haven, CT: Yale University Press, 2013), 205.

85. Margo Schlanger, "Infiltrate the NSA," *Democracy: A Journal of Ideas,* no. 35 (Winter 2015), www.democracyjournal.org/35/infiltrate-the-nsa.php?page=all.

86. Moglen, "Privacy under Attack."

87. Michael Scaturro, "Meet Germany's 'Average' Data Protesters," Deutsche Welle, September 22, 2014, www.dw.de/meet-germanys-average-data-protesters /av-17940098.

88. See, e.g., Benjamin Weiser and Nathaniel Popperjan, "Focus Is on Bitcoin Trader at Trial over Silk Road Black Market," *New York Times,* January 15, 2015.

89. danah boyd, *It's Complicated: The Social Lives of Networked Teens* (New Haven, CT: Yale University Press, 2014), 45.

90. Ibid.

91. McCahill and Finn, "The Social Impact of Surveillance," 283–284.

92. Ibid., 288.

93. See Daniel Ellsberg, *Secrets: A Memoir of Vietnam and the Pentagon Papers* (New York: Penguin, 2003), 199–214.

94. Quoted in *"The Most Dangerous Man in America: Daniel Ellsberg and the Pentagon Papers:* Interview," PBS, www.pbs.org/pov/mostdangerousman/inter view.php.

95. See Grant Robertson, "Edward Snowden and Daniel Ellsberg Speak Truth to Power at HOPE X," *Daily Dot,* July 20, 2014, www.dailydot.com/politics/hopex -edward-snowden-daniel-ellsberg.

96. See Michel Foucault, *The Courage of Truth: The Government of Self and Others II. Lectures at the College de France, 1983–1984* (New York: Picador, 2012); see also the forthcoming book by the French philosopher Frédéric Gros on disobedience and the courage of truth. John Rajchman at Columbia University is exploring these issues in a brilliant paper, "Citizen Four: Edward Snowden and the Question of Truth and Power," presented at the Columbia Center for Contemporary Critical Thought, April 6, 2015, and at the IX Colóquio Internacional Michel Foucault, Rio

de Janeiro, April 14–17, 2015. Charleyne Biondi at Sciences Po is writing as well; see her "Edward Snowden: le courage de la vérité à l'âge de la surveillance de masse," http://web.law.columbia.edu/sites/default/files/microsites/academic-fellows/images/journee_detudes_ehess_dec_16_2014.pdf. Austin Sarfan at the University of Chicago is also writing on this topic.

97. Matt Apuzzo, "C.I.A. Officer Is Found Guilty in Leak Tied to Times Reporter," *New York Times,* January 26, 2015.

98. Ibid. For instance, in the case of Jeffrey A. Sterling, a former CIA agent who was accused and convicted of leaking information to the *New York Times* about a covert operation to disrupt Iran's nuclear program, there was no direct evidence of a leak.

99. Snowden's first interview, Sunday, June 9, 2013.

Chapter Twelve: POLITICAL DISOBEDIENCE

1. Taken from http://goodsecurityquestions.com/examples.htm.

2. Eben Moglen, "Privacy under Attack: The NSA Files Revealed New Threats to Democracy," *Guardian,* May 27, 2014.

3. See generally Bernard E. Harcourt, "Political Disobedience," in W. J. T. Mitchell, Michael Taussig, and Bernard E. Harcourt, *Occupy: Three Inquiries in Disobedience* (Chicago: University of Chicago Press, 2013); David Graeber, *The Democracy Project: A History, a Crisis, a Movement* (New York: Spiegel and Grau, 2013), 87–98.

4. George Orwell, *1984* (New York: Random House, 2009), 263.

5. See Fabienne Brion and Bernard E. Harcourt, "The Louvain Lectures in Context," in Michel Foucault, *Wrong-Doing, Truth-Telling: The Function of Avowal in Justice* (Chicago: University of Chicago Press, 2014), 274–283; Daniel Defert, *Une vie politique: Entretiens avec Philippe Artières et Eric Favereau, avec la collaboration de Joséphine Gross* (Paris: Seuil, 2014), 36–76; Audrey Kiéfer, *Michel Foucault: Le GIP, l'histoire et l'action* (PhD diss., Université de Picardie Jules Verne d'Amiens, 2009), 169–172; Philippe Artières, Laurent Quéro, and Michelle Zancarini-Fournel, eds., *Le Groupe d'information sur les prisons: Archives d'une lutte, 1970–1972* (Paris: Éditions de l'IMEC, 2003).

6. Kevin Thompson, "Foucault and the Legacy of the Prisons Information Group," paper presented at DePaul University, May 8, 2015.

7. See generally Bernard E. Harcourt, "The Dialectic of Theory and Practice," *Carceral Notebooks,* vol. 12, ed. Perry Zurn and Andrew Dilts (forthcoming 2015).

8. Ibid.

9. Graeber, *The Democracy Project,* 89.

10. See Jodi Dean, "Occupy Wall Street: After the Anarchist Moment," *Socialist Register* 49 (2013): 54–55 ("Occupy Wall Street began as a left politics for a neoliberal age, an age that has championed individuality and excluded collectivity"); Doug Henwood, "The Occupy Wall Street Non-Agenda," *LBO News from Doug Henwood,* September 29, 2011, http://lbo-news.com/2011/09/29/the-occupy-wall -street-non-agenda.

11. Gilles Deleuze and Félix Guattari, *Anti-Oedipus: Capitalism and Schizophrenia* (Minneapolis: University of Minnesota Press, 1983), 344.

ACKNOWLEDGMENTS

I cannot express fully how much this work owes to countless inspiring and enriching conversations with Mia Ruyter, Daniel Defert, François Ewald, Frédéric Gros, David Pozen, Jesús Rodríguez-Velasco, W. J. T. Mitchell, Isadora Ruyter-Harcourt, Léonard Ruyter-Harcourt, and Renata Salecl. It has benefited immensely from rich exchanges, conversation, and seminars with friends and colleagues, and I would especially like to thank and acknowledge Etienne Balibar, Wendy Brown, Elizabeth Emens, Didier Fassin, Katherine Franke, Bob Gooding-Williams, Stathis Gourgouris, David Halperin, Todd May, Keally McBride, Eben Moglen, Calvin Morrill, Martha Nussbaum, Frank Pasquale, Jonathan Simon, Ann Stoler, Mick Taussig, Kendall Thomas, Loïc Wacquant, and Tim Wu. The book was reworked, thankfully, as a result of the prodding and pushing of three anonymous reviewers for Harvard University Press. I thank them each for their insights, patience, and generosity of spirit and intellect.

For consistently excellent and extensive research, I am indebted to Daniel Henry at the University of Chicago and Julian Azran at Columbia University. Léonard Ruyter-Harcourt and Henry Utset conducted probing research on surveillance in the early American and French republics, and Léonard also conducted excellent research on WikiLeaks. Isadora Ruyter-Harcourt provided outstanding guidance throughout on all aspects of the digital age, for which I am extremely grateful. Chathan Vemuri provided marvelous research on humanism in the twentieth century, and Alexander de la Paz continuously offered feedback and material. I am also indebted to students in my research seminar at the École des hautes études en sciences sociales in Paris, at Columbia University, and at the University of Chicago.

Once again, Lindsay Waters proved to be insightful, encouraging, and demanding, for which I am deeply grateful. I am delighted as well to have worked so closely with Amanda Peery at Harvard University Press, with Deborah Grahame-Smith and Sue Warga at Westchester Publishing Services, and with Claire Merrill at Columbia Law School.

INDEX

acausal, 162–163

Accurate Resistance (Germany), 277

ACLU v. Clapper, 295n7, 321n31

Actuarial: actuarial instruments, 23, 147; actuarial logic, 147–149, 156

Acxiom, 14, 160, 199, 202, 204

addiction, 50, 110, 122, 200

adolescent access to digital technology, 23, 288n68

advertisements, 2, 6, 20, 39, 48, 98, 126–127, 160–161, 210, 272

Agamben, Giorgio, 287n50

Alexander, Michelle, 260

algorithms, 14–17, 22, 25, 45, 61, 117, 157–163, 195, 218, 230, 271

All American Cable Company, 68

Allen, Anita, 286n42

al-Mindhar, Khalid, 59

Almond, Gabriel, 211

al-Qaeda, 55

AlterNet.com, 222–223, 285n15

Alteveer, Ian, 109

Althusser, Louis, 214

Amadae, S.M., 317n45

Amatriain, Xavier, 158, 318n54

Amazon.com, 1, 48, 73, 114–116, 126, 139, 188, 215, 268; Amazon books, 31, 32; Amazon recommendations, 20–22, 162–163, 207; workplace monitoring, 195–196

American Civil Liberties Union (ACLU), 34, 54, 210, 294n4, 295n7, 321n31

amoral, 162–163

amusement park, 93–97

analog age, 13, 20–21, 28, 35, 132–136, 166, 181, 188, 223–224, 234–237, 250

ancien régime, 17, 62, 288n66

Android, 20, 220, 271

ankle bracelet, 20, 124, 236–243, 248

anonymity, 2, 57, 240, 242; defined, 319n5; loss of anonymity, 13–14, 19, 229–231; anonymous messaging, 117, 129–131; enhancing anonymity, 270–271, 273, 277; privacy, 166–168, 176; whistle-blowing, 266–268

Anonymous (hacker collaborative), 268, 271, 277

AOL, 10, 65–66, 114

apathy, 254, 258, 281

Apple Inc., 34, 48, 73, 271; Apple
   Watch, 20, 52, 122–129, 198, 237, 243,
   248, 310n67; governing (Begley),
   189–191; PRISM program, 10, 66;
   revenue, 210–211

App Store, 189, 192

Arce, Alberto, 286n16, 286n20

archaic data, 143–145

archaic forms of punishment, 235–236

Arendt, Hannah, 171, 287n50, 321n18

Assange, Julian, 266–268, 270, 339n21,
   340n29

Assassin's Creed Unity, 3

Associated Press, 7–10, 112, 286n16,
   286n20

asylums, 223–224, 226, 229, 330n17,
   337n17

AT&T, 65, 71–72, 76, 79, 132, 245;
   AT&T Folsom Street facility, 65,
   71, 76

Atlas platform, 6–7, 285n8

Australia, 75, 113

autonomy, 26, 131, 175–176, 178–179;
   defined, 319n5; loss of, 19, 166–168,
   181, 277

Baehr, Peter, 248, 333n7

Balkin, Jack M., 54, 276

Bamford, James, 27, 66, 68, 71, 98, 121,
   246–247

Barnett, James, 74

Baudrillard, Jean, 94, 304n63

Bauman, Zygmunt, 73, 287n50, 299n96

Beaumont, Gustave de, 84, 259, 302n18.
   See also Tocqueville, Alexis de

Becker, Gary, 176, 297n59. *See also*
   University of Chicago

Begley, Josh, 188, 190, 323n5, 324n9,
   342n76

Bejar, Arturo, 4

Belgacom, 241–242

Benedictine monasteries, 224, 229

Benhabib, Seyla, 321n18

Benjamin, Walter, 140

Bennett, Tony, 110, 307n6

Bentham, Jeremy, 27, 80–93, 118, 120,
   197; acoustic surveillance, 115–116;
   invention of panopticon, 85; and
   Orwell, 38. *See also* panopticon

Bentham, Samuel, 85

Biden, Vice-President Joe, 115

Big Brother, 26–28, 31–48, 53, 55–56,
   229, 275

Big Data, 60, 62, 80, 115, 127, 162–163;
   benefits of, 21–22, 231, 245–246;
   categories of, 137–140; defined, 132.
   See also data

Bigo, Didier, 287n50

Bill of Rights, 173

Bing, 2, 22

Binney, William, 76

biometric data, 218

Biondi, Charleyne, 343n96

biopower, 17, 92, 103

Black Chamber, 67–68

Blackphone, 271

black sites, 248, 275, 342n76

Blackstone, William, 37, 291n21

blogging, 3, 39, 41, 103, 110, 256

Bloomberg, 221, 327n76

Boeing Corp., 65, 68

Boltanski, Luc, *The New Spirit of
   Capitalism*, 48, 293n53

Booz Allen Hamilton, 65, 68, 98, 101, 269

BOUNDLESS INFORMANT
   program, 209

Bourdieu, Pierre, 64, 211, 297n60, 327n79, 328nn80–81

boyd, danah, 18, 277, 287n52, 288n68, 343n89

Braman, Sarah, 119

Brandeis, Louis, 321n29

Brazil, 269

Breton, André, 314n135

Bridle, James, 275, 342n74

Brooks, John, 271

Brown, Michael, 16

Brown, Wendy, 306n89, 322n33, 329n96, 347

Bryman, Alan, 95, 304n66

bulk telephony metadata, 27, 32, 55–56, 213, 321n31

Bunyan, John, 248

bureaucracy, 57, 181

Burgess, Ernest W., 149, 152, 163, 316n29, 316n33

Burnham, Walter Dean, 256, 337n14

Bush, President George H. W., 191

Bush, President George W., 32, 56–57, 177, 191

Butler, Desmond, 286n16, 286n20

BuzzFeed, 97, 305n85, 327n73, 335n37

Campbell, Colin, 96, 305n72

Canada, 75, 113, 216

Candy Crush, 19, 21

Cantillon, Richard, 165, 319n75

Carter, President Jimmy, 34

causality, 162

CCTV, 20, 94, 144, 206, 218, 221, 229–231, 236–237, 265, 275, 278, 331n37

Celebration (Florida), 95–96

cellphones, 1, 109, 131–139, 183, 206, 244–245, 278; capture of cell phone

data, 7–10; onboarding, 159–160; tracking by cell phone, 15–16, 92, 253

censorship, 14, 68, 191, 268, 298n75

census data, 132, 139

Center for Constitutional Rights, 210

Centers for Disease Control and Prevention, 162

Chamayou, Grégoire, 296n56, 308n40

chat rooms, 130–131

chemical castration, 237

Cheney, Vice President Dick, 177–178, 322n35

Chertoff, Michael, 74

Chiapello, Eve, 48, 293n53

Chicago School, 99, 178, 322n36

Chilling Effect, 218–219, 232, 330nn8–9

China, 188, 269–270, 297n62; Chinese prisoner camps, 226

Chomsky, Noam, 177, 322n34

Chrome, 234

Church Commission, 298n75

CIA, 24, 57, 344nn97–98

circulation of power, 15, 17, 22–23, 25, 64, 84–85, 91, 97, 118, 213, 281

Cisco, 210, 242

Clapper, James, 34

Clinton, President Bill, 55, 57

code breakers, 59

Coen, Deborah, 315n22

Cohen, Julie, 24

COINTELPRO, 188

Cold War, 25, 59, 153, 156, 169, 172, 178, 311n79, 317n45, 321n19, 322n41

Cole, David, 77, 289n72, 300n118, 308n33

Colin, Nicolas, 274, 342n70

Collateral Murder video, 265–266

Collin, Pierre, 274, 342n70

Colls, Robert, 33, 291n16

Combined DNA Index System (CODIS), 144

consumption, 18, 65, 94–97, 103, 107, 126, 139, 145, 162, 187, 191–192, 194, 197, 207, 230; consumer data, 198, 293n49, 327n67; consumer scores, 204–205, 326n57; consumerism, 51, 96, 127, 305n72; consuming subject, 197

convergence of digital life and correctional supervision, 25, 236–240

cookies, 4–7, 52, 159–160, 181

Cooper, David, 37, 223, 291n24, 330n18

Coras, Jean de, 142, 145, 314n1, 314n4

Cornell University, 42, 101, 331n28

corporate espionage, 73, 216

correctional supervision, 20, 235–236, 239

cost-benefit analysis, 57–58, 100, 167, 176, 179

courage, 40, 171, 211, 279, 281, 283; courage of truth, 279, 343n96

Cox, Chris, 45

credit scores, 204–205

Crouch, Ian, 32, 37, 289n79, 290n11

Crowell, William P., 68–69

crystal palace, 25, 107, 110

Cuba, 7–9

Cukier, Kenneth Neil, 162, 245, 288n63, 318n72, 335n43

dangerous individual, 23, 36, 148, 315n25

data: identity data, 137–138; machine data, 138; payload data, 193–194, 324; people data, 137–138; smart data, 137; social data, 138–139; structured data, 139; transactional data, 138; unstructured data, 139; video data, 3, 6, 22, 75, 111–114, 132, 139–140

data brokers, 2, 14, 159–160, 187, 198–199, 202, 204, 265, 272, 325n41, 326n44, 326n54

data control, 62, 146, 153

data double, 18, 157, 228

data exhaust, 138–139

datafication, 21–22, 245

Datalogix, 204

data markets, 187–188, 198, 207, 215

data mining, 3, 22, 26, 64, 91, 121, 124, 129, 154, 159, 162–164, 194, 254; data skimming, 241–242

Davidson, Amy, 113, 291n13, 308n25

Davis, Natalie Zemon, 142, 144, 314n2, 314n6

Dean, Jodi, 283, 287n49, 345n10

Debord, Guy, 19, 88–90, 287n54, 303nn38–39

Defense Advanced Research Projects Agency (DARPA), 101

Defert, Daniel, 282, 344n5, 347

Deleuze, Gilles, 19, 47–52, 227–228, 262, 282–283; *Anti-Oedipus*, 48–49, 51, 227, 262, 283, 293n54, 294n66, 331n26, 345n11; "Postscript on the Societies of Control," 19, 262–263, 317n48

Dell Corporation, 242

democratic theory, 258

Der Spiegel, 15, 34, 240–241, 266, 287n45, 310n56, 334n23

desiring machines, 48–52, 228, 281

despotism, 254–255, 258–260, 281–283; ancient despotism, 254; democratic despotism, 255, 281

Diaspora (social network), 272

Didi-Huberman, Georges, 91, 304n49

digital avowals, 99–100, 129

digital devices, 50, 127

digital economy, 187, 207, 274

digital lust, 13, 18, 20, 47–50, 110

digital monitoring, 20, 22, 219, 237, 243

digital narratives, 128

digital security, 97, 229, 270, 305n86

digital self, 1, 10–16, 102–104, 109–110, 128, 157, 162, 227–231, 236, 253

digital sovereignty, 188

digital things, 135–136

Dilts, Andrew, 306n91, 337n19, 344n7

Direction Générale de la Sécurité Extérieure (DGSE), 2, 167, 297n61

disciplinary power, 82, 85–87, 223

discipline, 17, 21, 36, 80, 84, 86–89, 91, 93, 148–149, 163–164, 275

dislike button, 41–42

Disney Corporation, 95–96, 305n69

Disneyland, 94–95

distributed denial of service, 268, 277

Dixon, Pam, 205, 326n57

DNA, 23, 56, 144, 179, 312n95

DNI Presenter, 10, 75

DonorTrends, 205

doppelgänger logic, 25, 141–165, 217, 230

Douglas, Justice William O., 173–175, 177–178, 321n29

drones, 21, 103, 115, 188–192, 248, 275

Dropbox, 15

Drug Enforcement Administration (DEA), 74, 244

Duhigg, Charles, 195, 310n69, 325n28, 326n49

duodividual, 157, 159

Dwyer, Jim, 341n58

Dzur, Albert, 258

Easton, David, 211, 327n78

eBay, 22, 114, 139, 215

Eco, Umberto, 95, 304n64

economic espionage, 65, 73, 216

Efrati, Amir, 285n1

Egypt, 269

Ehrlich, Judith, 278

Eisenhower, President Dwight, 66, 297n63

electronic bracelet, 20, 237. *See also* ankle bracelet

Electronic Frontier Foundation, 270, 341n47

Ellsberg, Daniel, 267, 278–279, 343n93

Emens, Elizabeth, 308n37, 347

emotional contagion, 43–44, 292n31, 293n47

empathy, 40

Ericson, Richard, 50, 157, 166, 228, 293n55

Espionage Act, 268

Étatlab, 276

Ewald, François, 90, 151, 297n59, 303n45, 316n39, 347

examination of the self, 100

exercise of power, 22, 86, 88, 115, 120, 191

exhibitionary complex, 110, 307n6

Experian, 198

experimentation, 42–46, 110, 118–119, 167, 234, 281; panoptic experiments, 88; scientific experiments, 138, 164, 219

Explorer, 234

expository power, 15, 21, 25, 50, 91–92, 124, 157; defined, 131

expository society, 1–28, 36, 95, 163, 281; architecture of, 107–111; as total institution, 232–233; identified, 19, 90, 117; narratives, 128–129; and neoliberalism, 97–100

E-ZPass, 1

Facebook, 3–7, 21–24; censorship, 191–192; digital advertising, 14; Diaspora alternative, 272; dislike

Facebook (*continued*)
  button, 41–42; emotional contagion
  research, 41–47; Millions March FB
  page, 16; neoliberalism, 99–100;
  NYPD, 90, 115, 243–245; onboarding,
  159–161; PRISM, 10–13, 66, 75; privacy
  as passé, 166; revenues, 210–211;
  sexual orientation, 131; size, 136, 138;
  statelike, 215; NSA Tailored Access
  Operations, 241; tracking, 3–7,
  222–223; UPSTREAM, 65
facial recognition, 22–23, 91, 112–113,
  144, 304n48
factory discipline, 85, 197, 241
FAIRVIEW program, 72
false consciousness, 227, 331n25
fascism, 49, 51, 99, 170
Fassin, Didier, 347
FBI, 65–67, 79, 91, 144, 188, 209, 211,
  219, 253, 272–273
Ferguson (Missouri), 16, 130
Ferguson, Thomas, 256, 337n14
fetish, 47, 110, 122. *See also* addiction
Financial Times, 206, 325n34, 327n63
Finlay, Robert, 314n6
Finn, Rachel, 329n7, 342n80, 343n91.
  See also McCahill, Michael
Firefox, 234, 272–273
Five Eyes, 75, 77, 113, 121, 269
Flickr.com, 14, 36, 342n74
Foreign Intelligence Surveillance Act
  (FISA), 11, 175, 210, 298n75
Foreign Intelligence Surveillance
  Court (FISC), 56, 58, 59–60, 71, 210,
  327n75
Fort Meade (MD), 70, 77
Foucault, Michel, 19, 27–28, 36–38, 51,
  80–93, 97, 100, 115, 119–120, 196,
  214–215, 223, 275, 282; *Discipline and*

Punish, 80, 84, 87–89, 91; *History
  of Madness*, 223; knots of statelike
  power, 214–216; *Penal Theories and
  Institutions*, 81, 214, 301n6; *Security,
  Territory, Population*, 92–93; *The
  Birth of Biopolitics*, 92, 297n59,
  316n39, 329n96; *The History of
  Sexuality*, 92, 306n101; *The Punitive
  Society*, 81, 89, 92, 97, 214; *Truth and
  Juridical Form*, 82; *Wrong-Doing,
  Truth-Telling*, 300n4, 306n93, 312n88,
  316n38, 320n15, 344n5
FreedomBox foundation, 271
French penal code, 20
Friedersdorf, Conor, 54, 295n6
Friedman, Milton, 177–178
Friedman, Thomas, 54, 294n5, 336n4
future dangerousness, 23. *See also*
  dangerous individual

Garner, Eric, 16, 130, 312n89, 342n79
Gaussian bell curve, 148
Geeksphone, 271
Gellman, Bart, 11, 286n36
Gellman, Robert, 205, 326n57
Gillespie, Tarleton, 159, 230, 318n60,
  331n34
Gilliom, John, 78, 300n123
Glaeser, Andreas, 336n2
glass house, 111, 118, 125, 140, 236, 314n135
Gmail, 7, 17, 50, 101–103, 188, 285n15,
  306n95
Goel, Vindu, 6, 285n7, 292n31, 293n48
Goffman, Erving, 217, 221, 223–229,
  232–233; structure of the self, 223;
  systems of punishments and
  privileges, 224–227
Google, 1–7; Android, 271; digital
  advertising, 47–48; forecasting flu

epidemics, 21; Gmail, 101–103; Google Calendar, 103; Google Checkout, 103; Google Docs, 103; Google Drive, 103, 198; Google Earth, 275; Google Groups, 103; Google Hangouts, 103; Google Inbox, 103; Google Maps, 189; Google Voice, 103, 220; Google+, 3–4; mission, 120; neoliberalism, 169; Optic Nerve, 114; Orkut, 103; PRISM program, 10–15, 66; recommendation algorithms, 157–159, 198; ScanOps segregation, 125; Daniel Soar, 220–221; statelike, 72, 79, 188, 210–211, 215; Street View, 131, 193–194, 222–223, 235; Tailored Access Operations, 241; tracking, 3–7; UPSTREAM program, 65, 76

Gottschalk, Marie, 333n6

Government Communications Headquarters (GCHQ), 2, 111–114, 121, 167, 241–242, 246

GPS, 17, 20, 127, 136, 139, 144, 197, 236–238, 243, 248; GPS surveillance, 237; GPS tracking, 20, 197, 236, 243, 248

Graeber, David, 255, 282, 344n3

Graham, Dan, 107–109, 140, 307n1

Graves, Michael, 96

Gray, Patrick, 272

Greenwald, Glenn, 11, 24, 32, 54, 65–67, 70, 75–76, 80, 122, 216, 240, 269

Greycoder, 270, 341n48

Griswold v. Connecticut, 174, 321n29

Gros, Frédéric, 343n96, 347

group-based prediction, 147, 149, 163

GrubHub, 183

Guantanamo Prison, 179, 248

Guardian (newspaper), 2, 31, 34; Josh Begley, 189–190; BOUNDLESS INFORMANT, 209; corporate-intelligence cooperation, 66–72; Optic Nerve, 111–114; PRISM, 11–13, 209; Edward Snowden, 268–269; UPSTREAM, 209; WikiLeaks, 266; XKeyscore, 62–64

Guattari, Félix, 47–52, 227–228, 283. *See also* Deleuze, Gilles

Guerre, Martin, 141–144, 314n1, 315n13

Guillauté, Jacques-François, 62–63, 164, 296n56, 297n57

habitual offender, 36, 148. *See also* dangerous individual

hackers, 2, 65, 72, 253, 273, 279

Hacking, Ian, 148, 150, 315n19, 316n26

Haggerty, Kevin, 24, 50, 157, 166, 228, 293n55

Halliburton, 177–178, 322n35

Haraway, Donna, 308n40

Hard Rock Café, 95

Harvard Business School, 153

Harvey, David, 177, 306n89, 322n34

hashing, 160, 318n65

Hayden, NSA Director Michael, 24, 70–71, 77, 115

Hayek, Friedrich, 178, 322n36

Head, Simon, 196–197, 325n35

"Hedge Two-Way Mirror Walkabout," 107–109, 140, 307n4. *See also* Graham, Dan

Hegel, G.W.F., 82, 85, 92, 116, 304n56

Henwood, Doug, 283, 345n10

Hewlett-Packard, 210–211

Hilbert, Martin, 133–135, 313n101

Hirschman, Albert, 293n51

Hitler, Adolf, 49, 227

Hobbes, Thomas, 17, 165, 337n19

Holder, Eric, 261, 338n30

home monitoring, 236

homeland security, 19, 74, 246

homo digitalis, 18

Hoover, J. Edgar, 187–188

Hoover, President Herbert, 68

HOPE X, 279, 343n95

hopelessness, 211

hospitals, 85, 197, 223, 258

Hotels.com, 116

Hotmail.com, 65, 67

Hotwire.com, 116

human capital, 99, 196

human essence, 168–169, 176–177, 179

humanism, 166–183, 319n5, 320n10;
American humanism, 171–172;
European post-war humanism,
170–172; humanist discourse,
171–172, 176, 179–180

humiliation/degradation, 224, 229

Huxley, Aldous, 46, 111

Hwang, Tim, 181, 323n42

hyperreality, 95, 304n64

identity theft, 141

ideology, 89, 94–95, 226–227, 260,
293n65

India, 241, 269–270

Indonesia, 269–270

Informants, 17, 62

Insecam, 24

Instagram, 1, 3, 13, 35–36, 41, 47, 50, 100,
110, 233; Biden's Instagram, 115;
NYPD monitoring, 243; people-
based marketing, 6

Integrated Automated Fingerprint
Identification System, 144, 315n16

Intel Corporation, 210

Internet of Things, 21, 135, 138

iPhone, 15, 20, 35, 52, 110, 127, 189,
291n17, 323n5, 341n50

Iraq, 177, 190–191, 265–266

iron cage, 236, 248–249, 333n7, 335n55,
336n59

IRS, 14, 116, 132, 181

Ive, Jony, 122–123

Jaspers, Karl, 170, 320n12

Jinman, Richard, 32, 290n10

Johnson, Philip, 96, 118, 125, 236, 249

Johnson, President Lyndon B., 154

jouissance, 40, 52

Julius, Nicolaus Heinrich, 82–85, 88,
92, 116, 301n9, 302n19; *Lectures on
Prisons*, 82, 85

Kafka, Franz, 26–27, 91, 221

Kantorowicz, Ernst, 13. *See also* two
bodies

Kardashian, Kim, 115, 307n9

Katz v. United States, 174, 321n28

Keller, Bill, 54, 289n81, 294n2, 336n4

Kelly, Ray, 244

Kennedy, Justice Anthony, 179

Kenya, 269

Kindle, 110, 253

King Jr., Rev. Martin Luther, 59

king's two bodies, 13. *See also* two
bodies

Klayman v. Obama, 32, 290n9, 295n7,
311n86

Klein, Naomi, 177, 322n34

Koopman, Colin, 308n40

Kramer, Adam D.I., 42–43, 45, 292n31,
293n47

Krauss, Rosalind, 309n41

Kristof, Nicholas, 31, 290n4

Kurgan, Laura, 119, 309n49
Kyllo v. United States, 179, 322n38

Laing, R.D., 37, 50, 223, 330n18
Laney, Doug, 132
Lanier, Jaron, 274, 342n69
leaderless social movements, 281–283
Leahy, Sen. Patrick, 7, 10, 34
Le Corbusier, 118
Lefebvre, Henri, 170, 320n13
Leon, Judge Richard, 32, 56–57, 59,
   60–61, 78, 167, 311n86
Lerman, Amy, 260, 338n29
Lessig, Larry, 24, 276
Leviathan, 17, 55, 299n107
Levine, Yasha, 102–103, 206, 285n15,
   306n100, 326n41, 327n69
Levy, Karen, 181, 308n38, 323n42,
   325n33, 330n10
Lexis-Nexis, 198
liberal democracy, 13, 17, 20, 22–23, 47;
   liberalism, 208–216; repression in,
   90–92, 115, 235–236, 287n50; virtual
   democracy, 254–260
Light, Jennifer S., 317n46
"Like" button, 1, 3–5, 35, 41, 50, 90, 115,
   138, 157, 217, 228
Lindsay, Mayor John, 154
Lindsay, Vachel, 173
linguistic analysis, 43, 102–103
LinkedIn, 41, 114, 242, 334n23
Liptak, Adam, 332n3
liquid modernity, 73, 299n96
literal transparence, 125, 309n41
local police departments, 72, 245
Lookalike Audiences, 160–161, 318n67
López, Priscila, 132–135, 313n101
Louis XIII, 93
Louis XIV, 93

Louis XV, 62
LOVEINT, 182
Lukes, Steven, 331n25
LUSTRE program, 65
Lynch, Judge Gerald E., 295n8,
   296n54
Lyon, David, 207

MacAskill, Ewen, 24, 269, 286n30,
   297n62, 323n43, 331n32, 335n45
Mall of America, 97, 107
malware, 216, 240–243, 273, 297n62
Manning, Chelsea, 125, 265–266, 268,
   270, 278
Maoists, 81
Marin, Louis, 95, 304n64
marketized subjects, 22, 26, 207
Martin, Reinhold, 309n47
Marx, Karl, 165, 170, 212, 319n77, 331n25
Masco, Joseph, 311n79
Mashable.com, 189–190, 323n6
mass incarceration, 19–20, 223, 236,
   248–249, 258, 333n6, 338n30
MasterCard, 268, 277, 340n33
matching logics, 145, 157–159, 162–163
Mauriac, François, 170, 320n14
May 1968, 81
Maycotte, Higinio, 137, 312n95
Mayer-Schoenberger, Viktor, 162, 245,
   288n63, 318n72, 335n43
McBride, Keally, 96, 305n70, 338n19,
   347
McCahill, Michael, 329n7, 342n80,
   343n91; UK Schools Study, 218–221,
   229–233, 278
McCarthyism, 188
McConnell, NSA Director Michael,
   68–69
McDonaldization, 95, 304n66

McKinsey and Company, 134, 136–137, 312n95

McNamara, Robert S., 154–156

mechanical calculator, 163

Medbase200, 199–200, 202–204

metaphors, 26–27, 46, 81, 118, 132, 249

Metasploit app, 273

MetroCards, 1, 17

Mexico, 216, 220, 270

Microsoft, 34, 72, 79, 167, 188, 233; cooperation with NSA and FBI, 65–67; PRISM, 10, 66; statelike power, 215; revenues, 210–211; Xbox privacy issues, 114

Miles, Steven, 77, 95–96, 240, 305n67

military discipline, 85, 197

military-industrial complex, 66, 227

Miller, D.A., 167, 287n47, 319n5, 320n6

Millions March NYC, 16, 130

Mills, Lindsay, 32

Mirowski, Phillip, 99, 306n90

mirrored glass pavilion, 25, 107–111, 120–140, 236, 249, 253, 279. *See also* glass house

Mitchell, Timothy, 212–213, 215, 328n85; "The Limits of the State," 213–214, 328n85

Mitchell, W. J. T., 89, 267, 303n42, 305n84, 319n5

Model Penal Code, 37, 291n21

Moglen, Eben, 24, 117, 211, 254, 272, 277, 281; on definition of privacy, 319n5

Monahan, Torin, 24, 78, 300n123, 325n33, 330n10

moral career, 224, 229

moralistic, 162–163

Morone, Jennifer Lyn, 263–264, 339n4

Morozov, Evgeny, 188, 256, 323n3, 337n15

mortification of the self, 19, 25, 217–225, 233, 329n1

mosaics, 129, 311n86

Mouse Trap, 265, 339n19

MySpace.com, 36, 65, 244

narratives of self, 128–131

Narus (Boeing), 65, 68

National Reconnaissance Office, 78

national security, 2, 56–58, 69, 73–74, 101, 112, 166, 175–178, 183, 190, 265

National Security Agency (NSA), 2, 10–13, 24, 66–72, 75–77, 120–122, 239–243; BOUNDLESS INFORMANT, 209; call chaining analysis, 121–122; corporate-intelligence cooperation, 66–77, 211; *doppelgänger* logic, 163; economic espionage, 65, 216, 241; Global Access Operations, 14, 121; harming reputations, 247–248; "We kill people based on metadata," 24, 115; LOVEINT, 182–183; LUSTRE program, 65; malware, 240–243; mission, 122; neoliberalism, 65, 70, 98, 213; Optic Nerve, 112–114; Orwell, 31–33; outsourcing, 65, 70, 73, 98, 101; PRISM, 10–15, 18, 65–66, 75–76, 180, 209, 239; QUANTUMINSERT, 240–242; RFID devices, 240; Section 215 program, 27–28, 32, 55–61, 115–116, 167, 213; sharing intimate images, 182, 230; size of, 77; Tailored Access Program, 126–127, 240–242; Treasure Map, 120–121, 131; UPSTREAM, 65, 76, 209, 239; Unit 8200 (Israel), 246–247; Utah Data

Center, 275; Xbox surveillance, 114;
XKeyscore, 24, 63–64, 75–76, 113, 134

Neiman Marcus, 65, 297n62

Nelson, Deborah, 172, 321n19

neoliberalism, 19, 64, 73, 92–93,
95–100, 98–100, 167–169, 176–178,
199, 215, 283; architecture of, 93–97;
defined, 98–99; distinct from
securitarian logics, 97–100; Facebook, 99–100; Foucault, 93, 215;
impact on privacy, 176–178

Netflix, 22, 48, 73, 79, 114, 145, 157–159,
188, 215, 227

Neustar, 73–74

New Deal, 55, 57–61, 78

New York City Police Department
(NYPD), 16, 79, 90, 243–244, 342n79

New York Times, 2, 34; art review, 109;
best-seller list, 260; editorializing,
31, 54, 130; exposing, 6–7, 266;
reporting, 74, 194, 202, 240, 242–243,
247

New Zealand, 75, 113

News Feed, 43–45, 292n31

Next Generation Identification, 91, 144

nonymous, 231

nothing to hide, 19, 178, 181, 183, 221,
232–233

Noyes, Andrew, 4

nudity, 112–113

Nussbaum, Martha, 40, 292n28, 347

NYU Brennan Center, 238

Obama, President Barack, 27, 31–33, 55,
57–61, 115, 191, 209–210, 216, 261, 279

objective self-awareness theory, 218,
329n5

Occupy Wall Street, 277, 282, 344n3,
345n10

Office of Tailored Access Operations
(TAO), 126, 240–242

onboarding, 159–160, 239

online shopping, 13, 189, 206

opacity, 25, 107, 119, 124–126, 205

Open Data movement, 276

Operation Crew Cut (NYPD), 243;
see also New York City Police
Department (NYPD)

Operation Shamrock, 68, 298n75

Operation Torpedo, 273

Optic Nerve, 111–113, 116, 230, 307n11,
308n22

Orbitz.com, 116

Orwell, George, 26, 31–39, 41, 43,
45–47, 49, 51–53, 56, 282, 289, 290n1,
344n4; *1984*, 26, 31–37, 39–40, 46–48,
282; Orwellian, 28, 32, 36, 52, 56, 221,
288n61, 290n7

Outlook.com, 65–67, 169

outsourcing, 64, 70, 73, 79, 98

packet sniffer, 73, 115

Paglen, Trevor, 275, 342nn75–76

Paltalk, 10, 66

panoptic power, 81, 91–92, 301n8

panopticon, 27, 38, 80–103, 120, 267;
acoustic surveillance, 115; contrast to
expository power, 90–92, 115–118,
126; employee supervision, 197;
positive view, 80, 275. *See also*
Bentham, Jeremy

Papachristou v. City of Jacksonville,
173–175, 321n21

paper-squeeze, 40–41, 164. *See also*
Guillauté, Jacques-François

Parole board, 152, 316n33

Parsons, Talcott, 236, 248–249,
336n57

Pasquale, Frank, 24, 124, 198, 275, 289n78, 310n71, 326n43, 342n72, 347; *The Black Box Society*, 124, 289n78, 310n71, 326n43

PatternTracer, 121

Pauley III, Judge William H., 55–56, 57, 59, 61, 78, 295n8, 296n54

PayPal, 268

penitence, 100, 129

Pentonville prison (London), 84

people-based marketing, 6–7, 159, 239

Petty, William, 164, 318n73

Pew Research Center, 166, 274, 313n112, 319n4, 329n2, 332n3, 342n68

phenomenal opacity, 25, 119, 124

Philosophy, 170–172; philosophy of risk, 152; philosophy of transparence, 266–267

Pidgin, 271

Pitch Interactive, 191

Planet Hollywood, 95

pockets of opacity, 124–126

Podesta, John, 60

Poindexter, Adm. John, 101, 120, 291n26

Poitras, Laura, 24, 66, 269, 275, 287n45, 297n62, 310n56, 325n30, 335n45

Polaroid, 13

political disobedience, 280–283

political disobedient, 46, 280–283

Poncela, Pierrette, 288n60

Posner, Richard, 175, 177

Pozen, David, 75, 299n107, 311n86, 347

PPBS analysis, 154

predilections, 13, 34, 47, 205. *See also* addiction; fetish

President's Review Group on Intelligence and Communications Technologies, 27, 57–60, 290n85

PRISM program, 10–12, 18, 65, 98, 100, 286n30, 297n61, 298n65

prison, 20, 38, 81–84, 149, 223–233, 235–236, 248–249, 259–261, 267–269, 282

Prisons Information Group (GIP), 282–283, 344n5

privacy, 32, 40, 211, 213, 218, 242–243, 274–275; defined, 319n5; eclipse of, 166–183; enhancing, 27, 58–60, 211, 269–272; Goffman, 229–233; historical, 68; loss of, 13–14, 19, 207, 277; Optic Nerve, 113–114; prospects, 274–275; risks to, 58

Privacy and Civil Liberties Oversight Board, 58, 60

privatization, 64, 73, 98–99, 167, 176, 274–275

Proceedings of the National Academy of Sciences, 44, 292n31

ProPublica, 223, 287n43, 300n109, 318n61, 326n54, 330n15, 340n34

Prosody, 271

Puritans, 86, 248

Pynchon, Thomas, 31

Quakers, 86, 248

quantified self, 13, 47, 128–129, 164–165

QUANTUMINSERT program, 241

Quetelet, Adolphe, 148

racialized overincarceration, 235, 248, 333n6, 338n30. *See also* mass incarceration

Radio Communications Act of 1912, 68

Radio Frequency/Pathway technology, 38, 219, 237, 240, 242, 291n26, 325n33, 334n20

Rajchman, John, 343n96

RAND Corporation, 153–155

rational choice approaches, 163, 168, 176–178, 180, 317n45

Reagan, President Ronald, 101

reasonable expectations of privacy, 172

recommendation algorithms, 14, 20, 22, 48, 60, 102, 157–160, 207, 217, 227, 318n54

regression analysis, 146, 150, 156, 162–163

Regulation of Investigatory Powers Act (UK), 113

Reich, Wilhelm, 47, 49–50, 228

relations of power, 15, 24, 64, 81–82, 89, 215, 262, 281

reputational harm, 112, 247

Reuters, 112, 114, 183, 265–266, 307n18, 323n46

Revere, Paul, 59

reversal of the spectacle into surveillance, 83–85, 88–89

RFID tags, 136

Riley, Terence, 309n43

risk, 19, 57–59, 61, 129, 141, 148–149, 151–152, 205, 238, 268, 279

Ritzer, George, 95, 304n66

Rockefeller IV, Sen. John D., 198–199, 202

Rodríguez-Velasco, Jesús, 347

Rohe, Mies van der, 118

Roosendaal, Arnold, 4

Rosa, Hartmut, 299n97

Rose, Nikolas, 148, 315n20

Rosenberg, Karen, 109, 307n4

Rothman, David, 223, 330n18

Rowe, Colin, 309n41

Rubenfeld, Jed, 320n5

Rumsfeld, Donald, 177

Rusbridger, Alan, 31, 290n3, 323n43, 331n32

Salecl, Renata, 347

Salon.com, 33, 291n16, 325n35

Samsung, 20, 188, 210, 215, 242, 288n61

Sanchez, Julian, 54, 294n3

Sartre, Jean-Paul, 170–172, 320n10

satellites, 78–79, 132, 172, 238

Sawyer, Stephen, 328n92

Scalia, Justice Antonin, 179

ScanOps (Google), 125

schizoanalysis, 50–51, 228

Schlanger, Margo, 277, 343n85

Schlesinger, James R., 155, 317n47

Schmidt, Eric, 169, 198, 288n58, 320n9

Schneier, Bruce, 54, 289n81, 294n1

Schnitt, Barry, 4–5

school discipline, 17, 85–87, 96, 197, 219–221, 278

Schopflin, Jaime, 5

Schumpeter, Joseph, 257

Scorelogix, 205

Secret Department of the Post Office, 68

securitarian logics, 17, 95, 101

securitarian power, 93, 95, 100, 116

sécurité, 2, 92–93, 96–97, 116–117, 304n59

security-commerce complex, 27, 65, 72–74, 77–79, 187, 215, 253

security-corporate collusion, 11, 26, 66–72

Sedwick, Eve Kosofsky, 311n75

Self, Jason, 270, 341n48

self-awareness, 218, 279

self-confidence, 26, 173, 217–219

self-esteem, 22, 218, 231

self-governance, 14, 86, 254

self-sacrifice, 279

selfie, 47, 52, 108–110, 115, 119, 122, 129–130

Semantic Traffic Analyzer STA 6400, 65

Sensenbrenner, Jim, 34

sensors, 32, 123, 132, 138

sexuality, 36, 49–50, 92, 214, 306n101

sexual orientation, 131, 247

Shanken, Ed, 275, 342n78

shell as hard as steel, 249–250, 283

Silent Circle, 271, 341n53

Silicon Valley, 65–66, 72, 157, 253

Simon, Jonathan, 315n20, 347

Simon, William, 80, 275

simulacrum, 94

Skinner, Quentin, 254, 336n3

Skocpol, Theda, 212, 328n82

Skydrive, 65, 67

Skype, 10–11, 14, 18, 22, 47, 66, 75, 79, 111, 188

Slutzky, Robert, 309n41

Smith, Adam, 165

Smith, Winston, 32–41, 46–47, 52–53, 282

Snapchat, 3, 36, 41, 110

Snowden, Edward, 2–3, 23–24, 75, 121, 256, 265–270, 278–279, 281; BOUND-LESS INFORMANT, 209; corporate-intelligence cooperation, 66–72; courage of truth, 278–279; economic espionage, 65, 216; "Five Eyes," 77; harming reputations, 247–248; LOVEINT, 182; Optic Nerve, 112–114; Orwell, 31–34; PRISM program, 10–11, 66, 209; reactions to revelations, 166–167, 180, 256; Section 215 program, 27; self-presentation, 98, 279; sharing nude pictures, 182, 230; Tailored Access Operations, 126–127, 240–243; Treasure Map program, 15;

UPSTREAM, 209; urgency, 281; XKeyscore, 64–66

Soar, Daniel, 220, 313n110, 330n11

social control, 148, 150, 153, 225

society: punitive society, 19, 81, 89–90, 92, 97, 214; society of the spectacle, 19, 88–90, 303n38; societies of control, 19, 262, 288n66, 317n48, 339n1; surveillance society, 78, 233, 300n123, 327n72

Solove, Daniel, 24, 221, 277, 287n51; on definition of privacy, 320n5

soma, 45, 111

Souter, Justice David, 179

sovereignty, 64, 93, 188, 262

speaking truth to power, 279

Sprint, 72

stahlhartes Gehäuse, 236, 249, 333n7

Stasi regime (GDR), 17, 126, 254, 336n2

state: "bring the state back in," 78, 212, 328n32; boundary of the state, 26, 187–216; state and society, 8, 208, 211–213; state security apparatus, 2, 22–23, 33, 81, 87–90, 101, 117–120, 198, 214–215, 275; state statistics, 132, 140, 146–150, 246; statelike, 79, 214–215; surveillance state, 27–28, 32, 54–79, 277

Stateville Prison (Joliet, IL), 90, 147, 149, 303n44

statistical control, 153–155, 162–163

Steel Mesh, 25, 234–235, 237, 239, 241, 243, 245, 247, 249, 332n1

Stefancik, Gregg, 5

Stewart, Jon, 97

stoic examination of conscience, 129

Stoler, Ann, 347

stop-and-frisk, 16

Storify.com, 128

Strauss, Leo, 178

Street View (Google), 15, 131, 193–194, 222, 235, 324n24

Studeman, William, 69

subjectivity, 14, 23, 26, 165, 217–221, 224–226, 233; humanist conception of, 172–175; rational choice approach to subjectivity, 178–179; shift in conceptions of, 180–181; subjectivation, 227. *See also* mortification of the self

Sunstein, Cass, 292n29

Supercomputer, 163

Supreme Court, 172, 174–176, 178–180, 193–194, 324

surveillance-industrial empire, 27, 66, 98

surveillant assemblage, 50, 228, 287n51, 293n55, 294n67, 317n49, 319n1, 331n31

symbolic interaction/power, 64, 79, 130, 229, 328n81

Szasz, Thomas, 37, 223, 291n24, 330n18

Target, 65, 124, 138, 162, 188, 194–195, 215; predicting pregnancy, 124, 194, 246

Taussig, Michael, 344n3

Taylorism, 196

TechAmerica Foundation, 139, 312n95, 313n116, 314n133

Telcordia, 74

telecommunications companies, 2, 27, 64–74, 178, 209, 213, 241, 245

telescreen, 21, 38

tenticular power, 78–79

themed space, 93, 96–97, 107; theme park, 93, 95–96; themed shopping mall, 93. *See also* amusement park

Thomas, Kendall, 347

Thompson, Kevin, 344n6

Thoreau, Henry David, 173, 321n26

thoughtcrime, 35, 37, 40

Tilh, Arnauld du, 141, 143–144

timetables, 17, 196

Tocqueville, Alexis de, 84, 254–255, 258–260, 281; *Democracy in America*, 254, 258, 260, 336n5; *On The Penitentiary System*, 84, 259–260

Tor, 271, 273, 277, 341n57

total awareness, 39, 62, 100, 115, 122, 131, 164–165

total institutions, 224, 226–227, 232–233

Travelocity.com, 116

Treasure Map program, 15, 120, 131, 287n45

trucking, 195, 325n33, 330n10

truth-telling, 100, 278–281. *See also* courage

Tumblr.com, 13, 75, 114

Twitter.com, 3–4, 31, 42, 75, 100, 114–115, 131, 138, 189, 227; Cuban Twitter, 7–10; NYPD, 243–244; onboarding, 160; recommend algorithms, 158

two bodies, 13, 124, 236, 253

Unit 8200 (Israel), 246, 335n44

United Kingdom, 65, 68, 75, 113, 195, 218, 276, 298n76

United States Agency for International Development (USAID), 7–10

University of California, Berkeley, 101

University of Chicago, 149, 176, 209, 221

UPS, 162, 168, 271

UPSTREAM program, 65

Uribe, Carlos Gomez, 158

USA FREEDOM Act, 27, 61, 167

USA PATRIOT Act, 295n8

Vaidhyanathan, Siva, 25, 47, 91, 126, 168

Venmo.com, 22

Venturi, Robert, 96

Verizon, 72, 245, 291n13

Vidler, Anthony, 118, 289n77, 309n44

Vimeo.com, 189, 297n59, 323n5, 339n17

Vine.com, 3, 47, 198

virtual authenticity, 127

virtual democracy, 253, 255–257, 259, 261

virtual seduction, 122

virtual transparence, 15–18, 22, 25, 107–109, 131, 249, 270; defined, 118–122; mortification, 233; radiation, 243; steel mesh, 234–237

Visa, 268

voter turnout, 16, 255–257

voyeuristic, 90, 115, 183, 229

Wacquant, Loïc, 333n6, 347

Wall Street Journal, 3–4, 202, 204, 207, 285n1

Walmart, 136, 162, 196

war on terror, 32, 56, 311n79

Warren, Samuel, 321n29

Weaver, Vesla, 260, 338n29

Weber, Max, 78, 188, 236, 248–249; *The Protestant Ethic and the Spirit of Capitalism*, 248, 333n7, 336n58

Western Union, 67–68, 298n75

WeTransfer.com, 15

whistle-blower, 266, 268, 271, 278, 294n5

Whitman, Walt, 173

Wi-Fi, 1, 15, 35, 47, 127, 193–194, 222

WikiLeaks, 24, 265–268, 277, 339n21, 340n26, 347

Wilson, Andrew Norman, 125, 310n74

Wired.com, 19, 52, 97, 137, 194, 243, 245, 273

Wiretap Act, 193–194

Wolfowitz, Paul, 177

Wolin, Sheldon S., 260, 337n18, 338n20

World Economic Forum, 139, 314n129

World Privacy Forum, 326n57

Wu, Tim, 125, 310n73, 347

Xbox, 20, 114, 234

XKeyscore Program, 24, 38, 62–64, 112–113, 122, 134; and PRISM, 10–13, 75–76

Yahoo!, 75, 210, 241, 293n49; and PRISM, 10, 66; and Optic Nerve, 111–114; Yahoo Messenger, 111–112

Yardley, Herbert, 67

Yoffe, Emily, 294n69, 307n10

youth gangs, 243–244, 292n30, 334n30

YouTube.com, 3, 10, 36, 103, 131, 136, 138, 241, 243; and PRISM, 65–66

Zappos.com, 116

Zerilli, Linda, 311n75

Zhao, Shanyang, 128, 287n51, 311n83, 329n1

Zimbardo, Philip, 219

Zuckerberg, Mark, 136, 166, 192, 286n41, 313n113

ZunZuneo, 8–10